SNAKES

Temple Pitviper (*Tropidolaemus wagleri*), southeastern Asia.

*The publisher gratefully acknowledges the generous contribution provided
by the General Endowment Fund and by the Director's Circle of the Associates of the
University of California Press, whose members are*

Evelyn Hemmings Chambers

June and Earl Cheit

Lloyd Cotsen

Phyllis K. Friedman

Susan and August Frugé

Harriett and Richard Gold

Ellina and Orville Golub

Florence and Leo Helzel

Yvonne Lenart

John T. Lescroart and Lisa Sawyer

Raymond Lifchez and Judith Stronach

Hannah and Thormund A. Miller

Mr. and Mrs. Richard C. Otter

Joan Palevsky

Lisa See and Richard Kendall

SNAKES   THE

HARRY W. GREENE

EVOLUTION OF MYSTERY IN NATURE

With photographs by Michael and Patricia Fogden

University of California Press   Berkeley   Los Angeles   London

University of California Press

Berkeley and Los Angeles, California

University of California Press, Ltd.

London, England

© 1997 by

The Regents of the University of California

Library of Congress Cataloging-in-Publication Data

Greene, Harry W., 1945–

   Snakes : the evolution of mystery in nature / Harry W.

Greene ; with photographs by Michael and Patricia Fogden.

      p.   cm.

   Includes bibliographic references (p.   ) and index.

   ISBN 0-520-20014-4 (alk. paper)

   1. Snakes.   I. Title.

   QL666.O6G69 1997

   597.96—dc 20   96-21928

Printed in Hong Kong

9 8 7 6 5 4 3 2 1

The epigraph for Part One is from Albert Camus, *The Myth of
Sisyphus and Other Essays*, translated by Justin O'Brien, © 1955
by Alfred A. Knopf and Random House; used by permission
of Random House. The epigraph for Part Two is from Pablo
Neruda, *Bestiary/Bestiario*, © 1958 by Editorial Losada, S.A.,
Buenos Aires, English translation by Elsa Neuberger, © 1965
and renewed 1993 by Harcourt Brace & Company; used by
permission of Harcourt Brace & Company.

CONTENTS

AUTHOR'S PREFACE

This book was provoked by a cantankerous challenge from Norman Maclean, late forest-fire fighter, fisherman, and professor of English literature. Norman's beautiful novella *A River Runs through It* is a lean, profound tale of families in a landscape. Two pages mention his wife and obliquely convey volumes about their relationship. A dear brother dies with no warning, abruptly in midparagraph, as tragedies so often intrude into our lives. The final line has a majestic and poetic solemnity, transcendent and reassuring: "I am haunted by waters." When we met for dinner in 1983, introduced by a mutual friend, Norman wanted to talk about snakes. As a young man he had visited dens with South Dakota's famous rattlesnake control man, A. M. Jackley, and once heard a rattler strike an ax handle: "Like someone sneaking up and hitting your bat with another bat." I wanted to ask him about writing, to talk about my having been an ambulance driver, a soldier, a teacher, and all the while a field biologist. Norman, who alternated between musing like a scholar and cursing like a cowboy, finally demanded, "Look, just tell me, why do you work with those damned old rattlesnakes? Go write about that!" It took several years to abandon my earlier pretensions, admit that I was unable to say much about more personal matters, and appreciate the wisdom of his advice.

Susan E. Abrams first suggested I write a book and instigated that dinner with Norman Maclean. My ideas for structuring the text were particularly influenced by Wade C. Sherbrooke's *Horned Lizards* (1981) and by our conversations about writing. Richard Shine's wonderful volume on Australian snakes (1991a) appeared while my book was in preparation, and I profited frequently from his insights. I also drew inspiration from Charles Bowden, Sgt. Jim Chee, Amarante Córdova, Henry S. Fitch, Glenn Frey, Carl Gans, Jim Harrison, Alden C. Hayes, Don Henley, Mark Knopfler, Jim Lehrer, Maya Lin, Barry Lopez, Paul S. Martin, Peter Matthiessen, Augustus McCrae, James McMurtry, Natalie Merchant, Karen Saeger, Jay M. Savage, George B. Schaller, Andy Sipowicz, Robert C. Stebbins, Alfred Russel Wallace, David Rains Wallace, and Edward O. Wilson. The Fogdens and I especially appreciate the enthusiastic support of James Clark, Valeurie Friedman, Steve Renick, and Rose Vekony of the University of California Press.

This book owes an enormous debt to hundreds of snake biologists, their published results, and their sources of financial support. My reliance on other naturalists and scientists is documented in the Notes, but here I must single out Richard A. Seigel and his colleagues for their fine overviews of snake biology (Seigel et al. 1987; Seigel and Collins 1993). For ideas, feedback on parts of the manuscript, and other help, I thank Kraig Adler, Aaron M. Bauer, Christopher Bell, Marcia Bonta, William R. Branch, Edmund D. Brodie III, Anne Brooke, Tina Brown, Gordon M. Burghardt, John E. Cadle, Jonathan A. Campbell, M. Brent Charland, Miyoko Chu, David B. Clark, Deborah A. Clark, Harold Cogger, Charles J. Cole, Roger Conant, Michael Cummings, David Cundall, Samuel Danon, Kevin deQueiroz, Benjamin E. Dial, C. Kenneth Dodd Jr., David Duvall, Blake Edgar, Michael Fogden, Patricia Fogden, Jacques A. Gauthier, Bethune Gibson, Linda Gibson, Mary Gleason, Krista Glickman, Anna Graybeal, Billie B. Hardy, David L. Hardy Sr., Bruce C. Jayne, Robert Kaplan, Mimi A. R. Koehl, Fred Kraus, William W. Lamar, Howard E. Lawler, Amy J. Lind, Jonathan B. Losos, Claudia Luke, Stephen P. Mackessy, Marcio Martins, Roy W. McDiarmid, Lyndon A. Mitchell, Julián Monge-Nájera, James B. Murphy, Robert W. Murphy, Göran Nilson, Mark O'Shea, Claire Phillips, Frank A. Pitelka, Thomas R. Powell, Devin A. Reese, Howard K. Reinert, Randall S. Reiserer, Heidi E. Robeck, Wendy E. Roberts, Javier A. Rodríguez-Robles, Carmen M. Rojas, Alan H. Savitzky, Ivan Sazima, Sarah L. Schmidt, Kurt Schwenk, Wade C. Sherbrooke, Richard Shine, Hobart M. Smith, Alejandro Solórzano, Demetri H. Theodoratus, Carol R. Townsend, Harold K. Voris, Van Wallach, Paul Weldon, and Wolfgang Wüster.

Several people merit special thanks. Marjorie G. Greene, the late Harry W. Greene, and William I. Greene nurtured my childhood interests at every turn. The late Paul Anderson and Charles C. Carpenter answered endless teenage questions, and a high school internship with Henry S. Fitch and Charles J. Cole set my course. William F. Pyburn and Gordon M. Burghardt sponsored my graduate work at a time when few people thought snakes worthy of serious study. Billie B. Hardy and David L. Hardy Sr. opened their homes to me and shared many joys of discovery, from Sonoran cuisine to exotic pitvipers. Above all I am indebted to Kelly R. Zamudio for reading the entire manuscript several times, encouraging—not just tolerating—my enthusiasm for creatures and places, and for her inspiring companionship.

PHOTOGRAPHERS' PREFACE

Over the past five years our quest for snakes has taken us to eighteen countries on six continents and twice around the world. We have memories of exciting finds and a continuing collection of images. But our success owes much to the generosity of friends who are fanatical snake enthusiasts. They helped us with advice and information and loaned us precious snakes, sometimes allowing us to disappear with them for days at a time in search of the perfect picture. For such favors we are immensely indebted to Bill Branch of the Port Elizabeth Museum, South Africa; Brent Breithaupt of the University of Wyoming Geology Museum; Greg Fyfe of the Arid Australian Reptile Display in Alice Springs; Alejandro Solórzano of the Serpentario, San José, Costa Rica; Barney Tomberlin and Tony Snell of Portal, Arizona; Romulus Whitaker and the Irula snake catchers of the Madras Crocodile Bank, India; and Anson Wong and staff of the Reptile Sanctuary, Penang, Malaysia. We are similarly indebted to Howard Reinert, who devoted two days to showing us the Timber Rattlesnakes (*Crotalus horridus*) that have been the subject of his pioneering telemetry studies in Pennsylvania; and to Harry Greene and David Hardy for showing us their equally impressive telemetry studies of Black-tailed Rattlesnakes (*C. molossus*) and Ridge-nosed Rattlesnakes (*C. willardi*) in Arizona and of Bushmasters (*Lachesis muta*) in Costa Rica.

Over the years a host of other people have taken us to see snakes, found us snakes to photograph, given us information and advice, or facilitated our work in other ways. We are grateful to all of the following and anyone else whose name we have inadvertently omitted: Andre Bartschi, Robin Chazdon, David and Deborah Clark, Ian Crabtree, Eladio Cruz, Maureen Donnelly, Mike Griffin, Tonio and Wilford Guindon, Craig Guyer, Marc Hayes, Matt Heydon, David and Carol Hughes, Robert Inger, Doug Levey, Milton and Diana Lieberman, Francis Lim, Jack Longino, Clive Marsh, Sandra and Geoff Nicholson, Chris Parkinson, Alan Pounds, Manuel Santana, Jay Savage, Mary Seeley, Wade Sherbrooke, Rick Shine, Rob Steubing, Alejandro Suárez, Walter and Karen Timmerman, Orlando Vargas, Ana Maria Velasco, John Visser, Doug Wechsler, Fred Wilson, Helen Young, and George Zug.

The majority of snakes in this book were photographed where we found them, and most of the rest were photographed in authentic habitat. For photographs of a few important snakes that live in countries we have never visited, we sought and received the cooperation of several zoos and dealers in the United States. For permission to photograph snakes in their collections, and for help in doing so, we are extremely grateful to Clay Garrett, Ardell Mitchell, Jim Murphy, and Dave Roberts at the Dallas Zoo; David Blody and Rick Reed at the Fort Worth Zoo; and Howard Lawler at the Arizona-Sonora Desert Museum. For generously allowing us to photograph two particularly rare and little-known species, we thank Bill Love and Rob MacInnes of Glades Herps, Fort Myers, Florida (for Madagascan Vinesnakes [*Langaha nasuta*]), and Louis Porras of Zooherp, Sandy, Utah (for Fea's Viper [*Azemiops feae*]). Photographing captive snakes in artificial surroundings is not as easy as it might seem; all of the above gave freely of their time to make the photographs look as natural as possible.

Finally, special thanks are due to Billie and David Hardy for outstanding hospitality while we were photographing rattlesnakes in the Chiricahuas; Peter Jenson for accommodations at Explornapo and Explorama Lodges in Peru; Pedro Proaño and Marco García of Metropolitan Touring for accommodations in the Oriente of Ecuador; *Geo Magazine* in Hamburg, and editorial staff Christiane Breustedt, Uta Henschel, and Venita Kaleps for assignments that produced many of our best snake pictures; John Eastcott for the unsurpassed duplicate transparencies that allow us to keep our originals safe; and Sue Fogden for keeping our affairs in order when we are in the field.

In 1766 Carl von Linné (also known as Carolus Linnaeus), a Swedish botanist and founder
of modern biological nomenclature, named the Bushmaster *Crotalus mutus,* literally, "the
silent rattlesnake." Almost forty years later, François-Marie Daudin placed this, the world's
largest viper, in a new genus, *Lachesis,* named for one of the three Fates. In Greek my-
thology, Clothos was the spinner of life's thread, and Lachesis, drawer of lots, chose the
length of one's strand; Atropos made the final cut. Perhaps the young French herpetologist's
choice reflected a distaste widespread among early naturalists (Linné called reptiles "foul,
loathsome beings"), as well as the legendary dangers associated with that snake.

Subsequent treatment of the Bushmaster generally has continued in the vein set by early
writers. Although rarely seen by naturalists, this great snake has inspired more than fifty,
often colorful, vernacular names in Latin America. Some Brazilians say the "Surucucú" ex-
tinguishes fires, others that it suckles milk from cows and sleeping women. Rural people on
the Atlantic versant of Costa Rica call the Bushmaster "Matabuey" (ox killer), while an endemic race of that snake on the Pacific side is known as "Plato Negro" (black plate; a popular beans-and-rice dish) for its dark-colored head. The image of Bushmasters in travelers'
accounts is often fanciful, as if monstrous serpents were slaying whole mule trains of
explorers.

Almost two centuries after the initial description of *Lachesis muta,* even herpetologists still
wrote florid prose about this famous inhabitant of neotropical lowland rain forest. In ele-
mentary school I read "Episode of the Bushmaster," in which Raymond L. Ditmars received
the giant viper from Trinidad and had trouble transferring the snake to a cage in his family's
New York City house. There is the unmistakable impression of danger, perhaps even ma-
levolence on the snake's part, as the young man maneuvered the "deliberately aggressive"
animal with a broom. Until recently few herpetologists had encountered Bushmasters, and
we knew little more than Ditmars had reported: they are huge, rare, and supposedly fero-
cious, and they lay eggs, a habit unique among New World vipers.

In 1982 several colleagues and I began an extended research project at La Selva Biological
Station in Costa Rica. La Selva has one of the richest vertebrate predator faunas on earth,
and we hoped to untangle dietary interactions among the more than one hundred species of

snakes, raptors, and mammalian carnivores at the site. Not surprisingly, giant vipers were on my mind from the start, so I asked a lifelong resident of the region to watch for them on his daily patrols of the reserve. At lunch a few days later, Edwin Paniagua told the station co-directors that he had indeed found a Matabuey to show me. Deborah and David Clark asked about its size, and Edwin replied, "Big enough, and you'd better tell that gringo this isn't a soccer game."

For almost an hour we hiked south at twice my normal speed, up and down mud-slick trails—and all on a full stomach. The surrounding forest was hot and humid, almost claustrophobically dark and fecund. Huge buttressed trees towered above us, obscuring the sky, and everywhere were the deep greens and rich browns of living plants or their decaying remnants. After a brief but torrential shower, the air reverberated with buzzes, screams, and croaks of countless insects, birds, and frogs, and a troop of howler monkeys roared in the distance. Slogging along, I mused half seriously that within minutes of dying in rain forest one would be overgrown by mosses, vines, and fungi, all the while devoured in tiny pieces by ants and fierce green katydids.

The terrain becomes more corrugated upslope from the entrance to La Selva, each ridge a little steeper and higher. When we finally veered off the path I was soaked in sweat, almost giddy with exhaustion and anticipation. Parting the leaves of understory palms and vines, we watched for "Balas" (bullets; *Paraponera clavata*), huge black ants with the most intensely painful and long-lasting sting of any hymenopteran. Edwin stopped fifty meters or so up a broad ravine and peered over an enormous fallen tree. Then, motioning caution with one hand, he pointed for Manuel Santana and me to lean over the chest-high log. Coiled in a mound on the forest floor, its calligraphic black and tan colors blending with surrounding debris, was the most magnificent snake I'd ever seen in nature. Thirty years after I'd read Ditmars's story, here was a live, wild Bushmaster—perhaps two and a half meters long, and thicker than my arm.

As we scrambled over to it, the Bushmaster's only responses were slight elevation and retraction of its head, then a slow, vertical sweep of the long black tongue, aimed directly at us. The snake's behavior was not exaggerated—no lunging strikes, no frenzied escape efforts—but there was a powerful sensation of measured readiness, like Clint Eastwood's squint in *High Plains Drifter*: "Don't come closer." With no experience handling really large vipers, I simply photographed that first *Lachesis muta* and watched it slowly crawl away. In the following decade our research group studied more than two dozen others, documenting a sedentary lifestyle, remarkably narrow diet of rodents, and dependence on undisturbed lowland tropical rain forest. Although much remains to be learned about Bushmasters, we now know more about this species than about many other vipers.

Bushmasters embody our cultural and scientific traditions about what it means to be a snake: extraordinarily cryptic, they obtain infrequent but large meals with minimal risk, depend heavily on chemical cues rather than vision and sound, and convey a certain inscrutability. Bushmasters invariably have symbolized grave danger, even though our fears are largely irrational. Their bite is extremely serious, yet accidents are so rare that we lack a clear picture of proper treatment. Perhaps no other serpent is such an icon for wilderness and the complex meaning of that word, including the profound uncertainties one lives with, and learns from, in remote places.

Green Parrotsnake (*Leptophis ahaetulla*), Costa Rica. Like many tropical rain forest snakes, this species climbs well.

Snakes might rank more nearly equivalent to birds and mammals if our formal classification of living creatures really reflected distinctive characteristics, richness of species, and variety of lifestyles. Biologically, these "limbless tetrapods" are highly specialized and remarkably diverse. More than 2,700 species of snakes are currently recognized, placed in about 420 genera and 18 families. Snakes inhabit all major ecosystems outside of the polar regions and are among the most common predators on other vertebrates. Despite a lack of functional limbs, they manage to ascend trees and burrow beneath the earth's surface. Big-eyed serpents glide through the tropical canopy, only rarely descending to the forest floor, and tiny blindsnakes raid the subterranean nests of ants by following the insects' pheromone trails. Some species with flattened, oarlike tails never emerge from the ocean, and Yellow-bellied Seasnakes (*Pelamis platurus*) are actually planktonic, passive travelers on surface currents. Diets range from snails and centipedes to bird eggs and porcupines, and some venomous snakes engulf prey more than one and a half times their own weight. Most serpents simply deposit their eggs in appropriate sites for incubation, the hatchlings capable of independent living, but many species give birth to fully formed young, and a few exhibit parental behavior.

Instead of taxonomic prominence and the subtle benefits it conveys in a world dominated by humans, anthropomorphic sympathies for fur, feathers, and visible facial gestures prevail: We bury snakes and lizards in Class Reptilia, together with turtles—a group with which they

Western Threadsnake (*Leptotyphlops occidentalis*), southern Africa. Threadsnakes and their kin are habitual burrowers.

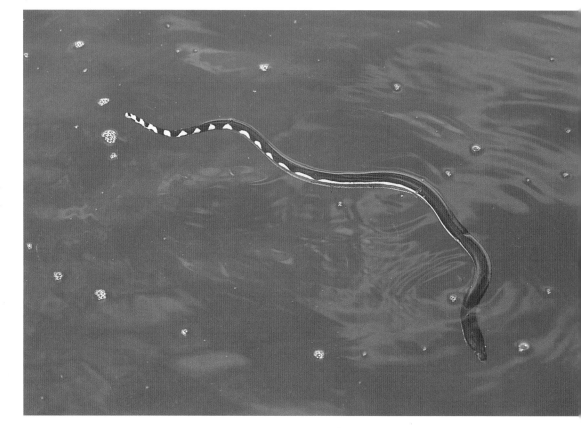

Yellow-bellied Seasnake (*Pelamis platurus*), Costa Rica. With numerous adaptations for marine life, this species feeds and breeds on the ocean surface.

haven't shared a common ancestor in more than two hundred million years. Birds are elevated to a formal group of their own, Class Aves, thereby obscuring the close relationship of our feathered friends with another group of "reptiles," the crocodilians! Nonetheless, despite deeply entrenched cultural and personal biases against them, snakes fascinate us. They represented healing to the ancient Greeks and knowledge to the Incas; a serpent tempted Eve in the biblical garden of Eden, and a giant cobra shaded Buddha. Snakes run the metaphorical gamut in aboriginal and modern lore, and in the past two decades these limbless creatures have even come into their own as subjects of scientific study.

This book aims to provide an overview of snakes that is both accessible to laypeople and scholarly in treatment, one that explores the beauty and intrigue of these animals against the backdrop of science. Several themes run through the text. The future of nature depends on rapid, widespread realization of its practical and aesthetic values. Appreciation of wild places and creatures comes in part from information garnered by biologists, from the facts that tell us why Giant Pandas (*Ailuropoda melanoleuca*) are unique and rain forests are special. I have tried to write a book that is accurate and up-to-date, and one that credits science for enriching our lives. I also believe that the past and present are reciprocally illuminating. Living creatures assume a historical dimension when we know something about their origins, and dry fossils come to life when we have a realistic context for ancient bones. The emphasis here is thus on evolutionary perspectives to understand a group of animals, both biologically and aesthetically. Indeed, I hope this book will exemplify the application of historical thinking in the appreciation of nature.

Science and art should be mutual pathways to understanding: We need something to confront sensually, something about which to feel, in order to place value on nature; we need the opportunity to interpret, to reflect on our own experiences and those of others in the context of new details and new perspectives. While describing the biology of snakes, I also wanted to explore their beauty and mystique, their roles in human experience. The introductory essays address each chapter's subject from a perspective other than traditional science and thereby portray these unusual, appealing organisms as more than facts. In this goal I am especially aided by the magnificent photographs of Michael and Patricia Fogden. I hope our joint efforts enhance appreciation for snakes and promote their conservation.

The uneven coverage of topics here reflects gaps in knowledge as well as our personal backgrounds, goals, and prejudices. My graduate degrees and subsequent research have emphasized snake behavior and ecology. Although I have done fieldwork in Europe, Africa, Latin America, the West Indies, and over much of the United States, for more than a decade my colleagues and I have concentrated on two regions. The tropical rain forests of Brazil and Costa Rica have exposed us to exceptionally rich snake assemblages—almost sixty species at La Selva—while allowing us to study them in the broader context of feeding biology. In Arizona and New Mexico, we focus on eight species of rattlesnakes that inhabit the arid grasslands, desert scrub, and forested mountain islands of that region. Opportunities to contemplate snakes in such different environments, as well as to observe close relatives at such extremes of geography, have been especially rewarding.

The emphasis on venomous serpents reflects my fascination with them, which transcends scientific interest. The details of natural history especially captivate me, and I have tried to generalize about snakes without losing sight of them as individual organisms in diverse environments; accordingly, I often describe real incidents from the field, identify particular prey

species, and so forth. I also am biased toward taxa and phenomena that are poorly emphasized elsewhere, such as Asian burrowing snakes (*Fimbrios*) with fringed lips and African sand-snakes (*Psammophis*) that polish their entire bodies with the secretions of special head glands. Michael and Patricia Fogden have worked primarily in southeastern Asia, southern Africa, Latin America, and the southwestern United States. In choosing and presenting photographs, we stressed artistic quality and congruence with the book's overall goals over exhaustiveness. Whenever possible, we sought images that accurately depict snakes in nature or doing something, rather than simple portraits.

The book is divided into three parts. The chapters in Part I summarize the classification and general biology of snakes, including relationships among major groups, structure and function, behavior, and ecology. Those in Part II survey all major lineages of snakes; the organization of those accounts varies for practical reasons. Small groups (e.g., blindsnakes) are discussed mainly in terms of anatomy, behavior, and other general topics; one especially large and perhaps artificial assemblage (the colubrids) is described in terms of major geo-

Above: Asian Green Vinesnake
(*Ahaetulla prasina*), Borneo, resting on a
leaf.

Opposite: Eyelash Pitviper (*Bothriechis
schlegelii*), Costa Rica, among palm
fruits.

graphic and taxonomic subdivisions. For reasons of special interest, venomous marine elapids (seakraits and seasnakes) are treated separately from the larger group to which they pertain. An overview of evolution and biogeography is postponed until Part III, because those topics build on details in the earlier taxonomic surveys for details. The final chapter, on snakes and other organisms, concludes with a discussion of the conservation status and future of snakes.

The essays that open each chapter recur throughout the main text and then fuse with it in the Epilogue. Most chapters also present a special topic, set apart from the main text, to address particular subjects and explain auxiliary concepts. References for general chapter topics as well as specific points in the text are provided at the end of the book, so as not to interrupt the narrative. Whenever possible I cite recent reviews, but if no summary exists (e.g., for blindsnakes), I provide numerous original sources. Readers can search readily for particular subjects by using the Contents, the List of Special Topics, and the Index. I hope the reader will also trust my dual agenda—facts and a feeling for the organisms in nature— and read this book from cover to cover.

At the final stage you tell me that this wondrous and multicolored

universe can be reduced to the atom and that the atom itself can be

reduced to the electron. All this is good and I wait for you to continue.

But you tell me of an invisible planetary system in which electrons

gravitate around a nucleus. You explain this world to me with an image.

I realize then that you have been reduced to poetry: I shall never know.

Have I the time to become indignant? You have changed theories. So

that science that was to teach me everything ends up in a hypothesis,

that lucidity founders in metaphor, that uncertainty is resolved in a

work of art.

Albert Camus, The Myth of Sisyphus

Peringuey's Adder (*Bitis peringueyi*)
sidewinding on dunes in Namibia.

CLASSIFICATION AND GENERAL BIOLOGY

Venerable, sometimes conflicting, traditions shape our views of nature, and humans were enthralled by slithering gaits, venomous bites, and various legendary attributes of snakelike animals long before recorded history. Some Native American petroglyphs depict fanciful horned organisms, only their elongate shapes suggestive of serpents, while others accurately portray rattlesnakes. Realistic Water Pythons (*Liasis fuscus*) and Common Death Adders (*Acanthophis antarcticus*) appear in the "dream time" paintings of Australian aborigines, whereas rural people in the southeastern United States still refer to certain lizards with long, fragile tails as "glass snakes"—focusing on limblessness rather than the external ear openings and eyelids never found on true serpents. Even biologists portray snakes mainly in terms of locomotion and feeding, although two thousand years ago Aristotle hinted at the truth when he linked forked tongues with a "twofold pleasure from savours, their gustatory sensation being as it were doubled." Perhaps this key characteristic is largely overlooked because of human antipathy for all but the most obvious and appealing odors, as well as fuzzy notions about what exactly is a snake.

The lives of serpents are exquisitely permeated by chemical phenomena. If newborn Queen Snakes (*Regina septemvittata*) are experimentally offered diverse odor samples on cotton swabs, they attack the molecular signature of freshly molted crayfish (the sole item in their natural diet) and ignore all others. Black-tailed Rattlesnakes (*Crotalus molossus*) use the distinctive spoor of woodrats (*Neotoma*) to choose strategic ambush sites near the runways and nest entrances of their favored prey. A male Plains Gartersnake (*Thamnophis radix*) deftly follows females by assessing pheromone trails on vegetation, and tropical snakes presumably distinguish their own kind by odor from among the five or six dozen species in a rain forest. By contrast, fresh-baked bread and barbecue smoke offer us pleasant but uninspiring hints of a chemical worldview. The fragrances of Kelly's hair when I first awake and of the desert after a thunderstorm are more impressive examples, flooding me with joy and hope before I consciously afford them meaning. And one hot afternoon, cruising mindlessly on a San Joaquin Valley freeway, my daydreams suddenly shifted from rattlesnakes to creamy tomato soup and crackers. Minutes later, passing a huge produce truck and still wondering why visions of childhood snacks had drifted in from nowhere, I realized I'd been traveling in the odor plume of an open load of tomatoes.

Early classifications of reptiles were based mainly on shapes and external structures as seen in museum specimens. Extensive attention to internal characteristics began in the 1800s with A.-M.-Constant Duméril, a Parisian savant, and blossomed thereafter with studies of lungs, reproductive organs, and vertebrae by Edward D. Cope in the United States. Nevertheless, until the turn of the century even herpetologists still considered some elongate lizards "snakes" and called limbless amphibians "pseudophidians," or false snakes. A fairly stable arrangement consistent with fossils emerged only with the work of French paleontologist Robert Hoffstetter in the 1950s, and the composition of snakes as a group continued to depend on whether certain extinct forms were included. Lists of definitive characteristics have appeared in the past few decades, but most of us couldn't recognize a snake as a snake from "lateral closure of the braincase wall" and other such features!

Novel ways of studying snakes were paralleled over the last two centuries by shifts in the philosophy of biological classification, from imposing order on God's Universe to portraying evolutionary history. Consensus is emerging, influenced by new data and new viewpoints, and relationships among major lineages are increasingly clear. For example, the formal category "Serpentes" includes the most recent common ancestor of extant snakes and all descendants of that ancestor—a definition that emphasizes living, intact organisms (about which more is known) and avoids the problem of whether certain poorly preserved fossils really are "snakes." The oldest evolutionary divergence was between blindsnakes and other serpents, although those two groups are not as different as once thought. Most important, our current concept coincides with neither complete limblessness nor eating really large prey, a fact that would surprise many biologists. Both of those traits actually arose after the origin of snakes, and both also occur in other groups of vertebrates.

With accurate notions of their membership, we can more confidently characterize the evolutionary chronicle of serpents. They arose from nearly limbless lizards, and although some snakes have expansive mouths, others still have the relatively inflexible jaws that typified their common ancestor. In addition to serpentine locomotion and feeding specializations, the origin of snakes foreshadowed an increased reliance on chemical cues—and innovations at both ends! A highly mobile tongue carries molecules to a receptor organ in the roof of the mouth, and the tongue's deeply forked shape facilitates directional localization of an odor's source. Snakes use those structures to explore a chemical world perhaps analogous in complexity and subliminal nuance to the textures and colors we perceive visually. Furthermore, all serpents have paired glands in the tail base from which foul substances are smeared about during encounters with predators. Twin innovations thus underlie a spectacular array of locomotor, feeding, and social activities: forked tongues facilitate intensive searching for mates, places to hide, and food, while tail glands protect otherwise occupied snakes from their enemies. Understanding these and other chemically mediated responses is an exciting challenge for science; meanwhile, the everyday lives of serpents are more easily imagined when I remember Kelly's hair, desert thunderstorms, and that truckload of tomatoes.

Beyond coining scientific names (an activity known as taxonomy), phylogenetic systematics is a branch of biology that discerns the history of evolutionary divergence among organisms and categorizes them accordingly. Within that framework, herpetologists divide and subdivide the 2,700 species of snakes into manageable groups, making sense of diversity by searching for similarities and dissimilarities among them. Throughout this book the numbers in those groups are approximate, both because of uncertainties about the status of some species and because new kinds of snakes are discovered each year.

This chapter introduces scientific nomenclature and common names. Even many biologists are unfamiliar with recent refinements, so some implications of phylogenetic systematics for understanding snakes are explained further in the Appendix. Here I also survey the major groups of snakes as background for the topics that follow, then summarize aspects of their general biology not covered elsewhere in the text.

SCIENTIFIC AND COMMON NAMES

For more than two hundred years biologists have used a binomial nomenclature pioneered by Carl von Linné, in which each species is known by a unique combination of italicized generic and specific names. The words themselves are derived from Latin or Greek or based on Latin endings. For example, the scientific name of the Western Rattlesnake is *Crotalus viridis*. Its species epithet, the second word, is Latin, meaning "green in color"; that species is one of thirty assigned to the genus *Crotalus*, based on the Greek word *krotalon*, meaning "a rattle." A formally named group of organisms, such as the genus *Crotalus* or the species *Crotalus viridis*, is called a "taxon." Subspecies, or geographical races within a species, have a third latinized name. In the first subspecies to be described, the subspecies name automatically repeats the species name: for example, the nominate subspecies of the Western Rattlesnake is *Crotalus viridis viridis* (Prairie Rattlesnake). In subsequently described subspecies a different name follows the specific name, as in *Crotalus viridis cerberus* (Arizona Black Rattlesnake). Genera and species often are abbreviated, as in *C. v. viridis* and *C. v. cerberus*. Definitions and some conceptual problems with species and subspecies are discussed in the Appendix.

Common names have no standing in biological nomenclature yet play an important role in communication among laypeople. I generally use English names in vogue for particular regions but follow personal preference in a few cases—I would as soon call a Black-headed Python (*Aspidites melanocephalus*) a "Teenage Pimple Serpent" as refer to colubrids of the genus *Tantilla* by their "standard" common name, blackhead snakes! I capitalize English names for plant and animal species (but not genera and other groups of species), to distinguish them from adjectives (e.g., dwarf, brown). In the interest of compactness, and with "rattlesnake" as a model, I combine words for groups of species (e.g., "seakraits" [*Laticauda*]), as opposed to single species (e.g., "Gopher Snake" [*Pituophis catenifer*]), but only when the root is one syllable with five or fewer letters and the modifier has no more than two syllables and six letters (e.g., "coralsnakes" versus "calico snakes" [*Oxyrhopus*]). English names for some higher taxa (e.g., "boas" for Boidae) are widely familiar, but I invented "dwarf pipesnakes" for *Anomochilus;* lacking popular alternatives, herpetologists call members of the largest snake family "colubrids" and refer to cobras and their relatives collectively as "elapids." Some laypeople find scientific names difficult, and many biologists are unfamiliar with common and scientific names outside their area of expertise, so I use English names whenever possible and provide scientific names for genera and species at the first mention in each introductory essay, text paragraph, special topic section, and photograph caption. Common and scientific names are cross-referenced in the Index.

SNAKES AND OTHER VERTEBRATES

In traditional Linnean taxonomy, groups above the species level (called "higher taxa") are assigned to ranked categories that supposedly represent older, more inclusive relationships. Genera are groups of similar, presumably related species (e.g., most rattlesnakes in *Crotalus*); families, denoted by the ending "-idae," are groups of similar, presumably related genera (e.g., cobras, seasnakes, and their relatives in Elapidae); orders are groups of similar, presumably related families (e.g., lizards and snakes in Squamata); and so on. Finer divisions in the ranked categories are sometimes used, most frequently subfamilies (denoted by the ending "-inae," as in Crotalinae for pitvipers, a subfamily within Viperidae). Tribes, denoted by the ending "-ini," are sometimes designated for groups of genera within subfamilies, (e.g., Sonorini for groundsnakes [*Sonora*] and their relatives within Colubrinae).

In point of fact, traditional classifications often fail to portray evolutionary relationships, as illustrated by the following familiar ranked system:

Class Amphibia (amphibians)
Class Reptilia (reptiles)
    Order Chelonia (turtles)
    Order Crocodylia (crocodilians)
    Order Sphenodontida (tuataras)
    Order Squamata (lizards, amphisbaenians, snakes)
Class Aves (birds)
Class Mammalia (mammals)

The above scheme omits the origins of four limbs in the common ancestor of terrestrial vertebrates (there is no named group for all four classes), fails to signify the origin of the shelled egg within that group (there is no named group for mammals, "reptiles," and birds), and ignores the fact that crocodilians and birds are more closely related to each other than either of them is to any of the others—all highly significant events in vertebrate evolution!

By contrast, phylogenetic classifications express evolutionary divergences as equally indented sister taxa (i.e., those most recently sharing a common ancestor; more than two taxa are indented equally when their phylogenetic relationships are uncertain). Each set of descendant, equally indented taxa is referred to by the next higher, less indented name. For example, amphibians and amniotes are each other's closest relatives (they are sister taxa), collectively known as tetrapods; conversely, the oldest evolutionary divergence within tetrapods was between amphibians and their sister taxon, the amniotes. The historical relationships of snakes with other major groups of living terrestrial vertebrates can be retrieved from the following phylogenetic classification (Fig. 1 expresses these relationships as a phylogenetic tree or cladogram; Chapter 15 discusses relationships within Squamata in more detail):

Tetrapoda (four-limbed vertebrates)
    Amphibia (frogs, salamanders, caecilians)
    Amniota (tetrapods with three embryonic membranes)
        Mammalia (mammals)
        Reptilia ("reptiles" and birds)
            Chelonia (turtles)
            Sauria (archosaurs and lepidosaurs)
                Archosauria (crocodilians and birds)
                    Crocodylia (crocodilians)
                    Aves (birds)
                Lepidosauria (tuataras and squamates)
                    Sphenodontida (tuataras)
                    Squamata (squamates)
                        Iguania (iguanas, chameleons, etc.)
                        Scleroglossa (other lizards, amphisbaenians, snakes)
                            Gekkota (geckos and flap-footed lizards)
                            Autarchoglossa (other lizards, amphisbaenians, snakes)

Snakes have diverged from ancestral conditions of the larger taxa to which they pertain, sometimes strikingly so, but also reflect their heritage as members of those more inclusive groups. For example, snakes have a segmented vertebral column, one among many characteristics of Vertebrata (a larger group including Tetrapoda and various fish). They share numerous features typical of tetrapods yet for the most part lack the girdles and limbs used for locomotion by most other members of that group. Like amniotes but unlike amphibians, the first snakes laid shelled eggs with three embryonic membranes (including the amnion, hence the larger group's name), although now many squamates and most mammals are viviparous. Snakes share numerous anatomical characteristics with other reptiles (including birds), by which they all differ from mammals (the sister taxon of reptiles within Amniota).

Among living Reptilia, the sister taxon of Squamata is Sphenodontida, encompassing two species of New Zealand tuataras (*Sphenodon*); together squamates and tuataras are named Lepidosauria. All squamates share numerous derived characteristics, including paired copulatory organs in males. Traditionally the approximately 6,700 living species of Squamata have been divided into lizards (Suborder Lacertilia, often inappropriately called Sauria), amphisbaenians (Suborder Amphisbaenia; see p. 48), and snakes (Suborder Serpentes, sometimes called Ophidia). Phylogenetic systematists reject that arrangement because Lacertilia is not monophyletic; monitors (Varanidae), alligator lizards (Anguidae), and their relatives probably are more closely related to snakes than to other lizards, while whiptails (Teiidae) and their

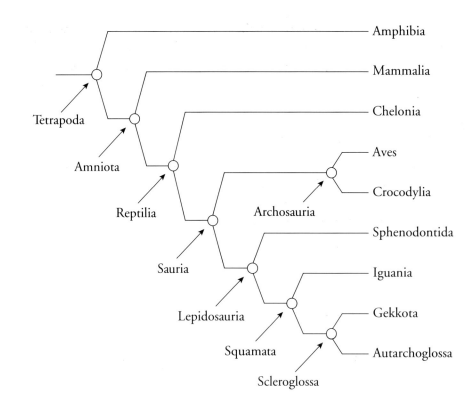

Fig. 1

Phylogenetic relationships among major groups of terrestrial vertebrates.

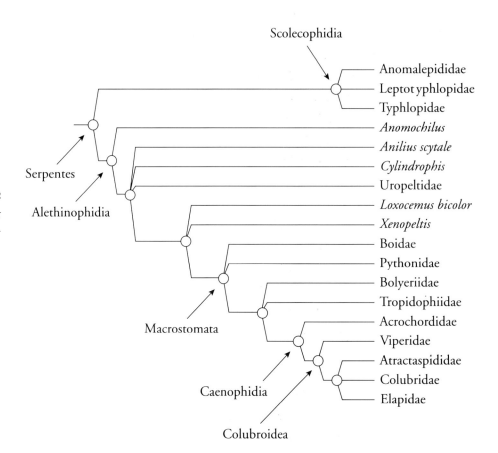

Fig. 2

Phylogenetic relationships among major groups of snakes.

relatives might be more closely related to amphisbaenians than to some other lizards (see the Appendix for definitions of monophyly, paraphyly, and polyphyly). Actually, amphisbaenians and snakes are lizards in exactly the same sense that humans are primates, primates are mammals, and so forth.

CLASSIFICATION AND THE DIVERGENCE OF HIGHER SNAKE TAXA

The arrangement of major snake groups herein reflects well-supported evolutionary divergence and monophyletic relationships. Rather than name them all formally, I designate one group by a bracketed list of included taxa; to facilitate discussion, two sets of paraphyletic basal taxa are indicated (framed in quotes; Fig. 2 expresses these relationships as a phylogenetic tree):

Serpentes
    Scolecophidia (blindsnakes)
        Anomalepididae
        Leptotyphlopidae
        Typhlopidae
    Alethinophidia
        Anomochilus (dwarf pipesnakes)
        ["basal alethinophidians," Macrostomata]
            "basal alethinophidians"
                Anilius scytale (Red Pipesnake)
                Cylindrophis (Asian pipesnakes)
                Uropeltidae (shield-tailed snakes)
                Loxocemus bicolor (Neotropical Sunbeam Snake)
                Xenopeltis (Asian sunbeam snakes)
            Macrostomata
                "basal macrostomatans"
                    Boidae (boas, sand boas, etc.)
                    Pythonidae (pythons)
                    Bolyeriidae (Round Island boas)
                    Tropidophiidae (dwarf boas)
             Caenophidia (advanced snakes)
                Acrochordidae (Australasian filesnakes)
                Colubroidea
                    Viperidae (vipers, pitvipers)
                    Atractaspididae (stiletto snakes, etc.)
                    Colubridae (ratsnakes, goo-eaters, etc.)
                    Elapidae (cobras, seasnakes, etc.)

Red Pipesnake (*Anilius scytale*), Brazil. In this species the eye is not covered by a spectacle but instead lies under a head scale. (Photograph by Marcio Martins.)

As indicated by this classification, the initial evolutionary divergence among snakes was between scolecophidians and all others, the latter as a group termed Alethinophidia (literally, "true snakes"). Scolecophidians are small creatures with eyes buried beneath the skin and highly polished, round scales of uniform size throughout the body. Called blindsnakes as a group, they are placed in the Anomalepididae (4 genera, 15 species), Leptotyphlopidae (2 genera, 80 species), and Typhlopidae (6 genera, 215 species).

Anomochilus (2 species of Indonesian dwarf pipesnakes) is structurally transitional between other alethinophidians and blindsnakes. Dwarf pipesnakes and five other small lineages of basal alethinophidians (once collectively called Anilioidea or anilioids) have undifferentiated or slightly enlarged ventral scales, stout jaws, and rather limited gapes. The other groups of living basal alethinophidians are Uropeltidae (9 genera, 46 species of shield-tailed snakes), *Cylindrophis* (9 species of Asian pipesnakes), *Anilius scytale* (Red Pipesnake), *Xenopeltis* (2 species of Asian sunbeam snakes), and *Loxocemus bicolor* (Neotropical Sunbeam Snake).

Basal macrostomatans (formerly called Booidea or booids) have moderately enlarged ventral scales and lighter, more mobile jaw elements than basal alethinophidians; in these and some other aspects, they are transitional between the latter and caenophidians. Basal macrostomatans include Bolyeriidae (2 genera, 2 species of Round Island boas) and Tropidophiidae (4 genera, 20 species of dwarf boas), as well as the better known Boidae (8 genera, 40 species of boas, sand boas, and their relatives), and Pythonidae (8 genera, 24 species of pythons). I

Carpet Python (*Morelia spilota*),
Australia, with labial scales modified
for infrared reception.

informally refer to basal alethinophidians, basal macrostomatans, and Acrochordidae together as basal snakes. Although highly specialized for aquatic life, Australasian filesnakes are phylogenetically and structurally transitional between basal macrostomatans and the Colubroidea; together Acrochordidae and Colubroidea are grouped as Caenophidia and informally called "advanced snakes."

Colubroidea encompass Atractaspididae (14 genera, 65 species of stiletto snakes [*Atractaspis*] and their associates), Colubridae (290 genera, almost 1,700 species of colubrids), Elapidae (63 genera, 272 species of cobras and their relatives), and Viperidae (30 genera, 230 species of vipers and pitvipers). Colubroids share modifications of the skull, such that the maxillary bones are freed from their primitive role for ingestion and available for other specializations. Venomous seakraits and seasnakes (ca. 15 genera and 55 species) sometimes are collectively separated as the Hydrophiidae, but that taxon includes at least two independent invasions of the marine environment by Australasian elapids and thus is polyphyletic.

The timing of major events in snake evolution is not well understood, owing in part to a relatively poor fossil record. It is discussed in more detail in Chapters 8 and 15, but here is a general outline: The group originated in the Mesozoic era, more than 65 million years ago (mya). Within the Cenozoic era that followed (up to the present), advanced snakes probably arose at least as long ago as the Oligocene epoch (35–25 mya). Numerous modern snake genera are known from the Miocene epoch (25–5 mya) and more recent deposits.

Three-striped Snake (*Dryocalamus tristrigatus*), Borneo. This poorly known species belongs to the Colubroidea, a large group characterized by diverse color patterns and lifestyles.

GENERAL ANATOMY

Snake heads vary from chunky to elongate and pointed; snakes' trunks may be cylindrical, tapered at each end, or compressed laterally or dorsoventrally. Some scolecophidians and basal snakes have rudimentary pelvic girdles and external hind limbs, but most serpents lack those structures. No snake has even traces of a pectoral girdle or front limbs. The musculoskeletal system of snakes is discussed later with respect to roles in locomotion (Chapter 2) and feeding (Chapter 3). In this book snake size usually is described in terms of the total length (snout-to-vent plus tail) of an average adult.

The internal anatomy of serpents resembles that of other vertebrates, but the viscera are modified to fit a tubular body. Major organs usually are staggered linearly and more elongate than in other squamates; in some snakes one member of a pair of organs is reduced or absent (see p. 49). Most basal snakes have two well-developed lungs, but in blindsnakes and colubroids the left lung is greatly reduced or lost. The right lung extends through much of a snake's body cavity and ends in a sacklike air storage area, especially well developed in seasnakes. A snake's esophagus extends from the mouth almost to midbody and is not sharply differentiated from the stomach; a pyloric valve separates the latter from a looped or coiled small intestine, which in turn empties into a short, straight, large intestine. The liver is elongate and bilobed. Paired kidneys and gonads (ovaries or testes) are staggered beside the midline in the posterior part of the body. There is no urinary bladder. Typically the right ovary is farther toward the head and the right oviduct is longer than the left. Some blindsnakes, some black-headed snakes (*Tantilla*), and at least one atractaspidid (*Polemon notatus*) have a single ovary or oviduct, the loss of one member of the pair probably associated with streamlined bodies in these slender, fossorial creatures. As in other reptiles (including birds), the cloaca is a common chamber that separately receives products of the kidneys, large intestine, and reproductive tracts; it opens through the vent.

Like other squamates, male snakes have paired copulatory organs (see Chapter 6 for other differences between the sexes). Each organ is called a hemipenis (literally, "half-penis"), a name that reflects an ancient, erroneous belief that the two grooved structures were pressed together to form a single mammal-like apparatus. Each squamate hemipenis is in fact a blind, inverted cylinder that can be everted through the vent by a combination of hydraulic pressure (from blood sinuses in the organ) and muscle action. Semen travels along a surface channel (the sulcus spermaticus, or sperm groove) into the female's cloaca, and a retractor muscle and reduced blood pressure invert the organ after mating. Hemipenes are stored in the base of the tail when not in use, giving the tails of male snakes a characteristically stouter shape and greater length than those of females.

The biological role of double hemipenes remains intriguing. Each copulatory organ receives sperm from only the testis on its side of the body. Common Kingsnakes (*Lampropeltis getula*) alternate the right and left hemipenes during successive copulations; Ottoman Vipers (*Vipera xanthina*) attempt to insert whichever organ is adjacent to a female's vent. Perhaps the squamate arrangement permits frequent copulation while ensuring recovery time for the alternating reproductive tracts. Hemipenes usually vary among species and higher taxa in shape (e.g., simple versus forked), ornamentation (e.g., presence or absence of spines, papillae, etc.), and size. As might be expected, the urogenital anatomy of males and females differs similarly in shape between two species of tree pitvipers (*Trimeresurus*); inexplicably, hemipenes and female cloacae each vary among populations of the Variable Reedsnake (*Calamaria lumbricoidea*), but not in ways that make sense in terms of their functional relationships with each other.

Cloacal scent glands are unique to and characteristic of all snakes, a fact that suggests they might have played an important role in the origin and successful radiation of the group. These glands are paired pouches, opening from the tail base into the posterior edge of the cloaca; their typically foul-smelling contents are ejected by muscles and probably deter predators. Some blindsnakes have an additional tail gland of unknown function.

Snake skin resembles that of most other reptiles, in that its scales are folded and thickened portions of the outermost layers, the dermis and epidermis. Each scale consists of an outer surface, an inner surface, a hinge zone where folding occurs, and a thin free margin that usually overlaps adjacent scales—all easily seen on a freshly shed skin. Snake scales are smooth or have longitudinal ridges (known as keels), and they usually overlap other scales posteriorly. Small pits and tubercles are sometimes visible with the naked eye, especially on the head, and the scale surfaces also are marked by microscopic ridges, tubercles, and other structures that differ among species and higher taxa.

Boas, pythons, and many vipers have small, irregularly arranged head scales, in contrast to the large, symmetrical head scales (also known as plates or shields) of most advanced snakes. Moving from front to back over the head, these large scales are the rostral (covering the snout), internasals (usually paired, behind the rostral), prefrontals (usually paired, behind the internasals), supraoculars (often fragmented, just over the eyes), frontal (between the supra-oculars), and parietals (paired, behind the frontal). Supralabial (upper) and infralabial (lower) lip scales border the mouths of snakes on each side. Behind each nasal scale (including or adjacent to the nostril) are loreal or preocular scales, or both, and behind each eye, between the supralabials and parietals, are postocular and temporal scales. On the underside of the head, an anterior mental scale generally is followed by large, paired chin shields and smaller gular scales. Most snakes have a longitudinal mental groove between rows of chin scales, but a few groups lack this expansion pleat (e.g., blindsnakes, dwarf pipesnakes [*Anomochilus*], and some colubroids that feed on slimy invertebrates).

Dorsal body scales typically overlap in diagonal, usually odd-numbered rows; these range from a dozen or so (e.g., ten in Tiger Ratsnake [*Spilotes pullatus*], thirteen in Asian coralsnakes [*Calliophis*]) to more than ninety in some seasnakes. In burrowers and other slender species, row counts of dorsal scales are low and relatively constant along the body, whereas in stocky serpents with tapered anterior and posterior parts the numbers are high at midbody and decrease at each end. Males of the Short Seasnake (*Lapemis curtus*) have spinelike, juxtaposed dorsal scales, and mid-dorsal scales of the bizarre Javan Mudsnake (*Xenodermis javanicus*) are large, non-overlapping knobs. Most snakes have enlarged, transverse ventral scales, but blindsnakes, most basal alethinophidians, and some seasnakes have scales that are uniform in size throughout the body. Ventral scale counts range from fewer than a hundred (e.g., in some African slug-eaters [*Duberria*]) to more than five hundred in some seasnakes. In most snakes a transversely enlarged anal scale covers the cloacal opening, marking the boundary between body and tail. Snakes with differentiated ventral scales also have enlarged subcaudal scales under the tail, often paired; these number from fewer than ten (e.g., in some reedsnakes [*Calamaria*]) to more than two hundred (e.g., in some wolfsnakes [*Lycodon*]).

Many lizards have crests, casques, and other fancy ornamentations, whereas—perhaps because they would impede locomotion—elaborations of the skin are rare in snakes. Scaly appendages occur on the snouts or over the eyes of some arboreal serpents (e.g., Madagascan Vinesnakes [*Langaha nasuta;* photos, pp. 124, 125]; also various vipers) and of the aquatic Tentacled Snake (*Erpeton tentaculatus;* photo, p. 183). Some snakes dig with an enlarged rostral scale, designed like a plow or spade (photo, p. 43). A few species have spikelike tail scales that also might aid in burrowing, and the modified tail tips of some others produce noise during antipredator displays.

Green Keeled Racer (*Chironius exoletus*), Costa Rica. Unlike most snakes, this species and its close relatives have an even number of scale rows.

Eyelash Bush Viper (*Atheris ceratophorus*), Tanzania. Fringed scales perhaps break up the outline of this snake's head, thereby concealing it from predators and prey.

Dorsal scales of Rhinoceros Adder (*Bitis nasicornis*), equatorial Africa.

Most snake skin colors result from pigments, although Asian sunbeam snakes (*Xenopeltis*) and some other species have microscopic structures that make their scales iridescent. Some boas, colubrids, and vipers become lighter at night and darken during the day. With age, melanin deposits in the skin often obscure the bright patterns of juvenile snakes; examples of such ontogenetic color shifts include several pitvipers and colubrids that are blotched or bright red as juveniles, then uniformly dark-colored as adults. Color changes sometimes vary geographically within a species, perhaps associated with changes in habitat and defensive responses: As adults, Black Ratsnakes (*Elaphe o. obsoleta*) are uniformly dark and Yellow Rat-snakes (*E. o. quadrivittata*) have black stripes on a light background, whereas all juveniles of that species are light brown or gray with a dorsal pattern of dark blotches. Adult Western Rattlesnakes (*Crotalus viridis*) are contrastingly blotched like their young over most of the range of that species, but some full-grown Arizona Black Rattlesnakes (*C. v. cerberus*) are sooty black with small golden flecks.

Adult color patterns are constant in most snake species, at least within a population. Well-known polymorphisms, however, include striped versus banded Common Kingsnakes (*Lampropeltis getula*); striped, banded, and unicolored Prairie Groundsnakes (*Sonora semiannulata*); and red versus gray or brown estuarine species in Asia (Dog-faced Watersnake [*Cerberus rynchops*]) and the United States (Salt Marsh Watersnake [*Nerodia clarkii*]). Although popu-

Arizona Black Rattlesnakes (*Crotalus viridis cerberus*). The juvenile (top) exhibits the short, tapered rattle string typical of young individuals; the adult male (bottom) has a long, parallel-sided rattle, typical of full-grown individuals. This subspecies undergoes a remarkable change in color with age.

Top: A female Black-tailed Rattlesnake (*Crotalus molossus*) with newborn young, Arizona. The baby rattlers are in pre-shed condition, indicated by their dull color patterns and bluish eyes. Bottom: Four days later, one of the baby Black-tailed Rattlesnakes sheds its skin in front of its mother. (Both photographs by David L. Hardy Sr.)

lar with hobbyists, free-living snakes with truly aberrant colors and patterns are easily visible to predators and thus rare. At least in Brazil, most naturally occurring albino snakes are either nocturnal or burrowers, therefore less likely to be seen by enemies, or are venomous and thus especially capable of defense.

Molting (shedding, ecdysis) in reptiles results from cyclical changes in the underlying skin structure; the end result—a shiny new skin—might facilitate growth, renew tissue abraded during locomotion or otherwise damaged, remove ectoparasites, and maximize chemical communication. Most squamates molt in small, ragged pieces over a period of days or weeks, but alligator lizards, amphisbaenians, and snakes typically shed their skins in one piece. Intervals between ecdysis range from a few weeks to several months and vary with temperature, health, growth, and feeding. For several days prior to molting the eyes are clouded gray or blue by fluid between the old and newly formed spectacles; snakes in that condition are usually inactive and sometimes remain hidden. Among those I found "in the blue," a Northern Cat-eyed Snake (*Leptodeira septentrionalis*) was hidden in a bromeliad, two Terciopelos (*Bothrops asper*) remained in burrows, several Bushmasters (*Lachesis muta*) were coiled on the forest floor, and a Lowland Bush Viper (*Atheris squamiger*) was resting on a branch.

A shedding snake rubs supralabial and infralabial scales free from the mouth's margin, then crawls forward; as a result of the inside-out nature of this process, the tail of the old skin points in the direction of the departing creature. Popular accounts imply that snakes shed against surface objects, but a free-living Indigo Snake (*Drymarchon corais*) was seen moving rectilinearly while spasmodic jerks loosened the skin and folded it backward; a large Terciopelo (*Bothrops asper*) simply crawled slowly over wet forest litter, emerging from its old skin in a matter of minutes. Chapter 13 discusses specialized molting behavior in marine snakes.

THE SENSES AND PHYSIOLOGICAL ECOLOGY

Behavior and physiology link an animal with its external environment. Like most other aspects of their biology, elongate shape and limblessness profoundly affect these responses in snakes. Vegetation and other environmental obstacles are more complex at ground level, so visual and auditory cues would be less effective there than they are for elevated organisms. Moreover, snakes evolved from a secretive, perhaps burrowing ancestor and thus relinquished an emphasis on visual and airborne sound cues early in their history.

Unlike most other amniotes, snakes lack external ears; able to perceive some low-frequency airborne sounds, they hear mainly vibrations conducted via the substrate. Ground-borne sounds are transmitted through a snake's body to the quadrate bone (a connection between the lower jaw and the skull) and thence through the columella (middle-ear bone) to the inner ear. Eyes vary from reduced or even absent in blindsnakes to birdlike globes in the Boomslang (*Dispholidus typus*) and some other colubrids. Usually the eye is protected by a convex spectacle, or brille, but head scales cover the eyes of blindsnakes and a few other taxa. Diurnal snakes have round pupils, and nocturnal species often have vertically elliptical pupils, which shut to a tight slit and thus protect especially sensitive retinas from daylight. Evolutionary history must also influence pupil shape, however, because many facultatively diurnal snakes (e.g., some boas, colubrids, and vipers) have "cat eyes."

Asian Green Vinesnake (*Ahaetulla prasina*), Malaysia. Note the horizontal pupil and groove on the side of the snout, both of which might facilitate binocularity in this visually hunting species.

Opposite: Yellow Blunt-headed Vinesnake (*Imantodes inornatus*), Costa Rica, with the vertical pupils typical of many nocturnal species.

Although humans might view ground-level environments as depauperate, the sensory world of snakes is surprisingly rich in tactile, thermal, and chemical cues. An elongate body predisposes heavy reliance on touch, because positional control mechanisms are integral to serpentine locomotion and because increased surface area enhances bodily contact during social encounters. Snake skin is indeed richly supplied with tactile receptors, and these animals seem especially responsive to touch. Pythons and boas also have highly sensitive thermal receptors in the supralabial and infralabial scales, and some vipers probably form infrared images with their facial pits (see p. 254). The Olive Seasnake (*Aipysurus laevis*) even has photoreceptors on its tail.

Chemical cues are of overriding importance in the lives of all serpents, processed by a sensory system with which we are largely unfamiliar (see the Special Topic in this chapter). Their nostrils, as in other terrestrial vertebrates, are openings for respiration; smell, however, plays a minor role in snake behavior. In contrast, they have elaborated ancestral mechanisms of autarchoglossan lizards to extremes found nowhere else among vertebrates. A deeply forked tongue gathers odor molecules, relative concentrations on the right and left tines indicating the proximity and direction of potential mates, prey, and enemies. Odor molecules are transferred from the tongue tips to the exquisitely sensitive vomeronasal (or Jacobson's) organ. Taste buds, concentrated in tissue along the tooth rows, function after objects have been grasped in the mouth. Later chapters detail the roles of chemical cues in the lives of serpents.

Like most other vertebrates and with the exception of brooding pythons, snakes do not produce enough excess metabolic heat to sustain high body temperatures irrespective of their surroundings; they are ectothermic ("cold-blooded" is best reserved for some humans!). Endothermic birds and mammals burn about ten times as much energy as ectotherms, and that, coupled with insulating feathers or fur, maintains their high body temperatures. Ectothermy does not, however, necessarily imply low or wildly fluctuating body temperatures; some reptiles behaviorally thermoregulate at fairly constant, high levels. Ectotherms often circumvent potential disadvantages of their low-energy lifestyles: snakes cannot sustain high aerobic activity for longer than a few minutes, as only endotherms can, but they escape predators by confrontation or sprinting to inaccessible retreats rather than by winning endurance races. Many endotherms can operate at relatively low temperatures, whereas serpents generally are inactive during harshly cold weather; some snakes, however, are active under surprisingly cold conditions, such as Desert Blacksnakes (*Walterinnesia aegyptia*) and European Adders (*Vipera berus*).

Temperature influences the speed at which bodily functions proceed in ectotherms, but extreme temperatures are dangerous for slender creatures, because high surface area/mass ratios dictate rapid heating and cooling rates. Snakes consequently prefer lower body temperatures (averaging around 30°C) than many lizards, and most desert serpents are nocturnal. Body elongation, however, has tremendous advantages: Snakes can rapidly warm and become

False Terciopelo (*Xenodon rabdocephalus*), Costa Rica, flicking its tongue.

Ethology has traditionally addressed four questions about behavior, codified by Nobel laureate Niko Tinbergen: its motivational and sensory control; the roles of genetics, maturation, and experience in shaping final form; its advantage in nature; and its evolutionary history. Gordon M. Burghardt confronted these issues by combining undergraduate training in chemistry, rigorous experiments, and an interest in the behavior of young animals. Using the literature on natural history and some everyday cotton swabs, Burghardt's doctoral research launched a generation of snake ethologists; thirty years later, he is bent on revising Tinbergen's manifesto.

As a graduate student, Burghardt presented naive, young snakes with surface washes from various potential prey, using horse meat and distilled water as experimental controls. Plains Gartersnakes (*Thamnophis radix*) briefly tongue-flicked at extracts of crickets, mice,

and the controls, whereas they rapidly attacked and tried to swallow swabs laced with the odors of leeches, worms, fish, or amphibians—all typical prey for that species. Several other natricine colubrids also preferred extracts that coincided with their natural diets; indeed, baby Queen Snakes (*Regina septemvittata*) reacted only to freshly molted crayfish. Not all species conformed so well to the cotton swab paradigm, but exceptions were still instructive: Wild Butler's Gartersnakes (*T. butleri*) eat only leeches and worms, but neonates also responded to fish and frog odors, so perhaps their chemosensory preferences have not yet diverged from those of the closely related Plains Gartersnake. Corn Snakes (*Elaphe guttata*) and Cottonmouths (*Agkistrodon piscivorus*) scarcely discriminated among the extracts, but both have broad diets that shift with age and thus might be hampered if they had preprogrammed perceptual constraints.

Overall, the feeding behavior of young snakes, evoked by specific stimuli and without the necessity of prior experience, nicely matches classical notions of an instinctive or inborn mechanism.

Other researchers have subsequently extended experimental approaches to many aspects of snake behavior, some discussed elsewhere in this book. Burghardt himself pursued a broad range of problems after graduate school, from chemical characterization of snake prey to behavioral development in hand-reared Black Bears (*Ursus americanus*) and free-living Green Iguanas (*Iguana iguana*). In his teaching and writing he often stressed Jacob von Uexküll's concept of *Umwelt*, or "outer world"—an animal's perceptions of its surroundings—and two projects sharpened his parallel concerns for the counterconcept of a perceptual inner world. Together we studied Eastern Hog-nosed Snakes (*Heterodon platirhinos*), establishing that threatened hatch-

Western Hog-nosed Snake (*Heterodon nasicus*), New Mexico, death-feigning.

lings consistently bluffed, death-feigned, or both; they had individually distinctive and stable behavioral profiles, although the responses could be modified by experience. Our most dramatic finding was that the little snakes would rapidly and repeatedly death-feign or right themselves and crawl, depending on whether a simulated predator (Gordon or a stuffed owl) watched them or averted its gaze. About the same time, Burghardt acquired a two-headed Black Ratsnake (*Elaphe o. obsoleta*), promptly named "IM" for the conflicting concepts of instinct and mind, and accumulated many fascinating observations on its behavior. The two heads often struggled over food, and dominance shifted repeatedly over periods of months; one head consistently preferred smaller prey than the other, although over the years each consumed roughly the same total mass.

After decades of studying reptiles and influenced by Donald R. Griffin,

Burghardt recently has raised a fifth and controversial question for ethologists: What are the private experiences of animals, and what role does an inner perceptual world play in their lives? Griffin, best known for discovering echolocation in bats, argued that mental events such as consciousness and awareness are indicated by surprising yet effective solutions to changing, unforeseen, and uncommon problems. In this regard the observations on snakes are inconclusive yet provide contrasting, tantalizing insights. IM's two heads seemed ludicrously irrational, with no reduction of conflict in the face of their common need to provision the same body! Hog-nosed snakes, however, meet Griffin's criteria rather well: they behave as if aware of their deceptive and dangerous relation to the predator—as if they rapidly, consciously assess the dynamic and alternative roles of bluffing, death-feigning, and escape.

But does a mere serpent have re-

flections and intentions? Can one dare speak of the mind of a snake? Burghardt advocates critical anthropomorphism, combining empirical natural history with our perceptions, intuition, and feelings to predict the outcomes of experiments and comparisons—in short, that we use human experience to forge novel, ultimately testable hypotheses about the inner worlds of animals. His approach steers clear of uncritical caricatures of other creatures as little more than poorly formed humans; in the spirit of "nothing ventured, nothing gained," it also rejects strict but stifling objectivism. Perhaps snakes hold special promise for answering ethology's fifth question, because they challenge us to go beyond the language and nonverbal gestures with which we readily identify mental events in fellow humans.

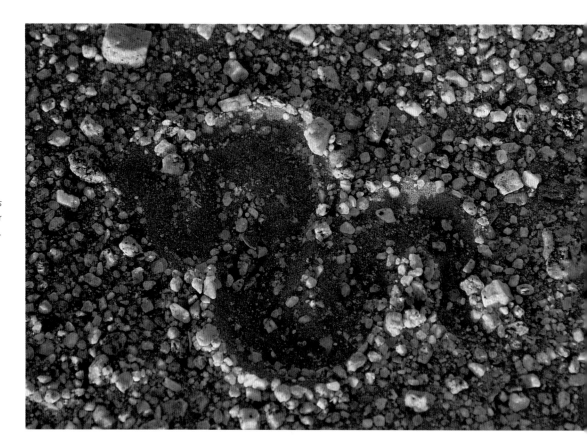

Imprint left by Horned Adder (*Bitis caudalis*) in Namib Desert gravel after basking in late afternoon sun.

Horned Adder (*Bitis caudalis*), southern Africa, drinking from its skin.

active by extending themselves in the sun or on a hot substrate; they can also conserve heat by coiling. Perhaps more important, differentially exposing small portions of the body lets snakes warm particular organs while remaining hidden and therefore safe from predators.

Snakes lose water through respiration, metabolic wastes, and—the more so because of high surface area/mass ratios—their skin. Like most other tetrapods, snakes conserve water by excreting uric acid (a white, semisolid slurry with the feces) instead of urine. Although some undoubtedly visit pools and streams, many species probably drink beads of dew or raindrops from their skin, and Peringuey's Adder (*Bitis peringueyi*) elevates and flattens its neck to condense coastal desert fog for drinking. Some marine snakes excrete excess salt via a gland under the tongue, whereas Salt Marsh Watersnakes (*Nerodia clarkii*) simply leave their estuarine habitats to drink and thus avoid the problem of consuming too much sodium. Desert serpents are more resistant to desiccation than species in humid regions, and some psammophiine colubrids may even condition their skin with nasal gland secretions to retard water loss (Chapter 9). Snakes seek moist, sheltered microhabitats during dry periods and prior to shedding, when they are especially sensitive to water loss.

turning to our camp for shade and water, still resolute, I made a new noose out of stronger material. Another hour later, we crept over the lava and repeated the entire operation, but the snake slipped out when the thick cord failed to close tightly. Game for a third try, Claudia wove a stout but pliant snare out of three strands of floss. Although she succeeded in approaching again, the snake dodged our noose, and this time we conceded defeat.

A few weeks later Claudia brought me a brick-red Coachwhip from Pisgah. As a professional biologist I sometimes remove animals from the wild for teaching and research, but I was uncomfortable about that snake. Now that it was coiled and subdued in a cage, the victim of some vague trophyism on my part, its colors weren't magical. Surely the one that got away was bigger, brighter, and faster, and I soon had the second snake released at its capture site. Looking back on that episode, I realize that something more primal than education or science motivated our breakneck pursuit over the volcanic rocks and sand. For a few hours we were more predators than scholars, engaged again in nature's rough-and-tumble chase. Big brains and long legs didn't carry the day, and we gained an appreciation for snakes far beyond the biomechanics of limbless locomotion.

Deserts inspire humans to look beyond immediate concerns, whether seeking eternal wisdom in rock tablets or simply the solace of wide open spaces. As I walk in the Mojave, looking for snakes and wondering about matters of life and death, small objects sometimes catch my eye: a bleached white skull of a kangaroo rat (*Dipodomys*), clean as wind, protruding from an old owl pellet. In an environment where the sun imposes unusually strict rules, desert creatures remind us of our own limitations—that "perfection" always depends on where you stand and that we are all transients. Back in Berkeley, while I pause for coffee on a gray winter day, incidents guide my memories in unexpected ways: I enjoy imagining that the first Pisgah Coachwhip really was fluorescent pink and two meters long, hoping it still lives in the wildest and most remote parts of the lava flow; I remember that my father was a decent man. He liked deserts too, and I wish I'd asked him why.

Locomotion incorporates support, propulsion, and steering. Each of those functions is strongly dependent on arms and legs in typical tetrapods, and all are accomplished by various amphibians and reptiles with elongate, limbless or near-limbless torsos. Compared with caecilians, amphisbaenians, and elongate lizards (see the Special Topic in this chapter), though, snakes travel with stunning finesse in a wide range of habitats and have speciated widely. Diverse environments confer both restrictions and opportunities on a moving organism, mainly because of the varying effects of gravity and friction; those influences are strongly evident in the details of serpentine travel, as well as in other modifications for particular lifestyles.

This chapter characterizes snake locomotion in relation to structure, ecology, and evolution. Two points warrant emphasis. First, "lateral undulation" and a few other widely used terms scarcely do justice to the locomotor versatility of these animals in nature. Additional movements and postures used in social interactions, hunting, resting, and defense are discussed in chapters on those topics and in the coverage of higher taxa. Second, many snakes do not fit neatly into habitat categories, so identifying particular adaptations for climbing, burrowing, and so forth can be problematic. Many terrestrially active tropical colubrids sleep

Orange-bellied Racer (*Mastigodryas melanolomus*), Costa Rica, sleeping at night on vegetation.

up in vegetation, and certain arboreal species also often crawl on the ground (e.g., neotropical blunt-headed vinesnakes [*Imantodes*]); even undeniably fossorial blindsnakes sometimes ascend trees (Chapter 7). Nevertheless, adaptations for swimming, burrowing, and climbing provide numerous, dramatic examples of convergent structural evolution in snakes. These are superimposed on locomotor mechanisms common to larger groups or even all serpents.

STRUCTURAL AND FUNCTIONAL VERSATILITY

Snake skeletons are simplified in overall design, particularly in reduction or loss of girdles and limbs, but individual bony elements are more complex than those of most other squamates. Elongation has been achieved primarily by increasing the number of trunk vertebrae, although some especially slender species (e.g., Asian vinesnakes [*Ahaetulla*]) also have lengthened the bones themselves. Snake vertebrae are connected by a strong ball-and-socket joint between each one, with two pairs of additional contact points that limit twisting; connecting ligaments further confine vertical and horizontal movements to about 25° between individual vertebrae. With few exceptions (e.g., stiletto snakes [*Atractaspis*]), the vertebral column is not strongly differentiated into regions anterior to the cloaca. There is no sternum, and because none of the long robust ribs join ventrally, the sides may more easily expand for breathing, feeding, reproduction, and defense. Snakes entirely lack an appendicular skeleton, but many basal snakes have remnants of a pelvic girdle and hind limbs. Although usually described as vestigial, the clawlike external legs are in fact used during male-to-male combat and in courtship by some boas and pythons.

As many as twenty repeated sets of discrete muscles on each side of a snake move its vertebrae, ribs, and skin. This bewilderingly complicated trunk structure reflects the need for

Scrub Python (*Morelia amethystina*),
Australia, descending on a tree buttress
by concertina locomotion.

at a constant rate. Rectilinear locomotion is impressive for its extraordinarily slow pace and seemingly instantaneous control, useful for ambushing prey by stealthy approach and avoiding detection by predators: Bushmasters (*Lachesis muta*) and some other stout snakes travel rod-straight, with the onset and cessation of movement almost imperceptible.

Concertina locomotion resembles the progression of an imaginary horizontal inchworm and is easily provoked by placing a snake in a plastic tube slightly wider than its body diameter. Body loops first expand laterally to form points of contact, by pushing against an adjacent substrate (e.g., the walls of a tunnel) or by enveloping it with coils (e.g., a branch); then either the posterior of the snake is pulled forward or the anterior is pushed ahead and new contacts are made. Because the body loops are stationary, momentum is lost with each repetition of

Ruthven's Earthsnake (*Geophis ruthveni*), Costa Rica, with preocular and loreal scales fused to reduce friction during tunneling.

these cycles; thus concertina locomotion is less efficient than lateral undulation. This mode prevails in some fossorial and arboreal snakes, perhaps because their substrates are ill suited for lateral undulation; flyingsnakes (*Chrysopelea*) even launch themselves in a concertina-like fashion, anchoring their posteriors on branches and rapidly straightening the foreparts into the air.

During sidewinding (illustrated on p. 9), two portions of the body are in passing contact with the ground while the remainder accelerates, a form of locomotion that is bewilderingly difficult to comprehend from casual observation. A sidewinding snake's anterior is rolled upward and forward off the substrate, then as that section is lowered and momentarily pauses, the posterior portions follow. Unlike concertina locomotion, sidewinding conserves momentum between contact cycles. Track ways thus consist of many discrete, parallel, roughly straight lines, and travel is at an angle to their long axes. Sidewinders (*Crotalus cerastes*), at least, move with impressive speed, and their tracks sometimes extend uninterrupted for hundreds of meters. Although many snakes can do so if placed on a smooth surface, well-coordinated and readily evoked sidewinding is restricted to the Pacific Viper Boa (*Candoia aspera*), the Dog-faced Watersnake (*Cerberus rynchops*) and a few other Asian colubrids, death adders (*Acanthophis*), and several species of New and Old World vipers. These are short-bodied snakes, less than a meter long, all found in sandy habitats or tropical mud flats. This ecological restriction suggests that sidewinding mainly facilitates locomotion on slick substrates, although it might also minimize contact with hot sand or mud during infrequent daytime movements.

Slide-pushing, striking, and jumping are distinctive but poorly understood locomotor modes. Slide-pushing superficially resembles lateral undulation and takes place on slick surfaces, such that a slipping, wriggling snake progresses slowly at an angle in the direction it faces. Striking is often used to capture prey: blindsnakes, Asian pipesnakes (*Cylindrophis*),

Mountain Patch-nosed Snakes (*Salvadora grahamiae*), Arizona, with rostral scale modified for digging.

and many colubroids simply approach and grasp prey, but boas, pythons, and some colubroids strike by retracting the head and neck before rapidly extending the foreparts and the opening mouth. Striking is similar to concertina locomotion (albeit much faster) in that a snake's anterior extends while the posterior remains anchored to the substrate. Most snakes coil and strike in a roughly horizontal plane, but stiletto snakes (*Atractaspis*) and some elapids fling their heads forward, downward, or backward to bite. Many species project less than half their body length, and most feeding and defensive strikes by pythons, boas, and some vipers, including large rattlesnakes (*Crotalus*), are considerably shorter. Terciopelos (*Bothrops asper*) and other neotropical lanceheads, however, sometimes extend surprisingly long portions of the body during defensive strikes. A blurry photograph published in 1936 even shows a heavy-bodied adult Puff Adder (*Bitis arietans*) in midstrike with only the posterior few centimeters of its body and tail still on the ground, and the much smaller African Horned Adders (*Bitis caudalis*) actually throw themselves into the air during strikes and frantic escape efforts.

HABITAT SPECIALIZATIONS

Snakes are somewhat predisposed to tunneling by virtue of their small girth, and they probably owe their elongation and limb loss to an initially fossorial existence. Although many rather typical-looking serpents are secretive and live beneath surface litter, true burrowers usually have modifications that decrease friction and allow them to penetrate compacted soil. Specialized fossorial snakes often possess a buttressed skull to resist and absorb the forces of tunneling; fused head and body scales (as few as thirteen dorsal rows in reedsnakes [*Calamaria*]) to minimize friction; and a short tail, often tipped by one or more spines and used as a lever for pushing off. Madagascan hog-nosed snakes (*Leioheterodon*), Shield-nosed Cobras (*Aspidelaps scutatus*), and other species with modified rostral scales use their snouts to open

Sonoran Mountain Kingsnake (*Lampro-peltis pyromelana*), Arizona, using its keeled ventral scales to climb by concertina locomotion.

and enlarge tunnels, then scoop soil out in a loop created by retracting the neck laterally. Fossorial serpents are typically less than 30 cm long, although Mole Snakes (*Pseudaspis cana*), Eastern Pinesnakes (*Pituophis melanoleucus*), and a few others exceed 2 m.

Sandy habitats provide special opportunities and problems for fossorial snakes in that those soils are easily penetrable but fail to form tunnels; lateral undulation is difficult because of slippage, at least for moderate to large surface-dwelling species. Shovel-nosed Snakes (*Chionactis occipitalis*) and other small species swim under the sand, while certain desert vipers resort to sidewinding above ground. Snakes that sand-swim have countersunk jaws and valvular nostrils, which prevent ingestion of sand grains; concave bellies in those species maintain an air space and permit them to breathe beneath a shifting overburden. The Arabian Sand Boa (*Eryx jayakari*) and Peringuey's Adder (*Bitis peringueyi*) ambush prey from under dune sand,

Temple Pitviper (*Tropidolaemus wagleri*), southeastern Asia, using its prehensile tail to grip vegetation.

since their dorsally placed eyes and nostrils can emerge while the rest of the head remains covered (photos, pp. 71–72).

Gravity and friction are especially problematic in vertical habitats, the more so when branches and foliage provide irregular, fragile, and discontinuous substrates. Nevertheless, creatures as different as slender Tiger Ratsnakes (*Spilotes pullatus*) and chunky Eyelash Pit-vipers (*Bothriechis schlegelii*) move readily between the ground and canopy. Attenuate climb-ers are often long-tailed and rely largely on lateral undulation, countering gravity by spreading their locomotor muscle mass widely over branches and foliage. Flyingsnakes (*Chrysopelea*) even parachute, their descent slowed by a concavity formed by longitudinal hinges on the ventral scales (Chapters 9, 15). Lacking limbs, climbing snakes rely on friction to grip by parts of the body, rather than suction or adhesion as in frogs and some other creatures with digits.

Terciopelo (*Bothrops asper*), Costa Rica, resting in the crotch of a rain forest tree—an unsettling surprise for night-time naturalists.

Slender arboreal snakes often are loaf-shaped in cross-section, with sharp keels on their ventral scales that press against surface irregularities, whereas some stout-bodied boas, pythons, vipers, and a few arboreal colubrids have prehensile or grasping tails that reduce the chance of falling when they extend their foreparts. Green Tree Pythons (*Morelia viridis*) and Emerald Tree Boas (*Corallus caninus*) drape a tight, symmetrical coil of the entire body over a tree limb during diurnal inactivity (photos, pp. 282–83).

Neotropical snail-eaters (*Dipsas*), Madagascan blunt-headed vinesnakes (*Lycodryas*), and a few other snakes seem especially well designed for bridging gaps in vegetation, their slender foreparts extended outward while the thicker hind end and tail remain attached to a branch. Among those species some have enlarged mid-dorsal scales underlain by dense connective tissue, which perhaps stiffen the body while it bridges gaps in vegetation. Pooling of blood is a potentially serious problem for elongate organisms in vertical environments; in arboreal snakes the heart is closer to the head than in terrestrial species, and various other anatomical and physiological mechanisms also counter the effects of gravity on circulation.

Body undulation converts readily to swimming, and some species in most snake families are further specialized for life in fresh water (the biology of marine elapids is detailed in Chapter 13). Aquatic forms range from stream-dwelling species less than 1 m long to the Green Anaconda (*Eunectes murinus*) at 8 m or more. North American gartersnakes (*Thamnophis*) and many other species enter fresh water only briefly to catch food, keeping their heads well above the surface to see and breathe while they swim. Green Anacondas, South American watersnakes (*Helicops*), and other species that routinely lurk under water have more dorsally placed eyes and nostrils. Many freshwater snakes are semiarboreal, basking in vegetation beside ponds or streams; they often have broad heads, perhaps for swallowing fat frogs and deep-bodied fish. Serpents that spend much time in water are generally confined to tropical climates, presumably because aquatic temperatures elsewhere would be prohibitively low.

THE EVOLUTION OF LIMBLESS LOCOMOTION

Limblessness or near limblessness has evolved more than a dozen times within terrestrial vertebrates (see the Special Topic in this chapter), and snakes are by far the most successful group in terms of number of species and variety of lifestyles. One plausible scenario for their origin is suggested by the fact that many elongate lizards, including several groups with spectacles instead of eyelids, are grass-swimmers. A burrowing or at least secretive, subsurface lifestyle is more likely primitive for serpents, however, based on its shared presence among blindsnakes and all basal alethinophidian lineages; moreover, the earliest snakes probably plowed through moist tropical forest soils, because today most basal alethinophidians and many blindsnakes are restricted to those habitats.

Although phylogenetic comparisons imply that the origin of snakes was associated with burrowing, body elongation might confer several locomotor advantages at any given mass—imagine, for example, how awkward lateral undulation would be for a primate! Slender organisms can enter cracks in or among hard substrates that would be impassable to stout bodies and can more easily crawl through dense vegetation. They are also more efficient tunnelers, because the force necessary for an object to penetrate a substrate is proportional to its cross-sectional area (consider the effort required to drive skinny versus fat nails of the same mass into a wall). And since an elongate animal can distribute its mass over a broader area, it more easily and safely traverses difficult substrates like soft mud, sand, and foliage.

Judging from the locomotor repertoires of blindsnakes, Asian pipesnakes (*Cylindrophis*), and some near-limbless lizards, the earliest snakes used primarily concertina locomotion to traverse crevices and perhaps tunnels formed largely by other organisms. Lateral undulation also is used by those taxa and probably played a role in serpent origins. Rectilinear locomotion seems to depend on muscular specializations and enlarged ventral scales; it evidently arose later, with the advent of macrostomatans. Sidewinding, jumping, parachuting, the use of prehensile tails in climbing, and highly efficient swimming also evolved (sometimes repeatedly) during the diversification of macrostomatans. Basal serpents range from the fossorial, pig-snouted Neotropical Sunbeam Snake (*Loxocemus bicolor*) to slender, arboreal Vine Boas (*Epicrates gracilis*), so the apparent diversity of locomotor modes among advanced snakes might reflect in part that there are many more species of them. Nevertheless, fast-moving, racerlike species are all colubroids.

Serpents turned one potential negative consequence of body elongation to their advantage. An object loses heat proportional to its relative surface area, so a cylinder will gain and lose heat much faster than a sphere of the same mass; perhaps this is why, at least in part, there are no really good mammalian "snakes" (weasels [*Mustela*] come closest; see the Special Topic in this chapter). Animals with attenuate bodies, however, can thermoregulate by protruding a loop from cover into the sun and thereby expose only small portions to predators. Other potential disadvantages of body elongation and their solutions in snakes are discussed in the chapters on feeding and behavior.

Purple Caecilian (*Gymnopis multiplicata*), Costa Rica. One of its two characteristic chemosensory tentacles is visible, rather like a white pimple, on the left side of the animal's face.

Given that the origin of tetrapods involved nothing less than the invention of limbs and walking locomotion by undulatory fish, it is remarkable that the subsequent loss of those features—body elongation and limb reduction—should be so common in terrestrial vertebrate evolution. A brief overview of other elongate tetrapods places the impressive adaptive radiation of snakes in a broader context.

Aïstopods and several other extinct amphibian groups were elongate and limbless, and among salamanders the large, eel-like amphiumas and sirens exhibit extreme limb reduction. Six families, 34 genera, and 154 species of caecilians, however, are the only fully limbless living amphibians. These wormlike, externally segmented creatures range in total length from 11 to 152 cm and live in tropical Africa, Asia, and Latin America. Most caecilians burrow and travel on the surface by lateral undulation and

modified concertina, but one small lineage is aquatic. With eyes reduced and sometimes buried in the skull, caecilians are heavily dependent on a special chemosensory tentacle on each side of the head. More than half of the species are viviparous, and those are highly unusual in that developing offspring use special fetal teeth to evoke the lipid-rich oviductal secretions on which they feed.

Amphisbaenians (from Greek words meaning "to go both ways") are burrowing squamate reptiles, perhaps related to whip-tailed lizards, and include 4 families, 18 genera, and 143 species found in arid and tropical parts of Africa, Eurasia, and the New World. They have quadrangular scales arranged in rings, very strong jaws, and total lengths of 12–72 cm. Amphisbaenians move by lateral undulation, concertina, a unique version of rectilinear locomotion, and jumping. Various modified snouts facilitate

tunneling in completely limbless forms, and three species of mole lizards (*Bipes,* the only amphisbaenians with legs) use their strong forelimbs during locomotion. Amphisbaenians feed on invertebrates, relying on chemical and vibrational cues to locate prey; the giant Yellow Amphisbaenian (*Amphisbaena alba*) follows trails of leaf-cutters to their nests, where it eats large beetles that live on fungus cultivated by the ants.

Body elongation and limb reduction have evolved within other scleroglossan squamate groups as well, including alligator lizards (Anguidae), girdled lizards (Cordylidae), snake-eyed lizards (Dibamidae), skinks (Scincidae), whiptails (Teiidae), and flap-footed lizards (Pygopodidae). Found throughout the world, these serpentine forms range from less than 10 cm to just over 1 m in total length. They are primarily grass-swimmers and surface burrowers, but some inhabit fine sand;

Burton's Flap-foot (*Lialis burtonis*), Australia. Among about 30 species of Australasian pygopodids, this one most resembles snakes, by virtue of its feeding on relatively large lizards; note the external ear opening, never found in true serpents.

most elongate squamates eat large quantities of small invertebrates. Other elongate lineages have not speciated widely, but Australasian pygopodids nicely parallel the serpents and have achieved a modest adaptive radiation. Among 36 species of flap-footed lizards, worm lizards (*Aprasia*) are like blindsnakes in general appearance and eat ant larvae and pupae; at the other extreme, Burton's Flap-footed Lizard (*Lialis burtonis*) resembles a more advanced snake in certain skull features that permit feeding on large skinks.

Variation among and within these groups implies that body elongation always preceded limblessness during the origins of snakelike creatures. Among flapfoots, girdle-tails, whiptails, skinks, and alligator lizards there are species with elongate bodies and four limbs, species with various degrees of limb reduction, and species with at least two limbs externally missing and the other pair highly reduced or absent; a recently discovered fossil caecilian even had fore- and hindlimbs. The structural basis for elongation of the trunk is usually more vertebrae, and evolutionarily limb loss is usually from front to back (sirenid salamanders and some amphisbaenians are exceptions, retaining only forelimbs). Most elongate amphibians and reptiles exhibit visceral simplification, although the precise manner varies: amphisbaenians have lost the right lung, whereas some snakes have lost the left.

Given the evolutionary frequency and success of elongation and limb reduction among squamates and certain amphibian lineages, why have these themes not been explored by other groups of vertebrates? Some specialized body forms are simply incompatible with a serpentine lifestyle, including jumping frogs, flying birds, and turtles. Large surface areas would cause rapid heat gain and loss in endotherms, and in fact among mammals the one that comes closest to a snake in shape lives on the edge physiologically: weasels (*Mustela*) capitalize on their slender builds to enter the burrows of prey, but they must eat very frequently to fuel a marginal lifestyle.

DIET AND FEEDING

An adult Hog-nosed Pitviper (*Porthidium nasutum*) is moderately stout and about 45 cm long when stretched out or 12 cm across in a tight coil; the tip of the snout is distinctly elevated, hence its name in English. Often common in tropical forests, this species occurs at low and moderate elevations from southern Mexico to Ecuador. Its color pattern consists of fifteen to twenty-three irregular dark blotches on a lighter background, and individual markings vary from shades of brown and gray through yellows and copper reds. These small creatures are so extraordinarily cryptic that we joked that their preferred microhabitat—ecological parlance for where an animal lives—is a place and a time: in the middle of a trail, fully exposed, just after three experienced herpetologists have passed by without seeing the snake! Twice I've found a Hog-nosed Pitviper coiled on a fallen tree limb, the proverbial bump on a log, perhaps waiting in ambush for passing rodents.

The first dozen or so Hog-nosed Pitvipers in our Costa Rican predator study had empty stomachs and thus shed little light on the natural history of that species, but on two successive days in 1984, fellow biologists brought in small ones that were bulging with food. I gently squeezed the prey forward and out of their mouths, a fairly routine procedure for studying diets without killing the snakes and only slightly trickier with venomous species. One of the pitvipers had eaten a 17-g Wedge-billed Woodcreeper (*Glyphorhynchus spiralus*), the other a 29-g Spiny Pocket Mouse (*Heteromys desmarestianus*); both meals had been swallowed headfirst. Minus its prey, each snake weighed 23 g. Subsequent observations from La Selva and other Costa Rican localities have fleshed out the natural history of this species. Females give birth to litters of six to nine fully formed snakes, each about 10 cm long. The diet changes with age, as it does in most other vipers. Juvenile Hog-nosed Pitvipers are born fully competent to hunt and kill their prey, mainly frogs and lizards. Adults of this species take mostly mammals, so a bird in the stomach of one of our first snakes was surprising. We also know that among snakes, only vipers can accomplish such truly spectacular feats of ingestion: at 129 percent of the Hog-nosed Pitviper's mass, that Spiny Pocket Mouse was roughly the equivalent of a 209-pound hamburger for me to swallow intact in less than an hour, without the benefit of hands or silverware!

3

Descriptions of color patterns, measurements of newborn young, and dramatic comparisons of prey items are the facts—our abstracted perceptions and evaluations of events in the forest—but I've often wondered about something more elusive: How did that woodcreeper fall victim to the Hog-nosed Pitviper, a strictly terrestrial predator? What was it like for a small snake, so accustomed to hunting lizards and rodents, to be faced with the abrupt prospect of rich rewards from an unlikely direction? How would that incident have played out if we hadn't intervened? In the woodcreeper's "real" world the pitviper must have existed only as vague uneasiness, a watchfulness and readiness to flee while the bird hopped over the tree trunk. Preoccupied with searching for insects among the bark flakes and crevices, propped upright against the gnarled wood on its spiny tail feathers, the woodcreeper kept an eye out for danger on the ground. It saw only leaves in all directions: irregular shades of browns, grays, reds, and pale yellows; here and there an occasionally dark stick or creamy stripe of fungus.

Lost in the monotony of organic cracks and cavities, unseen among the countless black shadows of leaf edges on the forest floor, were a pair of vertical slit pupils, a pair of small nostrils, and a pair of unique infrared imaging pits (for which pitvipers are named). The coordinated turret of senses on the snake's face was aimed at the base of the tree, and as the woodcreeper moved within range, they bombarded the Hog-nosed Pitviper's brain with such a cacophony of information that its body tensed uncontrollably. A warm object, large but not too large, was looming fast! The only outward signs of recognition on the serpent's part were a single, smooth oscillation of the forked, black tongue and an almost imperceptible shifting of the leaves. So fast that it would have been no more than a flicker to the human eye, an S-shaped coil coalesced and struck from the forest floor. The bird flexed both legs, pushed away from the trunk, and began a frantic wing stroke. In that same, terror-filled fraction of a second the pitviper veered in midstrike and snared the feathered thorax with curved, hollow teeth. Both creatures lay still after a brief, awkward struggle, as an amber fluid flowed through the translucent, needlelike fangs. After a couple of minutes the bird's heart failed, its final beats having circulated the venom that would also ensure rapid digestion once the snake had ingested its prey.

Soon the pitviper disengaged its long front teeth, maneuvered to just above the woodcreeper's beak, and began swallowing. For the next half hour, what had been the front end of a snake was transformed into a surrealistic contraption of distensible tissues and movable, toothy struts; with membranous blue skin stretched among the scales of its chin and neck, its jaws working in laborious side-to-side cycles, the Hog-nosed Pitviper literally pulled its head parts onto, around, and over the dead bird. Finally, sinuous body curves moved the protruding lump of a meal from throat to stomach. After yawning widely, the bulging serpent settled against a root, its head surrounded by compact coils, and bird tissue slowly converted to invisible reptile in the endless carpet of leaves. During the next week the fat little snake moved only twice, once when grazed by a windblown twig and the other time when a mosquito's bite caused it to flinch. For a few days, the Hog-nosed Pitviper's six small sensory apertures were attuned only to the unlikely approach of a predator on reptiles.

Snakes have solved a potentially serious problem—how to nourish a heavy body with a small mouth—that could otherwise offset the advantages of serpentine locomotion described in Chapter 2. Lizard-shaped vertebrates generally have wider heads than cylindrical animals of the same mass, so equivalent nutrition of an elongate creature might require more frequent search-and-capture operations, each of them risky and energetically expensive. (An enormous head for a given mass would necessitate a trunk so slender it could not support the giant front end; not surprisingly, no such tadpole-shaped terrestrial organisms exist.) Elongate amphibians and reptiles solve the small-head dilemma in one of three ways: by "eating lots," "eating chunks," or "eating big." Blindsnakes and most limbless lizards simply consume many tiny prey; such predators often contain food—sometimes several hundred items in a single stomach. A diet of small, highly clumped arthropods is feasible because large overall quantities of prey require few search-and-capture operations (e.g., traveling overland between ant colonies). Some caecilians and amphisbaenians shear chunks out of earthworms and other large prey with their exceptionally stout jaws, compensating for missing limbs by anchoring themselves and their victims against tunnel walls. Serpents have explored the third option—infrequent feeding on big prey—with anatomical modifications such that their skulls can essentially walk over and around large objects; perhaps not coincidentally, snakes also exceed other groups of elongate vertebrates in number of species by at least tenfold, and they occupy an unprecedented variety of habitats.

All snakes are carnivorous, but beyond that their diets run the gamut from meringue-like tropical frog eggs to porcupines. Some have very narrow tastes, like quill-snouted snakes (*Xenocalamus*) that consume only one species of amphisbaenian, whereas Coachwhips (*Masticophis flagellum*) and many others might take dozens of types of prey during a lifetime. Most snakes locate relatively large food items one-by-one and eat infrequently. Venomous elapids and vipers can consume other animals exceeding their own mass, and even many relatively unspecialized colubrids commonly eat prey weighing roughly 20 percent of their mass. This chapter describes the means by which snakes accomplish such remarkable feats. It then summarizes their tactics for finding, subduing, and digesting a wide variety of prey and finally explores the major role feeding specializations have played in the adaptive radiation of these limbless reptiles.

THE BASIC FEEDING MECHANISM OF MACROSTOMATE SNAKES

The skulls of many snakes are simplified and loosely constructed compared to those of lizards, encompassing major structural and functional shifts that facilitate swallowing large prey. I will first describe feeding specializations in terms of their extreme deployment in advanced snakes; stages in the early evolution of these mechanisms and subsequent modifications are discussed later, with respect to particular taxa.

Anatomical peculiarities of the lower jaws are primarily responsible for large gape in snakes. First, in other amniotes the two mandibles, or halves of the lower jaw, fuse anteriorly during embryonic development as a fairly rigid, bony joint, or symphysis (easily felt at the front of the human chin as a groove between two bumps). That mandibular symphysis never forms in most snakes. Instead, the two lower jaw tips are connected by ligaments and can move independently, a condition known as mandibular liberation. Flexibility is further enhanced

Hog-nosed Pitviper (*Porthidium nasu-tum*), Costa Rica, yawning with fangs extended. Note the inner rows of teeth in the roof of the mouth, the forward-placed opening to the windpipe, and the separate tips of the lower jaws, all key aspects of a snake's ability to swallow enormous prey.

Black Halloween Snake (*Pliocercus eury-zonus*), Costa Rica, eating a rain frog (*Eleutherodactylus*). The snake's mouth is opening as its throat and trunk muscles pull the frog into the esophagus.

by a joint in the middle of each mandible, which enables each lower jaw to bow outward and adjust to the shape of a prey item. Second, the lower jaws of other squamates are suspended from the skull proper, articulating at joints with quadrate bones that are themselves rather stiffly connected to the rear of the braincase on either side. In most snakes the quadrates are enlarged, slanting backward and outward, and form widely free-swinging struts on each side of the skull to which the lower jaws are independently attached. (It is interesting to note that during mammalian evolution the quadrate lost its role in the jaw joint and became a tiny inner-ear bone, the incus.) The quadrates of most snakes articulate with supratemporal bones on the rear of the skull, and in some species the supratemporals also are movable and serve as extra links in the jaw suspension (see the Special Topic in this chapter).

Most snakes engulf prey with alternating, side-to-side head movements, made possible by mandibular liberation, free-swinging quadrates, and highly mobile upper jaws. Bones of the snout and palate are movably attached to the braincase, rather than firmly articulated with it as in other vertebrates. Sharp, curved teeth are arranged on right and left palatomaxillary arches (the two upper-jaw segments), each arch shaped like a two-pronged fork and composed of four bones that are loosely attached to each other and the braincase. Starting at the rear of the roof of the mouth, near the midline, an elongate pterygoid bone (the handle of the fork) connects forward and along the midline of the head with a shorter palatine bone (the inner prong); together these two bones are called a palatopterygoid bar. Shortly before its junction with the palatine, each pterygoid also connects to the side through an ectopterygoid with the maxilla (together these two latter bones make the outer prong of the fork). Typically each palatomaxillary arch has an inner row of teeth on the pterygoid and palatine bones and an outer row on the maxilla, yielding a total of four longitudinal rows of upper teeth.

Compared to that of lizards, the jaw musculature of snakes is elaborate and elongate, characteristics that allow for increased gape and greater mobility of the tooth-bearing bones. Several pairs of muscles originate on the braincase and insert along the two palatopterygoid bars, enabling each upper-jaw unit to be moved forward and backward as well as up and down by pulling from various directions. During unilateral swallowing, a snake advances all the tooth-bearing bones on one side of its head out, over, and then against a food item, thereby engaging the teeth. As those teeth are applied to the prey, the opposite mandible and palatomaxillary arch disengage, then sweep outward and forward to initiate another cycle of jaw advance and retraction. Consequently, a snake doesn't actually pull prey into its mouth, as do most predators, but rather "walks" its head around and over its prey. Once prey has entered the throat, peristaltic contractions of the esophagus and sinuous curves in the trunk push it backward to the snake's stomach.

Unilateral feeding provides a spectacular example for teaching functional anatomy, because, with some prior explanation and a live snake that feeds readily, students can easily see movements of the quadrates beneath skin at the back of the head as well as alternating, side-to-side cycles of the mandibles and upper-jaw arches. Some of the modifications that permit snakes to swallow large prey can first be visualized with a simple model. Consider your forearms as mandibles, upper arms as quadrates, and torso as a skull. With hands clasped (like a mandibular symphysis) against your chin and elbows locked into your sides (such that the quadrates are built into the skull), swing your forearms down and then back up to their original position, simulating jaw opening and closing in a typical tetrapod. A triangular hole

Common Egg-eater (*Dasypeltis scabra*), southern Africa. The snake is shown first pushing its mouth onto an egg, then passing its head around the egg. Next it moves the egg into its throat, where vertebral spines will crush it. Finally, it ejects shell pieces of the crushed egg after having drained its contents.

A slender brown snake with an unusually viperlike color pattern crawls up into a colony of weaver birds (*Ploceus*), enters one of the dozens of pendulous nests, then swallows an egg three or four times the diameter of its own head. For a few minutes this Common Egg-eater (*Dasypeltis scabra*) is so misshapen that its displaced eyes and head scales, perched high atop the distended throat skin, vaguely recall the cockpit of a jumbo passenger jet. Unlike more typical serpents, the straining egg-eater pushes rather than pulls its head over the egg. As the enormous meal proceeds posteriorly and the snake's head regains normal appearance, its backbone arches over the egg. There is visible tensing of dorsal muscles, a sharp cracking sound, and then front-to-back wriggling movements as the contents of the fractured and compressed egg are squeezed back into the stomach. Eventually the body undulations reverse direction, and the serpent gapes strangely; a slimy, boat-shaped object is ejected from its mouth—crushed pieces of the egg's upper surface still connected by shell membrane like tiny, irregular tiles. First described by the French naturalist Jourdan in 1834, this astonishing spectacle was not well understood until it was studied by Carl Gans, a remarkable engineer-turned-biologist, during the latter half of this century.

Nondeformable and slippery food items pose special problems for ser-

Several attributes allow a *Dasypeltis* to lubricate, stabilize, ingest, and extract the contents of bird eggs. An enormous Harderian gland (behind each eye) perhaps supplies extra fluids to the mouth cavity. The palatomaxillary arches are rigid and solidly connected to the braincase, thus incapable of unilateral movements, and the few tiny teeth play no role in feeding. In most serpents the supratemporals are loosely attached to the skull and movably jointed with the quadrates, but in egg-eaters those two bones are fused on each side as elongate, swinging struts; unusually long mandibles also increase gape. An egg-eater's oral cavity is incredibly expansive, owing to folds of soft tissue so deep they form pockets between the lower jaws and lips when the mouth is closed. The supralabial scales are tightly attached along the upper jaws, but the skin near the back of the mouth is particularly elastic. There is no mental groove, and the enlarged chin scales form a continuous plate across the bottom of the mouth. After swallowing, the egg is crushed against a highly modified backbone: ventral surfaces of roughly the seventeenth to the thirty-eighth vertebrae are thickened and elongated as blunt spines, and those on the eighteenth to the twenty-sixth vertebrae actually pierce the esophagus through loosely bunched folds; nearby dorsal muscles are enlarged for extra compressive force. Those hypapophyses anterior to the most modified spines initially perforate the egg, after which it shifts backward for further

pents, the more so if those objects are relatively large. Common Kingsnakes (*Lampropeltis getula*), Bird Snakes (*Pseustes poecilonotus*), and some other colubrids occasionally eat small bird eggs, although those species possess no specializations for this unusual task. Japanese Ratsnakes (*Elaphe climacophora*) swallow and crush somewhat bigger eggs, aided by spines on the ventral surface of a few anterior vertebrae, then digest them shell and all. Only the six species of African egg-eaters

(and maybe the poorly known Indian Egg-eater [*Elachistodon westermanni*]) engulf relatively enormous eggs, then crush and eject the shells; by internally processing only the contents, egg-eaters increase their food storage capacity by 15–30 percent. The lack of anything comparable on other continents to African and Indian weaver birds in terms of their small size, extreme local abundance, and the accessibility of their nests might explain the unique origin and geographical re-breaking and cutting of the shell membrane. The most modified spines also prevent the shell from passing further posterior.

Snakes typically have sacrificed strength for flexibility in jaw structure, but most species still have sharp teeth, and some inflict painful defensive bites. African egg-eaters, although especially preoccupied and vulnerable during feeding, are almost toothless; instead of using dental weaponry, they thwart predators by closely resembling sympatric, dangerous vipers in appearance and behavior (see p. 226). Many egg-eaters have color patterns like those of saw-scaled vipers (*Echis*; photo, p. 251), and, again like those venomous snakes, they make rasping sounds by rubbing serrated, obliquely keeled scales against each other. Some Common Egg-eaters instead mimic the appearance of Lowland Swamp Adders (*Atheris superciliaris*), whereas still other highland populations of the same species look like Rhombic Night Adders (*Causus rhombeatus*; photo, p. 248).

Precisely how *Dasypeltis* eats eggs was revealed in a series of wide-ranging studies by Carl Gans, a pioneering figure in modern functional and evolutionary morphology. Fascinated by live animals since childhood, while in the military he collaborated with Masamitsu Oshima on a study of egg-eating by Japanese Ratsnakes (also called Awodaicho, or the Blue General). This was followed by a detailed investigation of feeding in African egg-eaters, and shortly thereafter, the onetime aspiring engineer having forsaken a career in boiler design for graduate work, by a Harvard dissertation on systematics of the genus *Dasypeltis* (published in 1959). He also soon analyzed defensive display behavior in egg-eaters and showed that they vary geographically so as to match the color patterns of local venomous snakes. Gans later capitalized on X-ray photography for a critical analysis of the mechanism of shell breakage during egg-eating, and his studies of *Dasypeltis* collectively led to a classic 1961 essay on the

feeding mechanism of advanced snakes.

Carl Gans's research on African egg-eaters has spanned dissection and description, high-speed film analysis, and X-rays of snakes feeding—all made the more insightful by his systematic studies and extensive field experience. Gans's research on those creatures has shed light on the feeding mechanisms of less-specialized serpents and motivated a generation of snake biologists; his investigation of color pattern variation in egg-eaters with respect to sympatric vipers directly inspired the solution of the coral-snake mimic problem (see p. 226). Of course much remains to be learned about egg-eating snakes, and electromyography, radiotelemetry, and other sophisticated techniques will no doubt reveal new twists. Eventually we may learn how the tiny hatchlings obtain food and how such an extremely narrow diet affects seasonal activity, reproduction, and other aspects of the biology of egg-eaters.

delimited by your extended forearms and torso thus determines what size item an animal of that head size could swallow. Now disengage your hands several centimeters for mandibular liberation, extend your forearms about halfway at the elbows, and swing your upper arms out from the shoulders like streptostylic quadrates; together they mark an opening roughly the size of an item that a snake could ingest and far larger than the maximum gape of a lizard with the same head size. This model lacks analogs for the upper jaws, but hand-over-hand movements of your arms simulate unilateral movements of the quadrates and mandibles.

Several other modifications also aid snakes in swallowing especially large prey. Whereas the skull of other squamates is open on its sides, with the upper surface and palate linked by bony struts, downgrowths of bone encase the serpent brain and protect it from struggling prey. Since snakes lack shoulder girdles and a sternum, the ends of their ribs are separated ventrally, and large food items can pass unobstructed. The long right lung stores quantities of air that are only partially reduced by adjacent food items in the stomach; moreover, snakes, because they have low metabolic rates, can wait for up to several minutes between respiratory cycles. The serpent's trachea opens anteriorly, rather than far back in the throat as in most other terrestrial vertebrates, and this enables a snake to breathe even while its mouth is stuffed with food. Snake skin is highly elastic and, all else being equal, those species with more scale rows probably can expand to a greater extent because of the increased amount of skin between their scales.

DIET CHARACTERISTICS AND CHOICE OF PREY

Predators encounter and choose, subdue, ingest, digest, and assimilate items in their diets. Each of those components has costs: while an animal feeds it expends energy, risks injury from the prey or other predators, and cannot perform other activities. Nutritional benefits of a diet must exceed all those costs if a predator is to survive and reproduce. Trade-offs among these cost-benefit components help us understand feeding in snakes—both their choices of particular items and their use of particular foraging tactics.

On a global basis, snake diets encompass an enormous range of prey types: earthworms; slugs and snails; crayfish, crabs, and other crustaceans; cuttlefish and squid; spiders, including big tarantulas; horrifically large centipedes; scorpions; ants, termites, and their eggs, larvae, and pupae; cicadas; grasshoppers and crickets; cockroaches; moth, beetle, and dragonfly larvae; diverse fish, including their buried eggs; caecilians; frogs and salamanders, including their eggs and larvae; turtles and their eggs; crocodilians; birds as different as swallows and egrets, and their eggs; lizards, amphisbaenians, snakes, and their eggs; and mammals as diverse as bats, antelope, and primates, including humans. Among those animals missing from the collective diets of snakes, elephants and other giants are too big for any serpents to eat; tuataras (*Sphenodon*) in snake-free New Zealand and many other kinds of organisms are simply never encountered. The overall biology of most snakes hinges on infrequent feeding on large items, so perhaps many invertebrates are too small for efficient use, and plants are nutritionally inadequate. Millipedes and a few other creatures might be insurmountably noxious, a surprising constraint given the highly toxic frogs and salamanders consumed by some serpents.

No single predator eats the remarkable range of items taken by snakes as a group, and even the total prey taken by a single species can be much broader than that of an individual. Diets of specialized predators can still vary with age or size, between the sexes or among individuals, and seasonally or geographically. For example, Striped Swampsnakes (*Regina alleni*) always

Northern Cat-eyed Snake (*Leptodeira septentrionalis*), Costa Rica, eating eggs of a Red-eyed Treefrog (*Agalychnis callidryas*). Like many other slender colubrids, these snakes cruise through their habitat in search of prey.

eat invertebrates, but juveniles take shrimp and dragonfly nymphs, whereas adults subsist entirely on crayfish. Female Arafura Filesnakes (*Acrochordus arafurae*), Diamond-backed Watersnakes (*Nerodia rhombifer*), and Yellow-lipped Seakraits (*Laticauda colubrina*) reach greater size than males and consequently add to their diets larger fish species not taken by the other sex. From the eastern United States to Mexico, Harlequin Coralsnakes (*Micrurus fulvius*) of all sizes feed primarily on small, secretive serpents, but adults also eat hatchlings of ratsnakes (*Elaphe*) and other larger snake species when appropriately small individuals are available. Food records for the Brown Treesnake (*Boiga irregularis;* see p. 179) consist of about one-third frogs and lizards, one-third birds, and one-third mammals, but that breakdown would not characterize an individual's stomach contents over a short time period anywhere in the species' geographic distribution; juveniles feed entirely on frogs and lizards, whereas adults eat mainly rodents in New Guinea and birds on smaller islands.

Snake prey also vary in size, ranging from tiny ant eggs to pigs and antelope weighing in excess of 50 kg. Beyond the obvious fact that some items are too large to eat, relative rather than absolute prey mass is crucial from the standpoint of the costs of handling and the nutritional value. Individual prey range from less than 1 percent of snake mass in adult Racers (*Coluber constrictor*) eating crickets and Western Aquatic Gartersnakes (*Thamnophis atratus*) gorging on tadpoles, to 137 percent in a Harlequin Coralsnake (*Micrurus fulvius*) that contained a glass lizard (*Ophisaurus*), and almost 156 percent in a small Common Lancehead (*Bothrops atrox*) that ate a whip-tailed lizard (*Cnemidophorus*). Of course, factors other than size also can influence the costs of handling and the nutritional value of prey. Feathers, fur, and scales are all indigestible for snakes, and a slug has only about one-fourth the energy content of a mouse of the same mass.

Insectivorous birds and other animals that must feed frequently on various small, common items have inspired optimal foraging theory, a branch of behavioral ecology that analyzes costs and benefits of feeding on an item-by-item basis. This approach is reasonable when the search costs, the handling costs, and the payoff per prey item are small and similar. For many snakes, however, the rates of prey encounter and the costs of handling are low relative to the predator's energy needs, whereas the payoff for each large food item is high. Thus a wild adult Bushmaster (*Lachesis muta*) might need as few as six to ten rats per year, each roughly 50–70 percent of the viper's mass, to repay the costs of searching for, handling, and digesting those meals. For a Western Terrestrial Gartersnake (*Thamnophis elegans*) the energetic costs of handling a salamander are less than 15 percent of the latter's caloric value, so even if most such prey escape following detection, they are still worth attacking. Perhaps the diets of many serpents are determined by intrinsic constraints (e.g., sensory preferences, gape) and relative availability of different species of prey, rather than by decisions to capture or ignore a particular item. For example, the feeding rule for most New World coralsnakes seems to be simply "If it is an elongate ectothermic vertebrate, small enough in diameter to seize, and does not struggle too much, eat it."

Not surprisingly, larger snakes sometimes eat larger prey than do smaller individuals of the same species, at least in part thanks to greater gape. Some serpents also pass up especially small and large items among acceptable types of prey, and this practice suggests that they do feed strategically on an item-by-item basis. The fact that large Viperine Watersnakes (*Natrix maura*), Beaked Seasnakes (*Enhydrina schistosa*), and Western Diamond-backed Rattlesnakes (*Crotalus atrox*) omit small prey from their diets might imply that larger, more profitable prey is encountered frequently (e.g., by predators on fish) or that camouflage and risk minimization are so important (e.g., for some vipers) that small items are ignored while the snake waits for larger prey. Conversely, large individuals of the Diamond Python (*Morelia s. spilota*), several species of terrestrial colubrids and elapids (e.g., Harlequin Coralsnake [*Micrurus fulvius*]), and some species of vipers also take prey almost as small as that which juveniles eat, presumably because for those adult snakes small prey are nutritious yet cheap to handle, without the additional costs of finding a better item.

Although chemical cues probably predominate in controlling all aspects of feeding for most species of snakes, visual, infrared, and tactile stimuli also can play important roles in foraging decisions. While still at a distance from prey, snakes use chemical and visual cues to hunt in only particular habitats or to follow the odor trails of specific animals; in proximity to a potential prey they use chemical, visual, and infrared cues; and after grasping an item in the jaws, they rely on taste and tactile information. For example, naive, young gartersnakes (*Thamnophis*) and Northern Watersnakes (*Nerodia sipedon*) begin foraging when they encounter water and may investigate fish prey solely on the basis of visual cues; chemical cues from the prey facilitate searching behavior and usually control the final attack and ingestion.

Some feeding preferences reflect tendencies present at birth or hatching, while others are influenced by maturation and experience; they vary enormously among snakes feeding on different prey and in different ecological contexts. At a certain body size, Plain-bellied Watersnakes (*Nerodia erythrogaster*) switch from preferring odors of fish to preferring those of frogs; that they do so regardless of experience suggests that an age-related shift is preprogrammed. Some adult Common Gartersnakes (*Thamnophis sirtalis*) attack frog extracts regardless of their experimental juvenile diet, so preferences also can be stable with aging. A species with a rather broad diet, the Plains Gartersnake (*T. radix*) increases its responsiveness to worms when raised exclusively on them; conversely, Butler's Gartersnake (*T. butleri*) specializes on worms, and its predilection for them is not diminished by an artificial juvenile diet of fish. Congenital differences in prey preferences among conspecific populations and between Common Gartersnakes and Western Terrestrial Gartersnakes (*T. elegans*) are consistent with geographic variation in their respective diets.

FINDING, CAPTURING, INGESTING, AND DIGESTING PREY

Predators often are portrayed as either "widely foraging" over large areas for appropriate victims or "sitting and waiting" in ambush for passing prey. The first of those two hunting modes is supposedly typical of slender, sometimes fast-moving colubrids and elapids; the second is said to characterize heavy-bodied pythons, boas, and vipers. Like many simple dichotomies, this one contains elements of truth but conceals important variation; it probably adequately describes the behavior of very few snakes.

Blindsnakes, Neotropical Sunbeam Snakes (*Loxocemus bicolor*), many colubrids, and most elapids of slender to moderately stout build do forage widely; they use chemical and visual cues to locate prey outright or to restrict their attention to specific microhabitats. Gopher Snakes (*Pituophis catenifer*) harvest nestling mammals, and Common Kingsnakes (*Lampropeltis getula*) sometimes dig up turtle eggs; individuals of both species travel considerable distances among clusters of sessile prey. Green Parrotsnakes (*Leptophis ahaetulla*) cruise through tropical forest, methodically exploring bromeliads that might shelter treefrogs, and Harlequin Coralsnakes (*Micrurus fulvius*) probe leaf litter for the burrowing snakes on which they feed. Common Gartersnakes (*Thamnophis sirtalis*) distinguish between the odors of recent and old earthworm feces to concentrate their hunting around the former, and Blackheaded Seasnakes (*Hydrophis melanocephalus*) swim quickly past inappropriate habitats while looking for sand-dwelling eels. Searchers do travel long distances, averaging about 185 m/day for Coachwhips (*Masticophis flagellum*) and up to 400 m/day for some terrestrial Australian elapids; an Olive Seasnake (*Aipysurus laevis*) moved almost 500 m during a single 26-minute submergence! Nonetheless, other attenuate serpents use ambush tactics. Orange-bellied Racers (*Mastigodryas melanolomus*) wait next to sunny spots on tropical forest trails, then seize tropical whiptails (*Ameiva*) when those lizards pause to bask. Green Mambas (*Dendroaspis angusticeps*) travel 5–20 m among trees, then remain at one site for days until they catch a bird or a mammal.

Most heavy-bodied boas, pythons, and vipers probably are less active than many colubrids and elapids; they typically strike prey from concealment, using chemical cues to choose good ambush sites (see p. 134 and the introductory essay to Chapter 8). Even the most sedentary snakes must occasionally forage widely for a good place to sit and wait, however, and they are thus more accurately called "mobile ambushers." Bushmasters (*Lachesis muta*) travel 5–20 m or more between hunting sites, next to trails and rodent runways, and remain at each location for days or even weeks. A Boa Constrictor (*Boa constrictor*) might crawl 20–85 m through tropical forest, enter the hole of a large mammal, and then coil there for two to four days, alertly facing the entrance. Sidewinders (*Crotalus cerastes*) move up to several hundred meters

Above: Ringed Slug-eater (*Sibon annulata*), Costa Rica, evaluating prospective prey.

Brown Blunt-headed Vinesnake (*Imantodes cenchoa*), Costa Rica, seizing an Understory Anole (*Anolis limifrons*). This nocturnal species methodically flicks its tongue over vegetation in search of sleeping lizards.

Terciopelo (*Bothrops asper*; lower left in photo), Costa Rica, in typical viper hunting posture, beside a rain forest trail.

Terciopelo (*Bothrops asper*), Costa Rica, swallowing a rain frog (*Eleutherodactylus*). This young viper still has a yellow tail tip, perhaps used as a lure, and has not entirely uncoiled from its ambush posture.

Calico Snake (*Oxyrhopus petola*), Costa Rica, constricting a Rain Forest Whip-tailed Lizard (*Ameiva festiva*).

each day, intermittently pausing to hunt lizards next to burrows and bushes. Some Striped Racers (*Elaphe taeniura*), Puerto Rican Boas (*Epicrates inornatus*), and Children's Pythons (*Antaresia childreni*) linger for months near cave entrances to capture flying bats. Several boas, pythons, vipers, and a few species of colubrids (Puerto Rican Racer [*Alsophis portoricensis*], the Brazilian *Tropidodryas striaticeps*) and elapids (death adders [*Acanthophis*]) enhance the efficiency of ambush tactics by wiggling their contrastingly colored, grublike tails to lure frogs and lizards into striking range.

Although all pythons, boas, and vipers perhaps hunt mainly by ambush, there are exceptions. Nocturnal Vine Boas (*Epicrates gracilis*) find diurnal lizards while crawling through vegetation on which their prey sleeps. Copperheads (*Agkistrodon contortrix*) sometimes encounter frogs by swimming along shorelines, and presumably those pitvipers locate cicada nymphs, moth pupae, and other insects in their diets by chemosensory searching.

Categorizing the overall feeding strategies of snakes is made even more difficult by their diverse techniques for capturing prey. Blindsnakes, Asian pipesnakes (*Cylindrophis*), and most colubrids and elapids simply approach and grasp food with their jaws. Boas, pythons, and vipers usually retract the head and neck, then rapidly strike their victims from ambush. Many serpents swallow small items that are alive and struggling, but they immobilize larger animals first by constriction (see Chapter 8) or venom injection. Among venomous snakes, colubrids and many elapids typically retain an initial grip until struggling ceases, whereas most vipers bite rapidly, release, and then relocate prey after it dies. Actual pursuit of prey by snakes is rare and probably confined to certain slender colubrids (e.g., some sandsnakes [*Psammophis*] chase elephant shrews, Green Parrotsnakes [*Leptophis ahaetulla*] chase treefrogs) and elapids (e.g., Australian whipsnakes [*Demansia*] chase lizards); failure rates for such chases are often high—about 88 percent for Four-striped Ratsnakes (*Elaphe quadrivirgata*) pursuing lizards and almost 100 percent for Western Ribbonsnakes (*Thamnophis proximus*) chasing frogs.

Now we can return to the supposedly dichotomous relationship between foraging modes and body shapes—widely searching by slender snakes and sit-and-wait hunting by heavy-bodied species. For all snakes, the likelihood of capture increases and predatory behavior changes when some prey characteristic (e.g., odor, microhabitat) is perceived—even though searching might then either continue or cease. Many serpents attempt to catch nearby prey soon after its detection; some come across an odor trail or other evidence of a particular animal, then search intensively until they find the animal; and others locate chemical traces of prey and then wait for it to pass by. Under the traditional dichotomy, the first and second of those three hunting modes would be lumped under widely foraging, and only the third qualifies as sitting and waiting. The second and third modes involve additional search time after the prey is detected, but the second entails continued movement before encountering it, and the third sometimes requires additional searching after the prey has been envenomed. Rather than just with overall feeding strategies, body shape in snakes is variously correlated with locomotor mode (itself related to habitat), defense mechanisms, and tactics of prey capture. Slender snakes undulate effectively and often flee when danger approaches, but they cannot be categorized as a group in terms of foraging biology. Heavy-bodied snakes are prone to rectilinear or concertina travel; as mobile ambushers, they rely on cryptic coloration and will pass up certain prey rather than risk detection by predators for a skimpy meal.

Coachwhips (*Masticophis flagellum*) and Sidewinders (*Crotalus cerastes*) in the Mojave Desert illustrate some correlates of radically different habits. The fast-moving colubrid is on the surface for about half as much time each day, spends about 2.6 times as much energy on activity, and eats about twice as much as the rattlesnake—although both species gain roughly the same net energy over the course of a year. Serpents that vary their behavior eventually might provide special insights into the costs and benefits of alternative foraging activities. Four-striped Ratsnakes (*Elaphe quadrivirgata*) are typically diurnal, wide-searching predators on lizards and rodents, but they may also lie on tree limbs in ambush for frogs in nocturnal breeding aggregations. Juvenile Viperine Watersnakes (*Natrix maura*) cruise for earthworms and tadpoles, whereas adults lie in wait for fish. Black-necked Gartersnakes (*Thamnophis cyrtopsis*) even change tactics seasonally for the same prey species; during spring they hunt along stream banks for clustered Lowland Leopard Frogs (*Rana yavapaiensis*), but in summer they ambush the same, by then more widely dispersed, prey from submerged algal mats. Some Black Ratsnakes (*E. o. obsoleta*) switch from widely searching for arboreal bird nests to seasonally residing in Bank Swallow (*Riparia riparia*) colonies; these individuals are 15–20 percent longer than those that do not switch, their extra growth an indirect measure of what search costs would have been for a more typical, less sedentary lifestyle.

Ingestion of relatively small prey is straightforward, whereas large or otherwise difficult items often require some maneuvering prior to swallowing. Constricting snakes sometimes stretch birds and mammals in their coils; some species push prey against the substrate during ingestion; and many snakes briefly pull prey backward and thereby arrange it linearly. Vertebrates and some invertebrates have protruding body parts that, if left unfolded, might impede passage into a snake's gut, and thus they often are eaten headfirst. Harlequin Coralsnakes (*Micrurus fulvius*) and King Cobras (*Ophiophagus hannah*) retain their hold until venom

Double-banded Coralsnake Mimic (*Erythrolamprus bizonus*), Costa Rica, swallowing a Northern Cat-eyed Snake (*Leptodeira septentrionalis*) tail-first.

causes a prey snake to cease moving; then they use the feel of its overlapping ventral scales to "jaw-walk" toward its front end. Conversely, some coralsnake mimics (*Erythrolamprus*) seize prey snakes as they crawl away and thereby ingest them tail-first. Neonate ratsnakes (*Elaphe*) and kingsnakes (*Lampropeltis*) learn by trial and error that swallowing a mouse pup from most directions is impossible; perhaps they use the lie of the hair, the shape of the snout, the texture of the underlying skull, and other cues to guide headfirst ingestion. Juvenile Striped Swampsnakes (*Regina alleni*) swallow dragonfly naiads headfirst, so as to direct backward the spines, hooks, and other protuberances on those insects; adults of that species ingest crayfish tail-first, thereby folding the crustaceans' appendages and immobilizing their dangerous claws. Among extreme examples of pre-ingestive food manipulation, brownsnakes (*Storeria*) wedge snails under rocks, then twist the mollusks free of their shells, and White-bellied Mangrove Snakes (*Fordonia leucobalia*) actually twist the legs off large crabs before eating them.

Effective digestion necessitates diverse behavioral, anatomical, and physiological specializations—all influenced by particular prey types and lifestyles. Large food items impede movement and hinder locomotor escape from predators, so satiated snakes often retire to hidden refuges to reduce the likelihood of meeting a predator. Some species attain higher temperatures and thereby speed the processing of food, either by choosing warm microhabitats (e.g., a Common Kingsnake [*Lampropeltis getula*] under a flat rock exposed to the sun) or by protruding a body coil containing the prey into sunlight. Some arboreal snakes feed and defecate

Fire-bellied Snake (*Liophis epinephalus*), Costa Rica, swallowing a Harlequin Frog (*Atelopus varius*). The snake is evidently immune to the prey's highly toxic skin secretions.

more frequently than terrestrial species, lightening their load for vertical transport over flimsy vegetation. New World coralsnakes accommodate freshly ingested serpents in an especially elongate esophagus; some cobras and North American hog-nosed snakes (*Heterodon*) might help counter the poisonous skin of toads they eat with their enlarged adrenal glands. Most vertebrates maintain their stomach and intestine in more or less continuous readiness for feeding, but Asian Rock Pythons (*Python molurus*), Sidewinders (*Crotalus cerastes*), and perhaps other snakes shrink their guts and turn off digestive functions between large, infrequent meals.

Energy needs vary with body size, temperature, and behavior; activity costs are relatively higher for small snakes, and energy needs are increased by growth in young snakes and by production of eggs and young in females. Generally snakes must ingest about six to thirty meals per year, totaling 55–300 percent of their body mass, to satisfy overall metabolic requirements. One meal of 40 percent of its mass could sustain a viper for several months, and many snakes probably do not eat regularly throughout their active periods; gravid females of some species might not feed for months. Conversely, many snakes forage frequently when prey is highly available; some Common Gartersnakes (*Thamnophis sirtalis*) catch up to fourteen earthworms per hour and might forage for several hours each day. Despite much variation in lifestyles, however, all snakes contrast dramatically with small endotherms: for a viper the capture and consumption of a single prey might take hours, days, or even weeks, whereas a hummingbird decides in seconds whether to continue withdrawing nectar from one flower or move on to another. These differences in metabolic machinery and foraging biology have profound consequences in terms of predation risks, reproduction, activity cycles, and distributional limits.

Eyelash Pitviper (*Bothriechis schlegelii*), Costa Rica, striking from ambush at a Rufous-tailed Hummingbird (*Amazilia tzacatl*).

THE EVOLUTION OF GENERALISTS AND SPECIALISTS

Although snakes are notoriously specialized hunters, some have surprisingly broad diets. Racers (*Coluber constrictor*), for example, take insects, frogs, lizards, other snakes, nestling birds, and small mammals. Despite their reputation as rattlesnake killers, Common Kingsnakes (*Lampropeltis getula*) also consume turtle and bird eggs, lizards, other colubrids, and nestling rodents. Eyelash Pitvipers (*Bothriechis schlegelii*) eat frogs, lizards, birds, mouse opossums, bats, and rodents. Of course, dietary variation can reflect individual preferences, but multiple items in a single stomach (as found, e.g., in Indigo Snakes [*Drymarchon corais*]; see p. 196) prove that some individual snakes consume a range of prey types. Body size also affects the breadth of the diet, since larger snakes might encounter more kinds of prey by searching over greater distances and subdue a greater range of items. In general, serpents with especially broad diets either eat small items, so as to minimize the impact of toxins, teeth, and other defensive qualities of the prey, or are venomous and thus can immobilize a wide variety of prey types, regardless of size. Species of snakes with narrow diets perhaps typically evolve from populations with broad diets that encompass those of the specialists (e.g., Butler's Gartersnake [*Thamnophis butleri*], see above; Bushmasters [*Lachesis muta*], Chapter 14).

Many snakes are dietary specialists, and some North American desert species illustrate the importance of behavioral and ecological contexts for understanding particular feeding adaptations. Western Whip-tailed Lizards (*Cnemidophorus tigris*) forage widely by day and retire to shallow burrows at night. Coachwhips (*Masticophis flagellum*) are large, slender diurnal snakes with metabolic specializations that permit them to chase and catch the relatively small, active whiptails. In contrast, Long-nosed Snakes (*Rhinocheilus lecontei*) are powerful constrictors with a pointed rostral scale that facilitates digging; they extract sleeping lizards from their holes. Sidewinders (*Crotalus cerastes*) are slow-moving and lack cranial specializations for burrowing; instead, those small rattlesnakes possess immobilizing, tissue-destructive venom and catch relatively enormous whiptails by striking them from ambush during the day. Simply knowing that Western Whip-tailed Lizards are major diet items for each of those predators belies the differences in how those snakes interact with their prey and thus fails to shed light on their feeding adaptations.

A few animals are so formidable that, regardless of habitat and relative size, only predators with behavioral or morphological counteradaptations can safely or reliably incorporate them in their diets. Brown Watersnakes (*Nerodia taxispilota*) ingest catfish headfirst, wait for the prey to relax its erect pectoral and dorsal spines during struggling, and then rapidly fold them by further swallowing movements. Centipedes would be ideal, elongate-shaped snake food, except that they possess powerful venoms and sometimes even eat snakes; worldwide, only three groups of serpents frequently consume those arthropods. Most species of New World black-headed snakes (*Tantilla*) and of African centipede-eaters (*Aparallactus*) rapidly immobilize centipedes with venom and are immune to bites from their victims (photo, p. 81). Juveniles of more than a dozen species of vipers occasionally eat centipedes weighing up to 50 percent or more of their own mass, after having minimized the likelihood of a retaliatory bite from the prey by using the strike-and-release envenomization typical of those snakes.

The immobilization of lizards of the family Scincidae poses a different problem, and few species of predators eat them regularly. Skinks have reduced limbs and strong trunk musculature, and their capacity for wriggling free is enhanced further by shiny, hard scales. Raptors and unspecialized serpents have difficulty holding such slippery, squirming prey, which simply rolls out of beaks and off rows of teeth. With more than a thousand species globally, skinks are especially diverse in the Old World, and several Asian and African colubrids (e.g., mock vipers [*Psammodynastes*] and wolfsnakes [*Lycodon*], respectively) are modified for eating them. Enlarged anterior maxillary and mandibular teeth and bow-shaped jaw bones of those snakes create a gap in the mouth (like bottle tongs) within which skinks are trapped crosswise (photo, p. 180); most such skink specialists also are rear-fanged and immobilize their prey with venom. In a similar fashion, movable joints in the upper jaws of Round Island boas actively tilt their large front teeth downward and entrap a struggling lizard. Still another mechanism in Asian sunbeam snakes (*Xenopeltis*) and some colubrids (e.g., the Neck-banded Snake [*Scaphiodontophis annulatus*]) involves equal-sized teeth that are attached only by ligamentous hinges on their posterior edges. Such folding teeth allow prey to pass into the esophagus but lock into place when a skink struggles backward; swallowing is exceptionally rapid, so perhaps the lizards actually crawl away from the ratchetlike teeth and down the snake's gullet!

BROAD EVOLUTIONARY PATTERNS

Despite much variation in feeding biology among individuals and species of snakes, several clear patterns are discernible among more inclusive phylogenetic groupings. Two different

Peringuey's Adder (*Bitis peringueyi*),
Namibia, luring with tail.

meanings of "large prey" are central to understanding these trends, because they bear on the problems posed by particular types of prey. An item can be large in mass compared to the predator's mass, large in diameter compared to the predator's head, large in both respects, or small in both measures of relative size. Relative prey mass and prey diameter make different but accurate predictions about how particular serpents are built and how they behave, a correlation suggesting that those two factors have played important roles in the evolution of snake feeding biology.

All else being equal, a heavier animal struggles more effectively than a lighter one, so predators that eat heavier prey should have specializations for subduing their victims. Some birds (e.g., Harris Hawks [*Parabuteo unicinctus*]) and mammals (e.g., African Lions [*Panthera leo*]) overwhelm relatively heavy prey by cooperative hunting, whereas snakes use constriction or venoms, or both, to immobilize dangerous adversaries. Items that are large in diameter still

Peringuey's Adder (*Bitis peringueyi*), Namibia, swallowing a Sand Lizard (*Meroles anchietae*).

must pass intact into the mouth, so snakes that eat bulky prey require modifications for increased gape, usually especially long quadrates and mandibles (see the Special Topic in this chapter).

With two measures of prey size (relative mass and relative girth), each varying from small to large, we can characterize four feeding types in terms of those extremes. For most lizards and blindsnakes, lightweight prey require no immobilization and are so small in diameter that they can be swallowed with ease regardless of variation in shape (Type I). The two measures interact, so snakes that eat heavy prey fall into two extremes with respect to the shape of their prey: those that take heavy, elongate items are constrictors or venomous but do not require large gapes (Type II; e.g., photo, p. 67), whereas those that eat heavy, bulky prey have specializations for immobilization and for gape (Type III; e.g., photo, p. 50). Snakes that eat prey that are not circular in cross-section (such as many fish) or otherwise have low densities for their diameter (such as birds, because of their feathers) require gape specializations but not venom or constriction (Type IV). Snakes that eat lightweight prey (Types I and IV) must feed more frequently than those who take heavy items (Types II and III) to obtain equivalent nutrition.

Most lizards (including elongate, limbless species) and almost all blindsnakes eat tiny insects, a fact that suggests that the earliest snakes also were Type I predators. Asian pipesnakes (*Cylindrophis*) and most other basal alethinophidians constrict elongate, heavy vertebrates

(Type II), thus mirroring an ancestral shift from frequent feeding on small prey to infrequent predation on heavy items, but without major changes to facilitate greater gape. Pythons and other basal macrostomatans have more expansive jaws than basal alethinophidians, such that the latter can swallow objects that are both heavy and bulky (Type III). Advanced snakes have an even more flexible jaw mechanism, so that inner rows of teeth on the palatomaxillary arches are primarily responsible for moving the mouth over prey; as a result, outer maxillary bones are available for the evolution of venom injection, enlarged teeth for snaring skinks, and various other specializations (encompassing Types I–IV). Historical implications of these feeding modes are discussed further in later chapters.

VENOMOUS SNAKES AND SNAKEBITE

4

Asked to name the most dangerous snake, herpetologists mention several candidates. Drop for toxic drop, the Inland Taipan (*Oxyuranus microlepidotus*) wins hands down: a bite from this Australian cobra relative contains enough venom to kill two hundred thousand mice—more than twice the punch of the closely related Taipan (*O. scutellatus*) and about seventy-five times that of an Eastern Diamond-backed Rattlesnake (*Crotalus adamanteus*). Of course taipans and large pitvipers are undeniably deadly, but simple comparisons are misleading because toxicity also varies from snake to snake and victim to victim—and effects on mice don't always translate to lethality in humans. As to which species can quickly place one in grave jeopardy, bites from Black Mambas (*Dendroaspis polylepis*) and King Cobras (*Ophiophagus hannah*) can be life-threatening within minutes. Which snake causes the most accidents varies with local abundance and habits as well as human behavior; Copperheads (*Agkistrodon contortrix*) are common in the eastern United States and difficult to see, although the many bites from this small species virtually never result in deaths. Which snake kills the most people is uncertain because many fatalities probably go unreported, but Asian Saw-scaled Vipers (*Echis carinatus*), cobras, and neotropical lanceheads (*Bothrops*) are major contenders. Those questions don't fully address the matter of danger, nor do they acknowledge that humans are at least largely irrelevant to the evolution of differences among venomous snakes.

Having worked mainly in western North America, where rattlesnakes are easily avoided and survival from snakebite is likely, I long believed that only Polar Bears (*Ursus maritimus*), Grizzlies (*U. horribilis*), and an occasional large crocodilian were really dangerous to New World field biologists. Some rain forest pitvipers, however, convinced me that they too might pose a serious threat to life and limb, and tropical snakes have inspired much of what we know about venoms and their therapeutic implications. The Terciopelo (*Bothrops asper*; its common name means "velvet" in Spanish) ranges from Mexico to South America and plays a distinctive role in local lore and economy. This species often is called Fer-de-Lance (French for "iron spear point," referring to the shape of its head), but I prefer the name widely used in Central America, which alludes to the lush-textured skin of large black-and-gold females. My visceral reactions were inspired by viewing the enormous fangs and copious venom flow when I handled these vipers, and by a close friend's bite from a previously unseen Terciopelo. Regarding the gruesome snakebite sequelae for this species, Clodomiro Picado, Costa Rica's first academic biologist and a pioneering expert on snakebites, introduced a 1931 book this way:

Moments after being bitten, the man feels a live fire germinating in the wound, as if red-hot tongs contorted his flesh; that which was mortified enlarges to monstrosity, and lividness invades him. The unfortunate victim witnesses his body becoming a corpse piece by piece; a chill of death invades all his being, and soon bloody threads fall from his gums; and his eyes, without intending to, will also cry blood, until, beaten by suffering and anguish, he loses the sense of reality. If we then ask the unlucky man something, he may still see us through blurred eyes, but we get no response; and perhaps a final sweat of red pearls or a mouthful of blackish blood warns of impending death . . . Those who read on in these pages will see how science is able to triumph over such desolation.

Born in 1887, Clodomiro Picado graduated from the Sorbonne in 1913, published 131 scientific papers and many popular articles on topics as diverse as bromeliads and culture, and died by his own hand in 1944. I have no hint as to whether he found snakes themselves interesting, but certainly at that time snakebite was a major public health problem in Latin America and inspired government support for science. Today the Instituto Clodomiro Picado is a major research center and produces high-quality antivenom, making Terciopelos effectively a cash crop for Costa Rica through exports of life-saving serum to other Central American countries. Similar institutions were founded in various other countries (Instituto Butantan in Brazil is among the oldest), and today, as a result of their efforts, many fewer deaths result from snakebites. The spin-off benefits of neotropical snake venom research have been far-reaching and often unexpected; captopril (Capoten), a billion-dollar drug for high blood pressure with few side effects, was inspired by a component in the venom of Jararacas (*Bothrops jararaca*).

Although Bushmasters (*Lachesis muta*) and Terciopelos are found in some of the same areas and both are extraordinarily cryptic in the wild, those two great pitvipers could scarcely be more different in many respects. For all their fame, Bushmasters are rarely seen, whereas almost everyone who works in the fields and forests of lowland Costa Rica sees Terciopelos. Bushmasters typically lay a few large eggs, and Terciopelos give birth to dozens of small young; Bushmasters feed on rodents throughout their lives, but Terciopelos switch from eating mainly frogs and lizards as juveniles to mammals as adults. Bushmasters are confined to primary forests and rarely strike at passersby, whereas Terciopelos are common in disturbed habitats and are perhaps more prone to bite defensively.

Up close and personal, Terciopelos at La Selva seemed unpredictable and nervous, at times almost explosive. The first large one I encountered turned back in a striking coil when its escape was blocked with a snakehook—an anxious moment, because by then I was knee-deep in swamp mud. Another shot between my legs as I attempted to catch it, and twice we saw escaping Terciopelos reverse direction under palm leaves, as if to ambush a pursuer. An adult took refuge in a hole on a riverbank, pointing out like a cannon from its vegetative ramparts. Even a 2-m pitviper, however, appears tranquil and almost benign through binoculars or a long camera lens. Thus protected by distance, for weeks our research team monitored a pregnant Terciopelo basking on a tree-fall and were rewarded by a rare sight, the snake drinking rain droplets off her own skin. We worked closely around another big female, unaware for several minutes of a smaller male lying next to her, and neither snake so much as twitched. In point of fact, both Bushmasters and Terciopelos avoid contact with humans; those at La Selva usually remained still or glided off when we approached them closely.

Injury and death influence our cultural traditions and individual feelings about snakes. Because so many campesinos suffered horribly from snakebite, Clodomiro Picado railed in vivid prose against "treacherous" pitvipers for robbing his countrymen of even the chance to struggle for survival. Sixty-five years later, despite the electric caution Terciopelos inspire (and perhaps also because of it), bites to biologists are actually exceedingly rare (about one per half million researcher-hours in the field); among nine recent cases, seven were serious, but there were no fatalities. With a realistic notion of danger, our attention turns to scientifically interesting, as yet unanswered questions. Do Inland Taipans and Eastern Diamond-backed Rattlesnakes really have such powerful venoms only for feeding on a mere rodent now and then, or might defense also have influenced the evolution of their toxins? Could temperaments differ between the two large Costa Rican pitvipers because Terciopelos have pursued a riskier lifestyle, consistent with their colonizing abilities in open, weedy habitats? Rational perspectives will further reduce the risk of snakebite and elucidate the ecological roles of venoms, but I'll bet laypeople will always wonder what is the most dangerous snake. So will some biologists.

Biological toxins are substances with deleterious effects on other organisms, through chemical damage to living tissues; toxins produced in specialized glands and injected into other organisms by spurs, stingers, spines, or teeth are called venoms. Among squamate reptiles, toxic glandular secretions are restricted to beaded lizards (*Heloderma*) and advanced snakes, and within that latter group perhaps half or more have venoms. Among the Colubroidea, roughly 250–500 species cause serious human accidents, but about 50 species produce the majority of problems. Bites from venomous snakes number several hundred thousand a year or more on a global basis, of which perhaps more than twenty thousand result in deaths; conversely, snake venoms are rich in substances with potentially beneficial medical applications. As a result of their various positive and negative roles in human welfare, venoms are the subject of extensive biochemical and pharmacological research. The snakes themselves exhibit complex adaptations for dramatic ecological roles and thus are of great interest to biologists.

The evolution of venoms was perhaps inevitable as snakes acquired the ability to ingest ever-larger prey, because at some point they would have been able to swallow other organisms too powerful to subdue by brute force and too bulky to digest. Evidently social hunting has not been feasible for serpents, perhaps because other aspects of their biology do not favor cooperative behavior and because ingesting intact items necessarily prevents sharing food. Until about two decades ago herpetologists believed that because most snakes are harmless to humans those species are also nonvenomous, and that venom injection mechanisms evolved as a simple progression with increasing efficiency in one clade. Now we know that many advanced snakes are venomous, that the evolution of sophisticated dentitions and toxic secretions was complex, and that venoms perhaps played important roles in the overall radiation of colubroids. This chapter describes the anatomical and functional bases for venom injection, some ways that snakebite affects human welfare, and the evolutionary and ecological roles of venoms in snake biology.

FUNCTIONAL ANATOMY OF VENOM INJECTION MECHANISMS

Our understanding of how venomous snakes harm their victims emerged in the seventeenth and eighteenth centuries, during a general blossoming of science and medicine. In 1664 Francisco Redi, physician to the duke of Tuscany, proved that head gland secretions of European Adders (*Vipera berus*) caused the effects of snakebite; his careful experiments contradicted a then-prevalent view that harm stemmed from "enraged spirits" of the serpents. Other Europeans also studied snakebite, and by 1727 Captain G. Hall of colonial South Carolina demonstrated that a rattlesnake exhausts its venom supply with successive bites. In 1887 Henry Sewall at the University of Michigan, following closely on Louis Pasteur's success with a rabies vaccine, immunized pigeons against pitviper venom. By the latter half of the nineteenth century, snake venom research, most of it focused on biomedical issues, flourished in Brazil, Europe, India, and the United States. Marie Phisalix, a French scientist, published the first masterful synthesis of venomous animals and their biology in 1922, and since then tens of thousands of articles have been published on the subject.

The venom injection mechanisms of snakes closely parallel a hypodermic syringe in basic design. The syringe has a storage vessel for the fluid to be injected (the barrel), a pump (the plunger), and a delivery channel (the needle); the respective analogs for venom injection in snakes are oral glands for manufacture and storage, jaw musculature for supplying pressure, and teeth for introducing toxic secretions into other organisms. Each of those components varies in structure and function among venomous serpents, and each of them might have arisen along separate evolutionary trajectories.

Head glands, found in all squamates, are evidently rich in potential for adaptive modification (e.g., the sublingual salt excretion gland of seasnakes, Chapter 13). Mucus-producing supralabial glands are found on either side of the head of all alethinophidians, often extending as a continuous strip from near the snout to below and well behind the eye; the secretions of these modified salivary glands are introduced into the mouth via several ducts that open at the bases of maxillary teeth. Duvernoy's gland is larger and more discrete than the supralabial gland, is composed of branched tubules rather than simply aggregations of cells, and can be mucous, serous, or both. Present in many colubrids and some atractaspidids, it is situated immediately under the skin, above and near the angle of the jaw; it is embryonically derived from the same tissue as tooth enamel. Duvernoy's gland opens by a duct at the base of one or more posterior, often enlarged and sometimes grooved, teeth; there usually is no central storage chamber or lumen. A few colubrids (e.g., Boomslangs [*Dispholidus typus*]) resemble three groups of highly venomous snakes (i.e., those with true venom glands, described below), in that their Duvernoy's gland is encased in a thin capsule and attached to a small portion of the jaw-closing musculature. Among "nonvenomous" serpents, only blindsnakes, primitive snakes, and a few groups of colubrids lack Duvernoy's glands. There are no serous cells (and thus presumably no toxins produced) in the Duvernoy's glands of constricting colubrids such as ratsnakes (*Elaphe*), kingsnakes (*Lampropeltis*), Eastern Pinesnakes (*Pituophis melanoleucus*), and their close relatives.

Many-banded Coralsnake (*Micrurus multifasciatus*), Costa Rica. Like many venomous snakes, this species uses its venom to subdue prey as well as defend against predators; it distracts attention from its head with a tail display.

True venom glands—with a thick connective tissue capsule, a lumen, a separate compressor musculature, and a duct connecting to a single fang—are found in some atractaspidids, all elapids, and all viperids. The venom glands of stiletto snakes (*Atractaspis*) have unbranched, radially arranged tubules that are surrounded by mucus-secreting cells near their openings into a lumenlike central duct. Elapid venom glands resemble Duvernoy's glands in having many secretory granules and only a narrow lumen; an elongate mucus accessory region surrounds and opens into each duct. Viperid venom glands have a large lumen and a globular accessory gland situated around the duct. A distinct compressor muscle is derived from different parts of the jaw-closing muscles in stiletto snakes, elapids, and vipers. A few species in each of those groups have exceptionally long venom glands, surrounded throughout by compressor muscle and extending backward either just under the skin (in certain atractaspidids and some night adders [*Causus*]) or within the body cavity (in certain elapids).

Classically four tooth types have been identified among serpents, but they inadequately describe morphological variation in snake dentitions and could falsely convey evolutionary progressions from simple to complex venom injection mechanisms. Aglyphous snakes (from Greek, meaning "without a groove") have undifferentiated maxillary teeth, at least with respect to conducting glandular secretions; this dentition characterizes blindsnakes and primitive snakes, as well as some colubrids (including some species with Duvernoy's glands). Opisthoglyphous snakes (from Greek, for "a groove at the back") have enlarged teeth on the posterior ends of their maxillary bones, through which flow Duvernoy's gland secretions; a groove, when present, is on the anterior or lateral surface of enlarged rear teeth. Here simply

False Terciopelo (*Xenodon rabdocephalus*), Costa Rica, with its enlarged rear teeth clearly visible.

called rear-fanged, whether or not grooves are present, those species chew to bring their posterior teeth into contact with prey or potential predators; secretions are infused slowly into the victim, facilitated by indirect pressure from nearby jaw-closing muscles.

Opisthoglyphous dentition is restricted to some atractaspidids and many colubrids, perhaps including one-third or more of species in the latter group, and is highly variable. Whereas many rear-fanged snakes have one enlarged tooth on each upper jaw, black-headed snakes (*Tantilla*) and some others often have two possibly functional fangs at the back of each maxilla. South American mock pitvipers (*Tomodon*) have rather short maxillae, with only a few solid teeth anterior to their "rear" fangs, which as a result are surprisingly close to the front of the mouth. Some aglyphous and opisthoglyphous serpents, such as those that snare skinks, also have large anterior maxillary, palatine, or dentary teeth that usually are not associated with venom injection. Only a few opisthoglyphs (or opisthomegadonts [from Greek, for "rear large tooth"], as those with enlarged, ungrooved rear fangs are sometimes called) cause serious accidents to humans (see the Special Topic in this chapter).

Proteroglyphous snakes (from Greek, meaning "a groove at the front") have enlarged anterior fangs; the fangs are deeply grooved or even tubular. This type of dentition characterizes the Elapidae as well as atractaspidid harlequin snakes (*Homoroselaps*). Most proteroglyphs move the maxillary bones relatively little during biting; their so-called fixed front fangs remain erect when the snake's mouth is shut and are rather short—only 11 mm long in a 3.83-m King Cobra (*Ophiophagus hannah*). The base of a fang is called the "pedicel," and each fang has an entrance lumen and a discharge orifice; often there is a cutting edge on one side near the tip. Yellow-bellied Seasnakes (*Pelamis platurus*) have many small, solid teeth posterior to small

Plains Black-headed Snake (*Tantilla nigriceps*), Arizona, envenoming a centipede. (Photograph by Harry W. Greene)

Central American Coralsnake (*Micrurus nigrocinctus*), Costa Rica. This elapid is swallowing a Common Slug-eater (*Sibon nebulata*) headfirst, having immobilized the prey by injecting its venom.

DEADLY COLUBRIDS AND FAMOUS HERPETOLOGISTS

Boomslang (*Dispholidus typus*), southern Africa.

Until the latter half of this century, herpetologists and physicians thought of colubrids as harmless snakes. Biologists had long known that some snakes had rear fangs, "inferior" venom injection mechanisms that might immobilize prey; although a few fatalities were on record, until 1957 the possibility that such snakes were deadly to humans seemed at most remote. The deaths of two prominent herpetologists from African colubrid bites changed that assessment, and recent events reveal that several other species of rear-fanged snakes have venoms that are potentially lethal to large vertebrates.

When Chicago's Lincoln Park Zoo sent a small Boomslang (*Dispholidus typus*) to the nearby Field Museum of Natural History for identification, the three experienced herpetologists who examined it were not especially cautious. At age sixty-eight, Karl P. Schmidt was the most senior

among them, as well as one of the world's most eminent and widely liked vertebrate biologists. Having studied African snakes as a young man, Schmidt immediately recognized the juvenile snake to be a Boomslang, despite some unusual scalation, and in the course of handling it he was promptly bitten. Although only one fang penetrated his thumb, Schmidt soon experienced nausea and some internal bleeding. He felt no great cause for concern and took careful notes on the symptoms. Schmidt felt better the next morning, but by midafternoon he was dead of a brain hemorrhage and respiratory collapse. As Clifford H. Pope, Schmidt's colleague and friend, concluded in an account of the incident, "A total lack of experience with Boomslang venom is largely to blame for the tragic events of September 25 and 26."

By 1972 several other deaths and

serious bites from Boomslangs and Savanna Twigsnakes (*Thelotornis capensis*) had been recorded. Following bites from each species, a prevalent symptom was chronic bleeding throughout the body. This precedent did not deter Robert Mertens, an extraordinarily accomplished German herpetologist (see the introductory essay to Chapter 11), from hand-feeding his pet Twigsnake. Bitten, he lingered for three weeks before dying. Ironically, Mertens generally was not fond of biologists from the United States, but one of the few for whom he had felt deep respect was Karl P. Schmidt; the two men were similar in their encyclopedic, uncommonly global knowledge of amphibians and reptiles. Among Robert Mertens's last words were something like "What a fitting death for a herpetologist."

More recently several fatalities have been caused by bites from a

Road Guarder (*Conophis lineatus*), Costa Rica.

Japanese natricine colubrid, the Ya-makagashi (*Rhabdophis tigrinus*), with symptoms including severe bleeding. Other possible candidates for deadly colubrids include Old World catsnakes and treesnakes (*Boiga;* Blanding's Treesnake [*B. blandingi*] has venom as toxic as that of some elapids), the Road Guarder (*Conophis lineatus*) of Mex-ico and Central America, and certain neotropical racers (*Alsophis* and *Philodryas*). Bites from Road Guard-ers have caused headache, nausea, pain, and—perhaps most signifi-cant, given the symptoms of Boom-slang, Twigsnake, and Yamakagashi bites—severe and prolonged bleeding. Bites from two species of South American green racers (*P. ol-fersi, P. viridissimus*) have resulted in severe symptoms in humans, and there are recent reports of a fatality caused by the latter species in Brazil.

Hog-nosed Pitviper (*Porthidium nasutum*), Costa Rica, eating a Rain Forest Whip-tailed Lizard (*Ameiva festiva*). Note how it uses its fangs in swallowing its prey, although vipers do not always do so.

fangs, so that their maxillae appear decidedly colubrid-like, whereas mambas (*Dendroaspis*) have no teeth posterior to their enlarged, somewhat mobile fangs. Proteroglyphs have well-developed true venom glands that connect through a duct on each side to openings at the base of the fangs; part of a jaw-closing muscle is attached to the outer sheath of each venom gland and expels secretions into the duct and through the fang. There is a small lumen in the venom gland, but most secretions are stored in the secretory cells themselves. Snakes with fixed front fangs typically seize prey and hang on until it has been immobilized by venom. Some cobras have their fangs modified for venom ejection in a defensive context (see p. 110).

Solenoglyphous snakes (from Greek, meaning "a channel or pipe" and "a groove") have a single, enlarged tubular fang on the anterior end of each short, highly movable maxillary bone. In vipers each maxilla pivots on a hinge with the prefrontal bone, itself hinged with the frontal; muscles attached to the pterygoid bones pull the ectopterygoids forward, thereby pushing the maxillae and erecting the fangs. Venom-conducting teeth of vipers and stiletto snakes (*Atractaspis*) are folded backward when not in use, owing to their relatively great length and mobility; a 1.8-m long Gaboon Adder (*Bitis gabonica*) has 29-mm fangs, about three times the length of those of a large King Cobra (*Ophiophagus hannah*). A viper can pull prey into its mouth without erecting the fangs, by partially extending the palatomaxillary arch, or it can erect the fangs completely for venom injection. At full fang extension the mouth is open almost 180°, and thus the snake can quickly stab and recoil from a victim. Solenoglyphs

Russell's Viper (*Daboia russelli*),
India, during venom extraction.

typically strike, release, and then relocate prey. In vipers most stored venom is contained in a large lumen; a substantial part of the jaw-closing musculature has been diverted to the large venom gland and forces secretions through a duct on each side of the head to the bases of the fangs. The overall consequences of movable front fangs in stiletto snakes and vipers are deep penetration and rapid infusion of toxins, although the precise mechanism differs between those two groups. Mambas (*Dendroaspis*) and death adders (*Acanthophis*) are somewhat similar, in that their fangs are more movable than those of other elapids; even some colubrids (e.g., false pitvipers [*Xenodon*]) have fairly mobile maxillae and thus considerable ability to erect their rear fangs.

Proteroglyphous and solenoglyphous snakes usually have one or more replacement fangs beside or behind the functional fangs; a fang is generally functional for about six to ten weeks in rattlesnakes (*Crotalus*) before it is lost. Because the fang sheath surrounds both fangs and the venom gland duct, transmission of venom is effectively maintained during replacement of one fang by another. Shed fangs often appear in the stomach contents and feces of vipers, and thus probably are swallowed with prey.

Most information on venom yields from diverse species has been obtained by forced milking, so we know little about venom production and injection that is ecologically relevant. The amount of stored venom expended in biting a mouse is about 55 percent in Brown Treesnakes (*Boiga irregularis*), 40–55 percent in Common Death Adders (*Acanthophis antarcticus*) and Tiger Snakes (*Notechis scutatus*), 6–10 percent in Habus (*Trimeresurus flavoviridis*) and Palestine Vipers (*Vipera palestinae*), and 20–55 percent in Western Rattlesnakes (*Crotalus viridis*). A typical bite from a Brown Treesnake contains three to eight times the lethal dose for a mouse, and almost half the venom is lodged in the skin instead of the viscera. By contrast, a Western Rattlesnake injects roughly three hundred times the amount required to kill a mouse, and most of that venom reaches the prey's interior. Cottonmouths (*Agkistrodon piscivorus*) inject more venom into rats than into mice, whereas Palestine Vipers do not. When venom glands are forcibly drained, replenishment rates peak at four to nine days in Egyptian Cobras (*Naja haje*) and Palestine Vipers, and glands are completely refilled in two to three weeks. About 20 percent of Palestine Viper bites to mice are "dry" (that is, they do not result in envenomization), as are that proportion or more of venomous snakebites to humans. However, the oft-mentioned possibility that venom injection is metered carefully during defensive encounters seems doubtful, given that many defensive bites to humans result in serious symptoms.

BIOCHEMISTRY, PHARMACOLOGY, AND BIOLOGICAL ROLES OF VENOMS

Classically snake venoms have been dichotomized as neurotoxic and hemotoxic (or hemo-lytic), and those two venom types are supposedly characteristic of elapids and viperids, re-spectively. The term "hemotoxic" implies effects on circulatory tissues, but I call such venoms "tissue-destructive," because they often affect far more than just the vascular system. Although the biochemistry and pharmacology of snake venoms have been investigated exten-sively, the functional properties usually are studied in terms of isolated compounds and with-out respect to other ingredients, in tissues or organisms and at dosages of questionable significance in nature. As a result, relatively little is known about the evolutionary and eco-logical significance of variation in these extremely complex and interesting mixtures.

The secretions of Duvernoy's glands are colorless, whereas those from the true venom glands of stiletto snakes (*Atractaspis*), elapids, and viperids are typically some shade of yellow. Some venoms are viscous and cloudy, whereas others are clear. They are often composed primarily of digestive enzymes, including proteases, cholinesterases, ribonucleases, and hya-luronidase, this last a spreading factor that increases absorption through connective tissue. About ten enzymes (mostly those that digest proteins and lipids) are found in almost all snake venoms, and as many as twenty or more may be found in that of a viper; other proteins, usually coupled with an enzyme, seem largely responsible for the toxic properties. Phospho-lipase A_2 (PLA_2) is a particularly widespread digestive enzyme in the venoms of colubrids, stiletto snakes, elapids, and viperids and combines synergistically with various toxic compo-nents. Major targets of venom action include the brain, nerves, heart and lungs, red blood cells and lining of blood vessels, clotting mechanisms, muscle cells, and cell membranes.

Viper venoms include several digestive enzymes with high molecular weight components; they spread slowly after injection and mainly through the lymphatic system (at least in hu-mans). Among their toxins, hemorrhagins are enzymes that primarily destroy the linings of blood vessels; hemorrhagins also are produced by some colubrids and by stiletto snakes (*Atractaspis*). Venoms of the latter, however, emphasize powerful cardiotoxins unique to that group. Elapid venoms generally have fewer components and ones with lower molecular weights than those of vipers; they spread rapidly in the bloodstream and contain substances that cause respiratory failure. Presynaptic neurotoxins block the release of transmitter mole-cules (acetylcholine) from nerve cells to muscles (as do some bacterial toxins; e.g., those that cause botulism), whereas postsynaptic neurotoxins prevent combination of transmitters with

Black Mamba (*Dendroaspis polylepis*), southern Africa. This species has an unusually toxic and fast-acting venom.

receptor sites on the muscles (as does curare, a plant-derived substance). The latter, which are more common in elapid venoms, include short-chain and long-chain neurotoxins, named for differences in the number of amino acids making up their protein molecules. A protein in mamba (*Dendroaspis*) venoms, derived from one of the short-chain neurotoxins of other elapids, is modified so that its destructive effects on cell membranes resemble those of structurally dissimilar toxins in viper venoms.

Dichotomous concepts of venom functions, such as tissue-destructive or neurotoxic, are based largely on clinical effects in humans. Bites from Boomslangs (*Dispholidus typus*) and Savanna Twigsnakes (*Thelotornis capensis*) cause serious coagulation problems, with victims hemorrhaging into various organs (see the Special Topic, this chapter). Viper venoms can cause massive tissue destruction and swelling, and in humans a third or more of the total blood volume may be lost into the tissues of an extremity within hours of a bite; death usually results from hypotensive shock, as blood pressure decreases to a point where the heart can no longer function. Crippling and limb loss, as well as some kidney and lung damage, are potential permanent effects in survivors of viper bites. By contrast, envenomization by an Indian Cobra (*Naja naja*) or a Black Mamba (*Dendroaspis polylepis*) leads to muscle paralysis and death by respiratory failure if untreated. Nevertheless, broad generalizations about venoms and their effects can be misleading. Bites from some Neotropical Rattlesnakes (*Crotalus durissus*) produce neurotoxic symptoms, such as drooping eyes and respiratory difficulties, whereas envenomization by Black-necked Spitting Cobras (*Naja nigricollis*) causes serious local tissue damage. Victims of marine elapid bites (e.g., by Beaked Seasnakes [*Enhydrina schistosa*]) exhibit clenched jaws and dark urine from muscle cell destruction, as well as respiratory failure. In fact, many snake venoms have both tissue-destructive and neurotoxic properties.

The long-term effects of venoms on lab mice might be meaningless with respect to the rapid capture, consumption, and digestion of prey in nature. Likewise, because humans are far larger than the prey of any advanced snakes, clinical studies are of uncertain relevance in understanding the biological roles of venoms. The Duvernoy's gland secretions of diverse rear-fanged atractaspidids and colubrids, however, immobilize victims and help the snakes digest them. For example, Northern Cat-eyed Snakes (*Leptodeira septentrionalis*) have difficulty swallowing sticky-toed Red-eyed Treefrogs (*Agalychnis callidryas*) until their prey have succumbed from prolonged chewing, and the bite of Puerto Rican Racers (*Alsophis portoricensis*) immobilizes and helps digest lizards. Duvernoy's gland secretions of the Montpellier Snake (*Malpolon monspessulanus*) contain various enzymes, including PLA_2 as part of a toxic component; even sublethal doses impair locomotion and respiration in mice and thus could help subdue natural prey. Head gland secretions from Brazilian Slug-eaters (*Sibynomorphus mikani*) immobilize prey and weaken its muscle tissues. The venoms of New World coralsnakes, cobras, seasnakes, and stiletto snakes (*Atractaspis*) primarily immobilize bitten prey and require up to several dozen minutes to take effect. Viper venoms induce clotting—perhaps the underlying cause of rapid death (often in a few seconds) from those snakes—when administered in high concentrations; their venoms also have pronounced digestive effects that are especially important at lower temperatures. Western Diamond-backed Rattlesnake (*Crotalus atrox*) venom digests the skin of a prey item, exposes the body cavity, and begins to break down the viscera within twenty-four hours.

Venom composition and toxicity may vary among related species, populations within a species, and individuals in a litter, as well as within the life of an individual—all differences that eventually might elucidate the ecological significance of venom biochemistry. The Aquatic Coralsnake (*Micrurus surinamensis*) feeds primarily on eels and has venom unlike that of the more than fifty other species of New World elapids, most of which feed primarily on snakes. Olive Seasnakes (*Aipysurus laevis*) have venom that rapidly incapacitates the fish on which they feed, whereas White-spotted Seasnakes (*A. eydouxi*) that eat only fish eggs have greatly reduced fangs and toxicity. The venoms of some Mojave Rattlesnakes (*Crotalus scutulatus*), some Neotropical Rattlesnakes (*C. durissus*), and the Midget Faded Rattlesnake (*C. viridis concolor*) exhibit increased toxicity and reduced tissue-destructive effects compared to the venoms of other populations of the same species, but the impact of those differences on their natural history is unknown. Inland Taipans (*Oxyuranus microlepidotus*), rodent-eating elapids with no age-related change in diet, show no venom differences between juveniles and adults. Conversely, Northern Pacific Rattlesnakes (*C. v. oreganus*) eat lizards as juveniles and mammals as adults; growth in that subspecies is accompanied by decreased concentrations of PLA_2 in the venom and diminished toxicity, but increased concentrations of proteases and digestive effects. An age-related dietary shift in Brazilian Lanceheads (*Bothrops moojeni*), from frogs and lizards to mammals, also is correlated with venom changes and loss of caudal luring behavior. Inexplicably, although Bushmasters (*Lachesis muta*) of all ages feed on rodents, the venom of juveniles shows very little toxicity in laboratory tests.

VENOMOUS SNAKES AND HUMAN WELFARE

The public health impact of venomous snakes is difficult to evaluate because many bites go unreported, but several patterns of variation are instructive. Fifty years ago—more than half

Indian Cobra (*Naja naja*), poised and ready to strike.

a century after the advent of antivenom—between 500,000 and 1 million people were bitten by venomous snakes annually throughout the world, most of them in tropical countries. At that time at least 40,000 people were killed per year by snakes, including 400–1,000 in Africa, more than 2,250 in South America, 25,000–35,000 in Asia, 50 in Europe, and 10–20 in the United States. Snakebites and fatalities are undoubtedly less frequent today, although accident rates are still roughly two to twenty times higher per capita in the tropics than in temperate countries; this disparity presumably reflects mainly the abundance of certain species in habitats where they are not easily seen by humans (e.g., Terciopelos [*Bothrops asper*] in Central American pastures). In one recent year there were more than 400 bites by pitvipers and 8 bites by coralsnakes in Costa Rica, of which about 8 percent of the former (all by Terciopelos) were fatal. Throughout Central America bites and deaths have been greatly reduced over the past few decades, mostly thanks to increased use of boots by fieldworkers and the availability of antivenom.

Today the mortality rate for properly treated bites in developed countries is far less than 1 percent; perhaps 1,000–2,000 venomous snakebites occur each year in the entire United States, most of them by rattlesnakes (*Crotalus*) and Copperheads (*Agkistrodon contortrix*), and they result in fewer than ten deaths. One can only imagine the public hysteria in Arizona if more than a hundred people per year died from snakebite—a figure comparable to the mortality rate of 3.3/100,000 for those accidents in Costa Rica! Epidemiologists make a useful distinction between legitimate bites—those accidentally incurred during normal activities— and illegitimate bites—those sustained while purposely interacting with snakes. More than half of all victims in both Costa Rica and the United States are under thirty years old; however, almost all Costa Rican bites are legitimate, and 68 percent are to legs and feet, whereas more than half the incidents with young men in the United States involve intentional handling of snakes and bites to the upper extremities.

First aid for snakebite encompasses numerous treatments, many of them fruitless or even dangerous. All are aimed at delaying the absorption of venom or neutralizing or removing it. Most tactics prevalent within this century in the United States were also used well before Europeans arrived in the New World, and some pre-date recorded history. Widely touted remedies include various concoctions and poultices (e.g., plants, alcohol), suction devices (e.g., a warm stone by Native Americans, a split chicken by rural people in the southeastern United States), and cutting or excision of the bitten part. Some well-known methods, each of them controversial and risky, include wrapping the limb in a tourniquet or ligature and then cutting and sucking the wound (an operation sometimes facilitated by spring-loaded blades); immersing the bitten limb in ice water; and applying electroshock to the wound (e.g., by a stun gun). Those practices skirt the prime rules of good first aid: to do no harm and to calm the victim (remaining cool-headed while someone whacks away at your hand with a pocket knife is difficult!). Under many circumstances, more appropriate responses to snakebite are simply to immobilize the bitten limb and safely transport the victim to a hospital. For elapid bites, tightly wrapping the injured limb with a bandage (not a tourniquet) slows the spread of venom until medical help is reached.

Hospital treatment of snakebite counteracts the injurious or deadly consequences of envenomization, primarily by antivenom therapy. Antivenom is produced by injecting minute but gradually increasing quantities of venom into other animals (often horses), then harvesting their serum with its venom-specific antibodies. Some antivenoms are monovalent and more effective against a single species; others are polyvalent and useful for several dangerous snakes in a particular region, especially when the identity of the snake involved is unknown. Antivenoms are best administered in hospitals, not only because they are most effective if given intravenously but also because some patients have severe allergic reactions to horse serum (antivenoms are now being developed based on only the purified antibodies, thus avoiding dangerous serum factors that are irrelevant to treatment). Moreover, because a major threat in many snakebites is hypotension or shock, physicians now anticipate and prevent this problem by bolstering the circulatory system with intravenous fluids. Victims of elapid bites in countries where antivenom is expensive or otherwise unavailable are sometimes saved by prolonged artificial respiration, until paralysis from the neurotoxins wears off.

A few points concerning snakebite are worth emphasizing. First, serpents avoid direct contact with humans, and bites occur because we unintentionally threaten them or deliberately seek an interaction. Illegitimate bites are largely a late twentieth-century problem and mainly a matter of testosterone tyranny, because young men disregard or do not appreciate the inherent hazards of handling venomous snakes. Many legitimate accidents could be prevented by simple precautions, such as wearing boots and using a flashlight outdoors at night, and almost all snakebites are now curable with proper medical care. Mortality today results from a lack of hospital facilities, still the case in many countries; from inadequate training of physicians, so that they do not anticipate such problems as shock following bites from large vipers; and from failure to seek medical care—three of nine recent fatalities from rattlesnake bites in Arizona were to persons who refused medical care. Finally, controversies about first aid and hospital treatment for snakebite are unusually widespread and persistent. Since snake-

Puff Adder (*Bitis arietans*), southern Africa. A hiker or bird-watcher walking over the log might step on this common, large venomous snake and be bitten.

bite is rare in most countries that have strong support for biomedical research and many incidents go unreported, those disagreements reflect in part the difficulties of assembling adequate case histories for comparisons among treatments. Uncertainties are also inevitable for events in which the outcome depends heavily on variation in snakes, victims, and treatment. And after all, if some defensive strikes result in dry bites and individual venomous serpents vary in toxicity, any "cure" may give the illusion of working.

THE EVOLUTION OF VENOM INJECTION MECHANISMS

Herpetologists usually have viewed fixed front fangs as unique to elapids, movable front fangs as unique to viperids, and rear fangs as having evolved several times within colubrids. Rear fangs would thus represent a transitional stage from nonvenomous snakes, either to elapids and thence to vipers or separately to each of the others. This scenario implies a simple trend toward increasingly efficient venom injection, with vipers at the pinnacle of snake evolution. Recently, the exclusion of stiletto snakes (*Atractapsis*) and harlequin snakes (*Homoroselaps*) from their traditional taxonomic placements has inspired an alternative hypothesis, namely, that proteroglyphy is primitive for colubroids and led independently to opisthoglyphy and solenoglyphy, as well as to secondarily nonvenomous aglyphous forms. That second view incorporates the independent evolution of true venom glands at least twice, as is likely, but evidence that vipers diverged before the origins of most other advanced snake lineages challenges its basic premise. Clearly venoms arose early in the evolution of Colubroidea, facilitated by the liberation of the maxillae from their primitive role of dragging a snake's mouth over prey (see Chapter 3). Probably the common ancestor (or ancestors) of all venomous snakes was aglyphous or opisthoglyphous, and various front-fanged conditions evolved several times from rear-fanged snakes.

Duvernoy's glands and true venom glands are derived embryonically from tissues that produce tooth enamel, and thus all are homologous at least to that extent. The more detailed evolutionary history of those glands is uncertain, but some general aspects of venom biochemistry hint at the origins of and changes in the secretions themselves. First, the head glands of many snakes make PLA_2s, and some of those molecules even work interchangeably with other toxic components among elapid and viperid venoms. These observations suggest that PLA_2s and enzymatic activity were characteristic of primitive venoms. Second, a prerequisite for venom evolution might have been the presence of circulating inhibitors against toxins. Pancreatic enzymes, likely precursors of some venom components in view of their structural similarities, are indeed associated with natural inhibitors, and the two could have converted in parallel to toxins and toxin inhibitors. As for venom-conducting teeth, the formation of a groove, or closed canal, obviously facilitates the flow of secretions into a victim's tissues. The initial evolutionary enlargement of rear teeth, however, could simply have aided in holding prey prior to the origin of specialized glands or even made it possible to puncture the puffed-up bodies of toads (as do false pitvipers [*Xenodon*]) as well as enhanced the inoculation of toxic secretions. In any case, two large adjacent teeth also might have formed together a crude groove (e.g., as in some littersnakes [*Rhadinaea*]), thus further exposing tissue to Duvernoy's gland secretions.

Regardless of their precise histories, similar venom injection mechanisms have evolved more than once among advanced serpents. Movable front fangs arose independently and with different functional properties in vipers and stiletto snakes (*Atractaspis,* Chapter 11), as well as to some extent in death adders (*Acanthophis*) and mambas (*Dendroaspis*), elapids that otherwise have fixed front fangs. Proteroglyphy perhaps evolved independently in elapids and harlequin snakes (*Homoroselaps*), although the phylogenetic status of the latter, and thus the number of times fixed front fangs have originated, is controversial. Long venom glands have evolved in some stiletto snakes and in another atractaspidid (*Micrelaps bicoloratus*), in long-glanded coralsnakes (*Maticora*) and Buerger's Forestsnake (*Toxicocalamus buergersi*)—both elapids—and in some night adders (*Causus*), among the viperids; those are all relatively slender burrowers, so perhaps long glands afford more storage space for venom in snakes with small heads. Reduction of the venom apparatus also has occurred, exemplified by a fangless African colubrid (*Aparallactus modestus*) that feeds on earthworms (its closest relatives are venomous centipede-eaters [other *Aparallactus*]) and by the Turtle-headed Seasnake (*Emydocephalus annulatus*), which preys entirely on fish eggs (its closest relatives are other marine elapids).

Venoms entail auxiliary adaptations beyond glands, teeth, and musculoskeletal modifications. For example, melanin deposits in the skin or the peritoneum of many vertebrates absorb ultraviolet radiation before it penetrates to deeper tissues; likewise, dark pigments over the head glands of some colubrids (e.g., Asian Green Vinesnakes [*Ahaetulla prasina*], Brown Vinesnakes [*Oxybelis aeneus*]) and many vipers might prevent detoxification of their venoms. Dark postocular stripes on the skin conceal a snake's eye and otherwise disrupt its outline (as

Desert Death Adder (*Acanthophis pyrrhus*), Australia. This species resembles a viper in appearance, and like some vipers it lures prey with its contrastingly colored tail tip.

Eyelash Pitviper (*Bothriechis schlegelii*), Costa Rica. The dark stripe behind each eye, typical of many vipers, protects venom in the head glands from ultraviolet radiation.

in nonvenomous Boa Constrictors [*Boa constrictor*]), but many of those snakes whose venoms emphasize digestive enzymes also have melanin in deeper layers of the skin, the compressor musculature, and even the glands themselves. Melanin is more extensively present in Ridge-nosed Rattlesnakes (*Crotalus willardi*) and other species at high altitudes, where ultraviolet radiation is stronger, than in Bushmasters (*Lachesis muta*) and other vipers that inhabit shady lowland forests. In some large African vipers (e.g., Puff Adders [*Bitis arietans*]), shiny irido-phores replace melanin, and ultraviolet radiation is reflected rather than absorbed.

Influenced by traditional concepts of venom composition and evolution, I long believed that immobilization of prey was the sole primitive function of snake venoms. According to that naive view, the biochemically simple neurotoxic venoms of elapids are sufficient for immobilizing organisms that are elongate and heavy (Type II prey) but do not require additional digestion; the more complex, tissue-destructive venoms of viperids are highly modified for immobilizing and digesting heavy, bulky animals (Type III prey; see Chapter 3). Size-related changes in viper diets and venoms (as seen above, for Western Rattlesnakes [*Crotalus viridis*]) would thus recapitulate a phylogenetic transition from immobilizing reptile prey to killing and digesting mammals.

In fact, the head gland secretions of all snakes might cleanse their teeth, lubricate prey before they swallow it, and even to some extent aid in digestion; moreover, even slight local tissue damage might briefly immobilize and thus help subdue prey. Recalling that both digestive and immobilizing effects characterize the venoms of diverse colubrids, it now seems more likely that vipers enhanced those primitive roles through radical changes in the anatomy of their heads as well as diversification in venom biochemistry. Equipped with spacious lumens to store venom and with long, folding fangs, most vipers immobilize their prey—often mammals with strong jaws and sturdy teeth—by injecting large quantities of toxins during brief contact, ensuring digestion with deep penetration of powerful, tissue-destructive enzymes. By contrast, the venoms of many elapids and some colubrids (e.g., Blanding's Treesnakes [*Boiga blandingi*]) emphasize non-enzymatic venom components, especially neurotoxins; perhaps acquisition of those characteristics and loss of enzymatic components were primitively associated in elapids with predation on other, relatively heavy, snakes (see Chapter 12)—items with high surface to mass relationships that would not require digestive venoms. An origin of neurotoxins coupled with a primitive diet of reptiles in early proteroglyphs also is consistent with the absence of inhibitors against those toxins in non-elapids and even some elapids. Under this view, a primitive elapid's feeding biology would mirror that of a viper in which the adult diet and venom composition had been lost.

Two more aspects of the evolution of venomous snakes deserve brief mention. First, where and how snakes feed, as well as the characteristics of their prey, have undoubtedly affected venom evolution. For example, heavy or otherwise dangerous prey can be subdued by some nonvenomous species in places where their victims cannot struggle effectively (e.g., Bull Snakes [*Pituophis catenifer sayi*] eating pocket gophers in tunnels). Furthermore, effective venom injection need not be rapid, given that some species feed under conditions where they probably are not vulnerable to predators, and minutes spent subduing prey are not otherwise critical (e.g., lyresnakes [*Trimorphodon*] seizing lizards in crevices). Finally, bulky prey might pose severe problems for digestion only in seasonally cold environments, which would explain why even nonvenomous lowland tropical snakes (e.g., some boas) can take such items, whereas only vipers do so at high altitudes and latitudes.

Second, the upper limits on prey mass for venomous snakes are set by stomach volume and locomotor capacity, rather than by limits to the immobilizing effects of the venom—after all, the juvenile Common Lancehead (*Bothrops atrox*) that contained a whip-tailed lizard (*Cnemidophorus*) 156 percent of its own weight could also have subdued me! There are no known feeding advantages associated with the especially powerful toxins of some snakes (e.g., certain populations of Mojave Rattlesnakes [*Crotalus scutulatus*], some species of saw-scaled

Terciopelo (*Bothrops asper*), Costa Rica, extending its fangs as a radio transmitter is carefully pushed into the snake's stomach.

vipers [*Echis*], taipans [*Oxyuranus*], and mambas [*Dendroaspis*]), raising the intriguing possibility that the irascibility and daunting threat displays of those snakes have co-evolved with their venoms. However, except in the case of spitting cobras (*Hemachatus,* some *Naja*; see pp. 109–10), the possibility that defensive roles have affected the evolution of snake venoms has not been explored.

OTHER VENOMOUS CREATURES

Venoms are relatively rare in nature, which suggests that they might be expensive in terms of energy to develop and maintain or that conditions favoring their evolution are uncommon. Furthermore, unlike at least most snake toxins, some other venoms probably have evolved primarily for defense against predators; likely examples of vertebrates with defensive venoms include blennies, scorpion fish, the Australian Platypus (*Ornithorhynchus anatinus*), and beaded lizards (*Heloderma*). Among other venoms that mainly function during feeding are those of centipedes, spiders, short-tailed shrews (*Blarina*), and the delightfully weird solenodons (*Solenodon*).

Fourteen species of deep-sea eels, collectively the Monognathidae, are stunningly reminiscent of some venomous snakes in their anatomy and capacity for feeding on large items. These fish are small, fragile-looking creatures that reach a maximum length of 160 mm; they lack upper jaws and have distensible lower jaws. One of the bones in the roof of a monognathid's mouth is modified into a sharp "rostral fang," a hollow spine divided into two ducts that open a short distance above its sharp, downward projecting tip. Basally these ducts connect with a secretory "rostral gland" in the roof of the mouth, through which the fang passes, and we infer that this is a venom apparatus. The diet of monognathids consists entirely of relatively enormous shrimp, and perhaps adductor muscles of the lower jaws force prey inward against the fang. Unlike vipers, after as little as one meal the eels lose the ability to feed (their lower jaws degenerate); they then reproduce once and die. Also unlike snakes, monognathid eels might use chemicals to attract their prey, rather than following chemical trails to locate food. Moreover, these delicate and strange little fish live as much as 5,400 m below the ocean's surface and never bite people.

Asian Long-nosed Vinesnake (*Ahaetulla nasuta*), India, in gaping threat display.

5

A beautiful Alaskan Malamute shared my life in graduate school and, immersed in a dissertation about behavioral evolution, I especially enjoyed her lupine personality. Layla's movements and postures seemed more wolflike than those of other dogs I'd watched; she'd often meet me with the "play bow," a stereotyped greeting that is widespread among wild canids, and she was intensely protective. Soon after moving to Berkeley, while I was jogging with Layla in a nearby park, we were attacked by a huge Doberman that escaped its owner's leash on a trail above us. There was no hope of outrunning the larger animal, so I stood my ground and moved Layla out of the way. She pulled back in front of me at the last instant, catching the full force of his charge, but the Doberman got only a mouthful of rump fur and a severely kicked head. I was angry and flushed with terror, my dog excited but not obviously frightened. When we encountered a hatchling Gopher Snake (*Pituophis catenifer*) on the same path a few days later, Layla poked her nose at the 30-cm snake and was rebuffed immediately by its exaggerated hissing and striking. Despite my urging, this 35-kg carnivore would not reapproach the 20-g reptile, so small she could have crushed it with a snap, its teeth so puny they could not even have pierced the delicate skin on her nose. In an equally intriguing standoff, another naturalist once saw an adult Eastern Pinesnake (*P. melanoleucus*) keep two Raccoons (*Procyon lotor*) at bay for an hour and then crawl away unharmed. Years after the Doberman attack, surveying antipredator mechanisms in reptiles, I returned to the puzzle of why large, competent predators shy away from small, harmless reptiles. "Layla's Paradox" led me to ponder additional aspects of snake biology as well as the limits to generalization in natural history.

Antipredator mechanisms are among the first things people notice about snakes, given our deeply entrenched fear of sinuous, limbless vertebrates and the usual unfortunate course of events. Each wave of humans invading the Americas encountered rattlesnakes, so dramatically unlike anything in the Old World that their silhouette is recognizable in ancient rock art, and their noisemaking *cascabeles* (jingle bells) appear in the early accounts of Spanish explorers. Today the widely familiar English names for the Puff Adder (*Bitis arietans*) and Cottonmouth (*Agkistrodon piscivorus*) reflect the snakes' reaction to threats, and a Thai name for the Blue Long-glanded Coralsnake (*Maticora bivirgata*) and other Asian species with defensive tail displays means "snake with two heads." Taxonomists have long been inspired by antipredator responses, using Greek words for "the liar" as the generic name of harmless neotropical birdsnakes with spectacular inflated throats (*Pseustes*), for "the spitter" as the species name of Javan Cobras that eject venom at an adversary's eyes (*Naja sputatrix*), and for "the bellower" as the subspecies name of the Florida Pinesnake with an especially loud hiss (*Pituophis melanoleucus mugitus*). *Ahaetulla,* the generic name for Asian vinesnakes, is an ancient Sri Lankan word meaning "eye picker," in reference to the tendency of these serpents to strike at the face of an adversary. Ironically, until recently snakes attracted little serious attention from behavioral ecologists, although roughly 20 percent of the world's snakes have had some aspect of their defensive reactions described.

The antipredator mechanisms of serpents are now known better than those of any other large group of terrestrial vertebrates, and some interesting patterns emerge from comparisons among them. Advanced snakes are essentially fragile creatures. Indigo Snakes (*Drymarchon corais*) and a few other species can bite hard, but most serpents have sacrificed jaw strength and big teeth—the usual vertebrate weaponry—for expansive swallowing ability. Not only do snakes have no limbs with which to grapple, but the fastest species are only about one-third as fleet as some fully limbed lizards. Physically engaging an adversary can prove lethal, as in the examples of an adult Neotropical Rattlesnake (*Crotalus durissus*) that was fatally injured by an Ocelot (*Felis pardalis*) and of a Terciopelo (*Bothrops asper*) that starved with a broken mandible. Given their overall frailty, snakes typically minimize risk and the expenditure of energy during encounters with predators. Even species as dangerous as Bushmasters (*Lachesis muta*) and King Cobras (*Ophiophagus hannah*) rely first on avoiding detection and then flee when discovered; once cornered, they resort to threat displays before actual combat.

Finally, only about 20 percent of advanced snakes can effectively defend themselves with venom; the rest depend mainly on camouflage, locomotion, and bluffing displays that are not backed up by retaliation. However, about 15–20 percent of New World snakes clearly resemble venomous coralsnakes, and 10–15 percent of the remainder look like pitvipers; if those trends hold worldwide, at least 25–35 percent of snakes that are not themselves effectively defended by venoms are protected by the presence of venomous species. South American Hog-nosed Snakes (*Lystrophis dorbignyi*) even combine mimicry of coralsnakes and of pitvipers, looking like the latter dorsally, when first discovered, and like the former ventrally, when their bright belly patterns are exposed during handling!

With those peculiarities of snake biology in mind, the solution to Layla's Paradox might be as follows. Most serpents are poor strugglers, so those lacking venom are essentially defenseless; confrontation might result in death, so even venomous species advertise their noxious qualities and avoid direct contact with enemies. Nevertheless, venomous snakes can kill large vertebrates; in the case of snakes, then—as opposed to many other animals—size is not a good predictor of whether a potential prey would be able to defend itself. Snakes also are unusual because, given their simplified body form, modest changes in appearance and behavior can make a harmless species resemble others that pose high risk for a predator (we owe this insight to the great nineteenth-century naturalist Alfred Russel Wallace, who noted the improbability "of a mimicry by which the elk could escape the wolf, or the buffalo from the tiger"). Accordingly, mistakes are easy to make, and they might be fatal. Most predators therefore should not hazard an attack unless facing death from starvation; those otherwise predisposed to attack venomous snakes would sometimes die young and leave no offspring. Layla had never been bitten by a venomous snake nor seen anyone express fear of one, but evidently she retained some relevant genetic baggage from wild canid ancestors. Attacking that little Gopher Snake just wasn't worth the risk.

Biologists seek comprehensive explanations for the diversity of life, but we might have to settle for fairly limited theories about antipredator mechanisms. Judging from species that already have been observed, we can make reasonably good predictions about an unstudied snake's response to danger based on its size, coloration, and habitat; beyond common principles of camouflage and a few other generalizations, however, we cannot predict much about a bird or a spider based on the behavior of snakes. This is because what works as defense depends on particular characteristics of the predator and the prey, as well as where and how they interact. Together, those variables and other uncertainties entail tremendous complexity, so that the possible number of outcomes across all animals is unfathomably diverse. Layla's Paradox reminds us that, because nature is so vast and idiosyncratic, sometimes truth and clarity are lurking in the details.

Because predation is the killing and consumption of another animal, defensive mechanisms against predators are best understood in terms of the costs and benefits of feeding discussed in Chapter 3. Those costs to the predator include the time and energy required to locate, capture, and digest an item, as well as any associated risk of injury or death. Antipredator tactics work when a potential prey actually raises any of those cost components to intolerably high levels or when it deceptively indicates higher costs or lower benefits than a predator could accept. A useful framework for exploring antipredator mechanisms also includes the four questions ethologists typically ask about behavior, namely, How does it develop within an individual? What sensory and motivational processes control it? What are its ecological consequences? and How did it evolve?

Snakes use a fantastic array of tactics to avoid detection and consumption by potential predators, running the gamut from open-mouth threat displays and amplified hissing to death-feigning and cloacal popping. Those tactics are the focus of this chapter, but first I briefly review predation on snakes and establish a natural context for understanding their responses.

PREDATION ON SNAKES

Large predatory invertebrates eat mainly other invertebrates, but some occasionally prey on small, secretive snakes and the juveniles of larger species. For instance, a diving beetle larva (*Dytiscus*) can capture and consume a young Western Terrestrial Gartersnake (*Thamnophis elegans*), and one crab even ate a hatchling King Cobra (*Ophiophagus hannah*). Western Blindsnakes (*Leptotyphlops humilis*), however, make up as much as 10 percent of the diet of large scorpions (*Hadrurus*) in arid western North America, and small tropical forest snakes may be especially vulnerable to the centipedes, spiders, and other large invertebrates common in such habitats.

Central American Coralsnake Mimic (*Erythrolamprus mimus*), Costa Rica, immobilizing a Forest Racer (*Dendrophidion vinitor*) by venom injection.

Fish occasionally eat serpents that enter the water; these prey range from blindsnakes (*Leptotyphlops macrolepis*) and Common Gartersnakes (*Thamnophis sirtalis*) to Black Mambas (*Dendroaspis polylepis*) and Puff Adders (*Bitis arietans*). Although amphibians typically eat invertebrates, large salamanders occasionally eat tiny snakes (e.g., a Lesser Siren [*Siren intermedia*] ate a Striped Swampsnake [*Regina alleni*]). Some Malaysian toads (*Bufo*) commonly feed on Flowerpot Blindsnakes (*Ramphotyphlops braminus*), and I found a Central American Bullfrog (*Leptodactylus pentadactylus*) swallowing an adult Brown Blunt-headed Vinesnake (*Imantodes cenchoa*). Large frogs occasionally even take highly venomous snakes; for instance, a North American Bullfrog (*Rana catesbeiana*) ate a Harlequin Coralsnake (*Micrurus fulvius*), and an African Bullfrog (*Pyxicephalus adsperus*) contained a litter of Rinkhals (*Hemachatus haemachatus*). Although turtles, crocodilians, and lizards rarely eat snakes, serpents themselves are well designed as predators on other elongate vertebrates; Asian pipesnakes (*Cylindrophis*), most New World coralsnakes, and a few colubrids (e.g., Short-tailed Snakes [*Stilosoma extenuatum*]) routinely prey on other snakes, even venomous species.

Many birds occasionally eat snakes, and some species prey heavily on them, even taking powerful constrictors or venomous elapids and viperids. For example, among the forty-four reptiles found in the nests of Brown Snake-Eagles (*Circaetus cinereus*) were forty-one snakes, including Puff Adders (*Bitis arietans*), Black Mambas (*Dendroaspis polylepis*), and Mozambique Spitting Cobras (*Naja mossambica*). Besides mammals as large as Kinkajous (*Potos flavus*), Guianan Crested Eagles (*Morphnus guianensis*) eat Rainbow Boas (*Epicrates cenchria*) and Emerald Tree Boas (*Corallus caninus*). In California, Red-tailed Hawks (*Buteo jamaicensis*) sometimes eat Western Rattlesnakes (*Crotalus viridis*), but they prefer Gopher Snakes (*Pituophis catenifer*) when those nonvenomous colubrids are available; in Kansas, however,

Blue-crowned Motmot (*Momotus momota*), Costa Rica, carrying a littersnake (*Rhadinaea*) to its nest.

venomous Copperheads (*Agkistrodon contortrix*) are among the most frequently taken prey by that raptor. Other well-known avian snake predators include Secretarybirds (*Sagittarius serpentarius*), Roadrunners (*Geococcyx californicus*), and various neotropical raptors (e.g., Laughing Falcons [*Herpetotheres cachinnans*], White Hawks [*Leucopternis alba*]).

Many mammals opportunistically eat snakes, and a few species specialize on them. Eastern Moles (*Scalopus aquaticus*) occasionally attack inactive juvenile Copperheads (*Agkistrodon contortrix*) under rocks. Among carnivores, Lions (*Panthera leo*) and Leopards (*P. pardus*) sometimes eat African Rock Pythons (*Python sebae*). In addition to insects, fruits, and other vertebrates, Ringtails (*Bassariscus astutus*) take small pitvipers, and fossilized droppings from

Central American Coralsnake (*Micrurus nigrocinctus*), Costa Rica, during a defensive tail display.

that raccoon relative contained bones of a Speckled Rattlesnake (*Crotalus mitchellii*). The stomach of a Selous Mongoose (*Paracynictis selousi*) contained a frog, rodents, crickets, twenty-one lizards, two Cape Wolfsnakes (*Lycophidion capense*), and a Southern Stiletto Snake (*Atractaspis bibronii*). Although no primates specialize on snakes, White-faced Monkeys (*Cebus capucinus*) sometimes eat harmless colubrids, and one, perhaps defensively, killed an adult Terciopelo (*Bothrops asper*) with a stick. One Patas Monkey (*Erythrocebus patas*) ate a meter-long West African Green Mamba (*Dendroaspis viridis*), and a Bornean Tarsier (*Tarsius bancanus*) consumed a Brown Long-glanded Coralsnake (*Maticora intestinalis*). Baboons are generalized terrestrial foragers in many arid parts of Africa, and they, along with birds, might be what drives viper mimicry by egg-eaters (*Dasypeltis*).

Clearly snakes are eaten by a wide range of enemies; even those that we might expect to be invulnerable are taken, including nocturnal, aquatic, fossorial, and highly venomous species. The handling abilities of predators depend on more than just relative prey size: Coatis (*Nasua narica*) are rebuffed by New World coralsnakes, but another small carnivore, the Grison (*Galictis vittata*), skillfully kills and eats those elapids. Most snakes are probably vulnerable to different predators when they are active than when they are inactive. As a result, some defense is effective in some cases, but no defense always works (except possibly that of Yellow-bellied Seasnakes [*Pelamis platurus*], Chapter 13). Predation on snakes is probably more intense and diverse in tropical regions throughout the world. At least twenty of eighty-nine diurnal African raptors feed wholly or in part on reptiles, including snakes, and many smaller carnivores on that continent also do so. Costa Rican snakes face a wide spectrum of predators, ranging from katydids and spiders to frogs, lizards, other snakes, birds, and mammals. Some predators hunt in particular microhabitats: for instance, the neotropical Crane-hawk (*Geranospiza caerulescens*) systematically searches bromeliads and holes in trees for large insects, amphibians, lizards, snakes, and nestling birds; New World coralsnakes search for smaller prey snakes in leaf litter. Such predators represent a constant threat to otherwise hidden, inactive animals, which implies that selection of a retreat site is important for avoiding predators.

Birds and mammals are perhaps especially significant for the evolution of antipredator mechanisms in snakes, because they are attentive to color and other visual details as well as clever at using their limbs and beaks or mouths to manipulate potentially dangerous objects. Both kinds of predators are thus especially likely to discover cryptic prey and effectively attack the head or other vulnerable parts, but at the same time they are more susceptible to deceptive displays. Moreover, parent birds and mammals often transport intact prey to their offspring, a practice that makes immobility and eventual escape from the nest a viable option for some snakes. Finally, both groups have high metabolic rates and therefore high consumption rates, and as a result their impact on prey populations is often much greater than that of ectothermic predators.

ANTIPREDATOR TACTICS OF SNAKES

Typically snakes exhibit a hierarchy of responses to predation, overlain by various specializations in particular lineages and even species. Most serpents first attempt to avoid detection; when that is impossible, they seek cover or resort to various displays, among other measures. Once discovered, blindsnakes and basal serpents generally rely on locomotor escape, struggling, discharging their cloacal contents, and, in the case of many boas and pythons, biting. The following descriptions of four species illustrate the wealth of responses to danger exhibited by advanced snakes.

Eastern Hog-nosed Snakes (*Heterodon platirhinos*) have one of the most complex defensive repertoires of any serpent. The dorsal pattern of this moderately stout species consists of dark blotches on a lighter ground color, although in some areas many individuals are melanistic. When initially threatened, an Eastern Hog-nosed Snake assumes an exaggerated S-coil, elevates and flattens its neck, and hisses loudly; when approached closely, it makes false strikes with its closed mouth. Those reactions are repeated if the snake is grasped, but after repeated harassment, the snake writhes violently with its mouth open limply and defecates. If placed on the ground, the hog-nosed snake continues to writhe for a short time, turns over, and lies limp—usually with its tongue hanging out of its open mouth. When placed upright, the hog-nosed snake immediately turns upside down again. If the threat is removed, the snake slowly rights itself and crawls away.

Another widespread North American colubrid, the Racer (*Coluber constrictor*), illustrates the responses of many slender, unicolored, fast-moving species of snakes. A startled adult Racer reacts by violent undulatory movement that quickly changes to rapid, graceful locomotion for thirty meters or more. If startled at very close range, a Racer sometimes coils with its head hidden and writhes, smearing scent gland secretions over its body. Alternatively, the

Black-tailed Rattlesnake (*Crotalus molossus*), Arizona, during a threat display and ready to strike.

Pink-bellied Littersnake (*Rhadinaea decorata*), Costa Rica. Its tail has recently been broken by a predator.

snake assumes an S-coil, vibrates its tail, and strikes repeatedly. If seized, a Racer continues to bite and discharge its cloacal contents, while twisting its entire body so strongly that sometimes the tail is broken off. Juvenile Racers have a boldly spotted color pattern and probably rely more heavily on camouflage for escape from predators than do adults.

If touched lightly, a Sonoran Coralsnake (*Micruroides euryxanthus*) quickly crawls away. When touched firmly or restrained, this bright red, black, and cream–banded species writhes rapidly and violently, with its head tucked laterally and its tail elevated in a coiled loop or spiral. As the coralsnake squirms and displays its tail, loops of the body move in vertical and horizontal planes. Its cloacal contents are expelled with a series of sharp popping sounds that a human can hear from at least one meter away. It continues to writhe and bite as long as it is restrained, and even a small adult of this species can envenom a large mammal, with serious consequences. If released, the snake will either crawl rapidly away or freeze for a moment, often with its head tucked under a loop of its body and with its tail elevated. Like some other New World elapids (e.g., Harlequin Coralsnakes [*Micrurus fulvius*]), this species readily regurgitates recently eaten prey when disturbed by a potential predator.

When approached from a distance, a Prairie Rattlesnake (*Crotalus v. viridis*) is either already immobile or freezes, its blotched dorsal color pattern difficult to perceive against natural backgrounds. If approached closely, the rattler's first response is usually to crawl rapidly away. If cornered or restrained, it assumes a pre-strike posture, its head raised and its foreparts drawn into an elevated S-coil. By that time the snake is typically rattling vigorously and continuously. If the predator remains in proximity or touches the rattler, it may strike and bite, hide its head under a coil of its body, or first hide its head and then suddenly strike from under a coil. If forcibly restrained, a Prairie Rattler bites, rattles, and discharges the cloacal scent glands in a fine spray.

Additional tactics used by other snakes include exuding blood from around the eyes, nostrils, and cloaca (e.g., Caribbean dwarf boas [*Tropidophis*], Long-nosed Snakes [*Rhinocheilus lecontei*]); rolling into a ball with the head hidden (e.g., Brown Sand Boas [*Eryx johnii*]; Viperine Watersnakes [*Natrix maura*]); actively breaking the tail (e.g., African watersnakes [*Natriciteres*]; some littersnakes [*Rhadinaea*]); ejecting a viscous, distasteful substance from the neck glands (some keelbacks [*Macropisthodon*]); spreading a hood and "spitting" venom (e.g., cobras; see the Special Topic in this chapter); and producing rasping or hissing sounds by rubbing body scales (e.g., saw-scaled vipers [*Echis*]) or vibrating the tail (e.g., some colubrids, most pitvipers). Other examples of defensive behavior for each of the major groups of snakes are given in later chapters.

SOURCES OF VARIATION: CONTROL, DEVELOPMENT, AND GENETICS

Snakes may vary widely in their responses to predators, among individuals in a population as well as geographically within a species. For example, Ring-necked Snakes (*Diadophis punctatus*) from most of the eastern United States have yellow bellies and lack a defensive tail display, whereas individuals from Florida and from western North America typically have scarlet venters and a startling inverted and coiled tail display. However, within a Kansas population roughly two-thirds of juvenile Ringnecks displayed their tails when handled but less than one-third of adults responded in kind. Several other species offer examples of geographic variation in defensive behavior. The Boomslangs (*Dispholidus typus*) that live in a treeless

Ring-necked Snake (*Diadophis punctatus*), Arizona, during a defensive tail display.

mountain area of southern Africa are more likely to bite than those found elsewhere. Coachwhips (*Masticophis flagellum*) usually escape rapidly and strike vigorously if cornered, but those in California's San Joaquin Valley form a stiff, irregular coil when restrained, concealing their heads. Prairie Rattlesnakes (*Crotalus v. viridis*) and some Northern Pacific Rattlesnakes (*C. v. oreganus*) are quick to rattle and strike, but Arizona Black Rattlesnakes (*C. v. cerberus*) seem rather mellow-tempered. Such variability could result from differences in stimulus, age, sex, temperature, and other factors, but they also might represent genetically based geographic differences.

Our views of defensive responses in snakes are heavily influenced by their reactions to humans, which often might be misleading. Rosy Boas (*Charina trivirgata*) react to gentle handling by forming a ball of coils, but when more vigorously threatened they release an extremely foul cloacal discharge. Violet Kukrisnakes (*Oligodon violaceus*) rarely defend themselves when picked up by humans, but they react to mongooses (*Herpestes*) by tail vibration and striking. In fact, snakes might perceive potential predators by visual, thermal, auditory, chemical, and tactile cues. Under experimental conditions, Plains Gartersnakes (*Thamnophis radix*) respond defensively (e.g., by head-hiding) if their heads are tapped and offensively if their tails are tapped. Prairie Rattlesnakes (*Crotalus v. viridis*) and Massasaugas (*Sistrurus catenatus*) react defensively to large, looming objects but not to eye marks and other precise stimuli on simulated predators. Pitvipers and some other species respond to the odor of Common Kingsnakes (*Lampropeltis getula*) and certain other snake-eating species with a stereotyped posture, in which the body is elevated and used as a shield or even a club against the attacker.

Response to a threatening stimulus can also be influenced by the context for an encounter. Indian Kraits (*Bungarus caeruleus*) hide their heads when threatened during daylight but crawl and bite when confronted at night. Whereas Queen Snakes (*Regina septemvittata*) initiate escape at greater distances when temperatures are relatively cool, European Adders (*Vipera berus*) have shorter evasion distances in cool spring and fall months than in the summer; those adders also flee for shorter distances when undernourished, during years when rodents are scarce. Solitary Prairie Rattlesnakes (*Crotalus v. viridis*) threatened in the open emphasize locomotor escape, whereas those near cover usually remain immobile; however, Prairie Rattlers grouped near cover frequently resort to escape and threat postures, which suggests social facilitation of individual responses to a predator.

An animal's characteristics reflect a complex interaction between genetic heritage and the environment within which it develops. Typical behavior at hatching or birth and stable variation among neonates indicate genetic influences on the defensive behavior of snakes. Of course, failure to exhibit appropriate responses during a first predator encounter might be fatal, and indeed naive young serpents often exhibit species-typical defensive behaviors. Examples include tail displays in Brown Sand Boas (*Eryx johnii*), cloacal popping in Western Hook-nosed Snakes (*Gyalopion canum*), hood spreading in Forest Cobras (*Naja melanoleuca*), tail vibration in Bushmasters (*Lachesis muta*), and diverse tactics in Eastern Hog-nosed Snakes (*Heterodon platirhinos;* see p. 30). Individually stable variation in defensive behavior occurs within and among litters of Eastern Hog-nosed Snakes and several species of gartersnakes (*Thamnophis*), and at least in the latter that variation reflects heritable variation among the parents.

Some defensive responses require no experience with predators yet are not elicited at hatching or birth, which suggests a maturational component to their normal expression. Many young Boa Constrictors (*Boa constrictor*) are mild-mannered, whereas adults in some populations hiss loudly and strike readily; conversely, juvenile Gopher Snakes (*Pituophis catenifer*) are more likely than adults to hiss, vibrate their tails, and strike. Mole Snakes (*Pseudaspis cana*) do not hiss and strike until eighteen hours after birth, and juvenile Copperheads (*Agkistrodon contortrix*) cannot eject a stream of cloacal scent gland liquid until they are several days old. Young Red Coffeesnakes (*Ninia sebae*) exhibit defensive head displays like those of adults, except that the hatchlings are wobbly, perhaps because of their lack of muscular coordination or their low trunk mass. Some age-related shifts in defensive responses are influenced by the fact that in proportion to the snake's mass, a juvenile crawls more slowly, struggles less effectively, and tires more readily than an adult. Consequently, some young serpents rely on seclusion and cryptic color patterns not present in adults (e.g., prominent blotches in Northern Watersnakes [*Nerodia sipedon*]), or they resort to behavior that does not involve locomotor escape (e.g., balling by Viperine Watersnakes [*Natrix maura*]).

How snakes adjust their defensive responses to experience with predators, full stomachs, pregnancy, and other factors is a fascinating problem about which we know very little. After repeated handling by humans, defensive behavior wanes in Eastern Hog-nosed Snakes (*Heterodon platirhinos*), Gopher Snakes (*Pituophis catenifer*), some gartersnakes (*Thamnophis*), and probably many other species. In some populations Timber Rattlesnakes (*Crotalus horridus*) abandon den sites when they are disturbed frequently, whereas in others they do not; it is possible, then, that differences in visitor behavior affect the snakes to varying degrees. Wild snakes with missing tails and other healed injuries illustrate that individuals can survive

Northern Death Adder (*Acanthophis praelongus*), Australia, with its body flattened in a defensive display.

Anatomical modifications that help snakes avoid discovery or defend themselves against predators include cryptic and warning colors; features that aid locomotor escape, sound production, or existing weaponry; and structures that are evidently only for defense. Examples of morphological specializations for defense include reinforced, clublike tail vertebrae in sand boas (*Eryx*) and their relatives, used to deflect attacks from the head (see p. 274); tracheal rings that enable throat inflation in neotropical birdsnakes (*Pseustes*) and Savanna Twigsnakes (*Thelotornis capensis*); glands in the neck skin of some keelbacks (*Macropisthodon*) that produce a noxious secretion; bright flash colors in North American Mudsnakes (*Farancia abacura*), Ring-necked Snakes (*Diadophis punctatus*), and many other species; and the cartilaginous preglottal keel that amplifies hissing in Eastern Pinesnakes (*Pituophis melanoleucus*) and their close relatives. As in most other animals, structures used for defense are often primarily adaptations for other roles and are either coopted outright or modified in minor ways. Usually their main role is movement and feeding, but male Common Kukrisnakes (*Oligodon cyclurus*) and Southern Coralsnakes (*Micrurus frontalis*) provide a notable exception, everting their hemipenes during defensive displays.

Afro-Asian Cobras (*Naja*) are particularly interesting because they confront enemies with multiple innovations, derived from both locomotor and feeding equipment. Cobras erect their hoods by extending elongated anterior ribs, such that the skin of their necks is stretched forward and to the side. Hooding typically occurs while the cobra's head is elevated, whether facing toward or away from an adversary. King Cobras (*Ophiophagus hannah*) sometimes growl during their erect defensive displays and, like most other cobras, strike from a hooded posture. When threatened for the first time, hatchlings of Forest Cobras (*Naja melanoleuca*), King Cobras, and probably other species readily spread their hoods. The hoods of most Asian cobras are shorter and wider than those of most African *Naja* and of King Cobras. Burrowing cobras (*Aspidelaps*), water cobras (*Boulengerina*), and some mambas (*Dendroaspis*) display less well developed hoods. Phylogenetic relationships among elapids are too poorly understood to determine whether the hood arose one or more times in those taxa, but plausible intermediate stages for the evolution of well-developed hoods are illustrated in elapids that flatten their entire body (e.g., death adders [*Acanthophis*]; New World coralsnakes) or their modestly enlarged neck region (e.g., Tiger Snakes [*Notechis scutatus*]).

Some cobras have a dark bar or other distinctive marking on the

Indian Cobra (*Naja naja*) with its hood spread.

ventral surface of the neck, easily visible from the front during defensive displays. Cobras in Africa, western Asia, Indonesia, and the Philippines do not have dorsal hood markings, although sometimes their body patterns are bright and contrasting (e.g., banded Egyptian Cobras [*Naja haje*]). Elsewhere in Asia, however, many Indian Cobras (*N. naja*) and Siamese Cobras (*N. siamensis*) have spectaclelike markings, whereas Monocled Cobras (*N. kaouthia*) have a single circular mark, and Chinese Cobras (*N. atra*) may be monocled or spectacled. In fact, the only snakes with dorsal hood marks are found in areas where cobras are sometimes associated with various religious beliefs and ceremonies, which suggests that favorable human attitudes toward these venomous snakes might have affected the evolution of their color patterns.

Perhaps some markings even reflect specific traditions, since Buddha's kiss reportedly left a monocled outline on the hood of a large cobra that shielded him from the sun. Of course, large eyelike markings might simply increase the fright value of a snake's defensive head display or give the impression of facing an adversary as the cobra retreats.

The Rinkhals (*Hemachatus hae-*

machatus) and eleven species of Afro-Asian cobras (*Naja*) have highly modified fangs and defensively eject, or "spit," venom at the eyes of their enemies. Nonspitters discharge their venom through large, elliptical orifices that end at the tip of the fang, whereas effective spitters have a smaller, circular and beveled aperture on the anterior surface of each fang above the tip. Spiral grooves inside the fangs of African spitters function like riflings in a gun barrel, forcing a spin on the emerging liquid and further enhancing accuracy. A spitting cobra lunges from an elevated posture, usually with an audible expulsion of air or a loud hiss, and directs a stream of well-aimed, coherent droplets at its victim. One Black-necked Spitting Cobra (*N. nigricollis*) expended all of its stored venom by spitting fifty-seven times in twenty minutes. The venom of Black-necked Spitters causes severe symptoms, including permanent blindness, if not treated; whereas the venoms of other African species and the Asian spitters mainly result in pain, swelling of the eyelids, and conjunctivitis. Cobras tend to spit and flee, but they do bite defensively, and Black-necked Spitters can cause extensive necrosis.

Four of seven species of African cobras spit, and seven of nine species in Asia spit to varying degrees. Within each region, species of *Naja* that spit are evidently monophyletic; therefore, spitting has arisen at least three times (i.e., in the Rinkhals and twice in other cobras). Within Asian snakes, variation in fang structure is not tightly correlated with spitting, which suggests that the behavior has been variably lost in some species. For example, the Monocled Cobra has spitter fangs but rarely spits; female Philippine Cobras (*N. philippinensis*) have larger discharge orifices than males, but there are no known behavioral differences in the two sexes. We might expect spitters to face an opponent more readily than nonspitters, and this seems to be true. Monocled Cobras are relatively placid, whereas Siamese Cobras and Sumatran Cobras (*N. sumatranus*)— both spitters—vigorously defend themselves.

Unfortunately, we have few details about the natural history of cobras with which to interpret their fascinating array of morphological and behavioral adaptations for defense. Possible ramifications for the rest of their biology are intriguing, especially since females of some

Asian *Naja,* some Shield-nosed Cobras (*Aspidelaps scutatus*), and King Cobras remain with and aggressively defend their incubating eggs—a parallel with the correlated origins of predator confrontation and parental care in pitvipers (Chapter 14). Recent systematic studies of cobras have greatly increased our understanding of variation in these snakes, and field studies of their behavior surely will provide important new insights.

predator encounters and thereby modify their future responses (photo, p. 104); I watched a Ring-necked Snake (*Diadophis punctatus*), besmeared with its own cloacal contents, endure more than five minutes of tentative mouthing by a captive Kit Fox (*Vulpes macrotis*) without apparent ill effects. In some cases hormonal factors might influence the expression of defensive behavior: males of three species of North American watersnakes (*Nerodia*) are more likely to bite than are females, and male Black Racers (*Coluber c. constrictor*), Black Mambas (*Dendroaspis polylepis*), and some seasnakes are reportedly more aggressive toward humans during the breeding season.

EVOLUTIONARY AND ECOLOGICAL CONSIDERATIONS

Cryptic coloration, locomotor escape, struggling, defecating, biting, and perhaps hissing are responses that probably go back to the earliest amniotes, judging from their use by representatives of all major living lineages. Many snakes retain those primitive defenses, and they are further characterized as a group by discharging their unique cloacal scent glands when threatened. Subsequent antipredator innovations within Serpentes often characterize several or all species in a clade. Defense mechanisms associated with such historical groupings include bluffing and then death-feigning in North American hog-nosed snakes (*Heterodon*), flattening of the body in certain xenodontine colubrids (e.g., Fire-bellied Snakes [*Liophis epinephalus*] and related taxa), bright warning colors and erratic displays in New World coralsnakes; see p. 226), hoods in cobras (e.g., *Naja;* see the Special Topic in this chapter), and the noisemaker of rattlesnakes (e.g., *Crotalus*).

Correlations with particular environmental variables and experimental studies provide insights into the biological roles of antipredator mechanisms. Unicolored and striped snakes tend to live in open habitats and respond to threat initially by rapid crawling; the absence of transverse markings makes it difficult to aim effectively for those species (some humans learn to "lead" their quarry by grabbing in front of a Racer [*Coluber constrictor*] or a gartersnake [*Thamnophis*]). Tail injuries seem particularly common in certain striped or unicolored species in aquatic habitats (e.g., African watersnakes [*Natriciteres*]) and open terrestrial habitats (e.g., some whipsnakes [*Masticophis*]), perhaps because predators often barely catch their relatively long tails, and the snakes twist free. Blotched and ringed snakes more often rely on immobility and confrontational defenses than on locomotor escape; the color patterns of rattlesnakes promote camouflage, whereas those of New World coralsnakes and the Bandy-bandy (*Vermicella annulata*) confuse or disorient predators during their thrashing displays. Defensive behavior and color pattern can vary concordantly among individuals, even within litters; for instance, striped Northwestern Gartersnakes (*Thamnophis ordinoides*) rely more on locomotor escape while spotted ones emphasize cryptic coloration and static displays. Such polymorphisms can also reflect selection pressures other than predation. Predators discover black European Adders (*Vipera berus*) more frequently than striped morphs, but although they may be more vulnerable to attack, black individuals thermoregulate better.

Behavioral responses have ecological correlates as well. Fossorial and terrestrial snakes often exhibit tail displays (e.g., Sonoran Coralsnakes [*Micruroides euryxanthus*]) and horizontal head displays (e.g., cobras), respectively, whereas vertical head displays (e.g., Savanna Twigsnakes [*Thelotornis capensis*]) and open-mouth threats (e.g., Asian Long-nosed Vinesnakes [*Ahaetulla nasuta*]) typify many arboreal species. Tail displays might direct attention away

Above: Savanna Twigsnake (*Thelotornis capensis*), southern Africa, hidden in vegetation.

Fire-bellied Snake (*Liophis epinephalus*), Costa Rica, spreading hood as a defensive display.

Bushmaster (*Lachesis muta*), Costa Rica,
lying camouflaged on the forest floor.

from the more vulnerable head, especially if that part is hidden under a coil of the body, to a dose of cloacal scent gland discharge; they also leave the potential prey's head free for a counterattacking bite. More than half of wild adult Brown Sand Boas (*Eryx johnii*) have heavily scarred posteriors, which indicates that displays of their heavily buttressed tails (see p. 274) have been effective against predators; one with a freshly bleeding tail, but otherwise uninjured, was found surrounded by tracks of a Golden Jackal (*Canis aureus*). Bright ventral flash colors characterize Ring-necked Snakes (*Diadophis punctatus*), mainly in areas that also have Scrub Jays (*Aphelocoma coerulescens*), sharp-eyed birds that feed on small animals in ground litter.

Head displays make terrestrial and arboreal snakes appear larger than life, thus too big for the predator to subdue; they probably also imply a willingness to bite, especially when accompanied by an open-mouth threat. Terrestrial snakes that orient upward toward enemies, such as North American hog-nosed snakes (*Heterodon*) and cobras, typically flatten their anteriors dorso-ventrally; arboreal species like birdsnakes (*Pseustes*) and Savanna Twigsnakes (*Thelotornis capensis*) that face adversaries laterally usually inflate their necks. Some arboreal species restrict their movements to periods when wind moves the surrounding vegetation, thereby

The False Terciopelo (*Xenodon rabdocephalus;* right), Costa Rica, closely resembles the Terciopelo (*Bothrops asper;* opposite) in its color pattern. The latter snake, however, is dangerously venomous.

remaining invisible against similarly colored backgrounds (e.g., Rough Greensnake [*Opheodrys aestivus*], Brown Vinesnake [*Oxybelis aeneus*]). Open-mouth threat displays are especially prevalent in colubrids and vipers, the latter including arboreal species (e.g., Lowland Bush Viper [*Atheris squamiger*]) as well as aquatic and terrestrial ones (e.g., Jumping Pitviper [*Atropoides nummifer*] and Cottonmouth [*Agkistrodon piscivorus*], respectively).

Charles Darwin was skeptical about death-feigning as a viable defense, and some modern-day zoologists also have doubted whether the spectacular writhing and inverted immobility of North American hog-nosed snakes (*Heterodon*), the Rinkhals (*Hemachatus haemachatus*), and occasionally other species could be effective against predators. An interesting alternative explanation was inspired by the fact that the most famous death-feigning snakes also have enlarged adrenal glands and so might temporarily collapse as a result of an imbalance in stress hormones. This, however, now seems unlikely, since hog-nosed snakes switch rapidly in and out of death-feigning in response to the presence or absence of a threat. A more plausible scenario is that death-feigning causes some predators to cease killing behavior, because the cues to elicit attack are absent. If the predator is a parent bird or mammal, the immobile snake can be left alive at a nest, from which it later might escape. Perhaps for the Rinkhals and other venomous species, the cessation of a predator's bites when death is feigned also gives the snake a chance to counterattack.

Venoms probably have influenced the overall evolutionary diversification of snakes, even beyond those groups that possess them. Toxic head gland secretions and elaborate defensive mechanisms (e.g., open-mouth threats, hoods, and sounds produced by rubbing scales and vibrating the tail) are nearly or entirely restricted to advanced snakes, which suggests that

those two attributes somehow co-evolved. Mimicry is far more widespread among serpents than among other terrestrial vertebrates, and most of the models are dangerously venomous elapids and vipers (see p. 226). New World mimics of coralsnakes include more than one hundred species of colubrids, such as Fire-bellied Snakes (*Liophis epinephalus*), some ground-snakes (*Sonora*), some calico snakes (*Oxyrhopus*), and some Milksnakes (*Lampropeltis triangulum;* see p. 198). Among possible New World pitviper mimics are three species of North American hog-nosed snakes (*Heterodon*), South American Hog-nosed Snakes (*Lystrophis dorbignyi*), the Eastern Pinesnake (*Pituophis melanoleucus*) and its relatives, two species of mock pitvipers (*Tomodon*), seven species of false pitvipers (*Xenodon, Waglerophis*), at least some of the roughly fifty species of New World slug- and snail-eaters (*Dipsas, Sibon*), and various diurnal arboreal colubrids (e.g., Green Parrotsnake [*Leptophis ahaetulla*]). In Australasia, some keelbacks (*Macropisthodon*) resemble cobras (*Naja*), several reedsnakes (*Calamaria*) are strikingly similar to long-glanded coralsnakes (*Maticora*), and some Viper Boas (*Candoia aspera*) look very much like Common Death Adders (*Acanthophis antarcticus*). Seemingly harmless species that particularly resemble Old World vipers in color pattern and often defensive behavior include six species of egg-eaters (*Dasypeltis*), several species of catsnakes (*Boiga, Telescopus*), two species of mock vipers (*Psammodynastes*), Diadem Ratsnakes (*Spalerosophis diadema*), and juvenile Mole Snakes (*Pseudaspis cana*).

Timber Rattlesnake (*Crotalus horridus*),
Pennsylvania.

BEHAVIOR, REPRODUCTION, AND POPULATION BIOLOGY

Timber Rattlesnakes (*Crotalus horridus*) were once prominent predators in the forests of eastern North America. John Smith mentioned them in "A Map of Virginia, with a Description of the Countrey . . . ," published in 1621, and rattler flags with the motto "Don't Tread on Me" were popular in the Revolutionary War. The *Pennsylvania Journal* for December 27, 1775, extolled this species, "found in no other quarter of the globe than America":

> Her eye exceed[s] in brilliance that of any other animal . . . and she has no eyelids. She may therefore be esteemed an emblem of vigilance. She never begins an attack, or, when once engaged, ever surrenders. She is therefore an emblem of magnanimity and true courage. . . . She never wounds until she has generously given notice even to her enemy, and cautioned . . . against the danger of treading on her. . . .
>
> The poison of her teeth is the necessary means of digesting her food, and at the same time is the certain destruction of her enemies. . . . The rattles . . . [are] just thirteen— exactly the number of colonies united in America. . . . One of these rattles, singly, is incapable of producing sound; but the ringing of thirteen together is sufficient to alarm the boldest man living. . . . She is beautiful in youth and her beauty increases with age; her tongue is also blue, and forked as lightning, and her abode is among the impenetrable rocks.

Scarcely more than two centuries later, these elegant, velvety creatures are endangered or already extinct throughout much of their former range. Their shockingly rapid decline stems from a collision between snake behavior, human customs, and the fundamental components of population biology.

Native Americans and early European colonists did not wipe out entire populations of Timber Rattlesnakes. Rather, it was the widespread persecution of those snakes in the late nineteenth and twentieth centuries that removed tens of thousands of them from a significant chunk of the continent, a saga that recalls the demise of Passenger Pigeons (*Ectopistes migratorius*), the near extinction of American Bison (*Bison bison*), and other larger-scale environmental travesties. During winter months Timber Rattlesnakes aggregate at rock outcrops, and thus entire local populations are vulnerable to mass destruction. Among countless incidents in Pennsylvania, a few hunters killed hundreds one day in 1886, and 250 were killed in a week in 1906 at another den. In the 1930s a state park superintendent covered an Illinois den with cement, and in the 1940s New York state wildlife personnel dynamited several dens. For decades bounties were offered, and local organizations promoted hunts as commercial activities. In New Jersey, New York, and Massachusetts about four thousand Timber Rattlers were sold over a thirty-year period. One bounty hunter alone took thousands from New York in the 1960s, and overall populations in that state have been reduced by 60 percent during this century. A majority of all historically active dens in the northeastern United States are now inactive, and populations have been reduced in many of those still harboring rattlers; small colonies persist in the Blue Mountains near Boston, but most major den complexes in New England have been destroyed or severely depleted.

Any healthy, stable population must contain enough reproducing females that new organisms reach maturity at roughly the rate they are lost. Number of offspring is not the key factor; rather, stability depends on immigration and the survival of offspring balancing emigration and mortality. Replacement thus entails growth rates, sex ratios, ages at maturity, litter size, and frequency of breeding—traits that combine to place new individuals in a population and counter local death rates. Although those life history traits vary in numerous ways, two extremes are illustrative: Some toads, flies, and other prolific organisms produce large numbers of small offspring, most of which die; the remainder grow rapidly and reach sexual maturity at a young age. Such creatures often do well in fragmented habitats, because even a few gravid females can establish or replenish a population. By contrast, elephants and some other species invest considerable energy in a few young, each with a high chance of survival.

Timber Rattlesnakes at cold northern latitudes exemplify the second strategy, at least in part because of their unusually short active season. Females require nine to ten years to reach maturity, breed every three to four years, and produce about ten large young; each female might reproduce only three to five times in a twenty-five-year lifetime. This strategy worked in the absence of widespread persecution by humans, because adult rattlers were largely safe from natural predators, and low renewal rates maintained viable populations. With low reproductive potential, however, a sustainable population requires at least forty-five snakes, including eight to ten mature females. Such groups recover slowly from abnormal depletion, and each adult female plays an important role in the long-term health of the population.

Other aspects of Timber Rattlesnake natural history also affect their fate. Individuals use a winter den, a summer hunting range, and transient areas, annually traveling hundreds of meters among those sites. A healthy population occupies at least 18 km² to encompass all three habitats, and even that does not assure dispersal, and thus gene flow, among populations. Urban and agricultural developments have drastically reduced wilderness, however, and the resulting fragmentation often slices across the total home range of a den population. In otherwise suitable terrain, encroaching vegetation might shade crucial summer basking sites and further reduce a female's chance of producing a litter. Saving the Timber Rattlesnake must therefore encompass, above all, protection of dens and preservation of large surrounding parcels. Among direct management measures, artificial basking clearings and other habitat enhancement might be appropriate in some cases; translocated rattlesnakes might repatriate extinct dens and augment genetic diversity in remaining small populations. Education programs must convince the public that even venomous serpents are worthy of respect and deserve a place in nature; the occasional "nuisance" rattler in a backyard or other places where it might pose a genuine threat to human welfare must be dealt with effectively.

With humans dominating the planet, extinction and conservation are inevitably cultural and political processes, played out against the backdrop of technology and our own burgeoning populations. Demographically the opposite of a good colonizing species, Timber Rattlesnakes are poorly adapted for highly fragmented forests and heavy adult mortality. Some of the roughly three million Europeans in North America at the time of the Revolutionary War disliked rattlesnakes, but there was no dynamite or gasoline with which to plunder dens. Colonists perhaps killed Timber Rattlers for food, as trophies, or out of spite—but so few humans could not exact a potentially irreversible decline. By 1979 there were more than 225 million people in the United States, and, widespread environmentalism notwithstanding, rattlesnakes were still not popular. That year the secretary of the interior fired C. Kenneth Dodd Jr., then employed by the Office of Endangered Species and now a leading conservation biologist, for protesting the sale of Timber Rattler meat in a fancy restaurant in Washington, D.C. As of 1992 Pennsylvania still permits sport hunting of the species, with full knowledge that 75 percent of the dens in the state are in jeopardy of total extinction.

The bottom line is that a formerly abundant predator is on the brink of disappearance in states where it once was emblematic of freedom and independence. Now Timber Rattlesnakes remind us that truly pristine landscapes no longer exist in the eastern United States, and that only strict protection and carefully planned intervention will maintain many vertebrate populations into the twenty-first century.

Replacing an individual in the following generation can be complicated, and routes to success vary greatly among serpents. Many species simply abandon their future progeny after laying eggs; others construct and guard nests; and some nourish developing young that are later born alive and fully competent. In keeping with heavy losses of juveniles to predation and starvation, Northern Watersnakes (*Nerodia sipedon*) and Puff Adders (*Bitis arietans*) produce enormous litters, from which most offspring die before they reproduce; well-defended and secretive, Louisiana Pinesnakes (*Pituophis ruthveni*) and Timber Rattlesnakes (*Crotalus horridus*) devote considerable energy to a few well-provisioned eggs or neonates, respectively. Those fundamental tradeoffs between mortality and reproduction are mediated by movement patterns, activity cycles, and social interactions—and snakes vary widely in this last respect as well. Although they were long regarded as asocial, recent field and laboratory studies reveal considerable complexity and diversity in the behavioral repertoires of serpents. This chapter surveys snake reproduction, with an emphasis on how social behavior, environmental characteristics, and evolutionary history influence their breeding biology.

Eyelash Pitviper (*Bothriechis schlegelii*), Costa Rica, with multicolored litter. Note the distinctively colored tail tips on the neonates, probably used as lures.

MOVEMENTS, ACTIVITY CYCLES, AND AGGREGATIONS

Snakes move to hunt prey, avoid predators, find mates, and locate favorable conditions for thermoregulation, egg-laying, and inactivity (e.g., ecdysis, hibernation). The fact that captive rattlesnakes flick their tongues more when cage props are displaced suggests they recognize the changes, while free-living individuals of many species repeatedly use particular sites for hunting, shedding, and hibernating. I have no doubt that wild snakes possess a detailed internal "map" of their surroundings. Serpents orient short-range movements (several centimeters to a few kilometers) with respect to local landmarks, using visual cues (e.g., Striped Whipsnakes [*Masticophis taeniatus*] alter their course in relation to nearby ridges), chemical pheromones (e.g., juvenile rattlesnakes follow adult trails to hibernacula), and perhaps other attributes of their surroundings. Some snakes also migrate seasonally over long distances and between distinct habitats, evidently navigating with chemical, visual, and even celestial cues. Common Gartersnakes (*Thamnophis sirtalis*) travel up to 18 km from their hibernacula to summer feeding grounds.

Activity is usually confined to either day or night, and these daily cycles are associated with temperature, the availability of prey, and the snake's own vulnerability to predators. Many species are exclusively diurnal in their movements, including Rough Greensnakes (*Opheodrys aestivus*) and Orange-bellied Racers (*Mastigodryas melanolomus*); nocturnal taxa include Round Island Ground Boas (*Casarea dussumieri*), catsnakes (*Boiga*), and cat-eyed snakes (*Leptodeira*). Others are not strictly diurnal or nocturnal but rather vary daily activity accord-

Horned Adder (*Bitis caudalis*), Namibia, with babies.

ing to local climatic conditions. For example, Gopher Snakes (*Pituophis catenifer*), Harlequin Coralsnakes (*Micrurus fulvius*), and Copperheads (*Agkistrodon contortrix*) are diurnal during cooler spring and fall months but nocturnal when the heat of summer makes daytime activity dangerous.

Inactive snakes often hide under cover or in the substrate, presumably to avoid detection by predators. Some diurnally active tropical colubrids sleep exposed on vegetation (e.g., Orange-bellied Racers [photo, p. 38], Green-striped Vinesnakes [*Xenoxybelis argenteus*]), thereby escaping ants and other ground predators; during the day, nocturnal tropical snakes usually sleep in hollow stems, between the tight-fitting bases of leaves (e.g., Brown Blunt-headed Vinesnakes [*Imantodes cenchoa*]), under surface objects (e.g., Montane Snail-eaters [*Dipsas oreas*]), or otherwise hidden. Whereas diurnal and fossorial snakes typically have round pupils, many nocturnal snakes close the iris to a tiny vertical slit during daylight exposure and thereby protect their especially sensitive retinas (photo, p. 125); Glossy Snakes (*Arizona elegans*) and some other nocturnal species can vary pupil shape from round in darkness to semi-elliptical in bright light.

Seasonal activity patterns are driven largely by weather, either directly, because of intolerable surface conditions, or indirectly, through its effects on the availability of prey and suitable incubation conditions. Rather than migrate or go into torpor, as do most birds and mammals, snakes, like other ectotherms, usually remain inactive during periods when prey is scarce,

Several males and a female Carpet
Python (*Morelia spilota*), Australia, in a
mating aggregation. (Photograph by
Richard Shine)

temperatures are extreme, or conditions are arid. Not surprisingly, two cues that usually influence the onset and cessation of activity in snakes are temperature and water. Species in temperate and subarctic regions are typically inactive during several cold months each year. Racers (*Coluber constrictor*), Four-striped Ratsnakes (*Elaphe quadrivirgata*), and many other temperate species have spring and fall activity peaks associated with movements to and from hibernacula as well as mating and feeding. By contrast, Crowned Snakes (*Tantilla coronata*) and some other fossorial species are on the surface only during the hottest or wettest months; their underground movements are unknown. Common Gartersnakes (*Thamnophis sirtalis*) and Harlequin Coralsnakes (*Micrurus fulvius*) are active during warm weather throughout the year in the southern United States, whereas the former species and many others hibernate at more extreme latitudes. Tropical species also vary their activities seasonally; Green Parrotsnakes (*Leptophis ahaetulla*), cat-eyed snakes (*Leptodeira*), and other predators on amphibians initiate hunting when heavy rains incite dense breeding aggregations of frogs.

Although individual snakes are not distributed randomly with respect to others in a population, most species are not territorial in the sense of actively defending an area; individuals of some taxa avoid each other, and some males actively defend a receptive female. Seasonal aggregations do occur for thermoregulation, feeding, oviposition, and mating, and such gatherings are an ancient aspect of the natural history of snakes (see p. 274). As many as eight thousand Common Gartersnakes (*Thamnophis sirtalis*) occupy some winter dens in Canada, and smaller aggregations occur in blindsnakes, boas and pythons, and many groups of advanced snakes. Although usually comprising one species, some winter aggregations include several taxa. Avoiding extreme weather might be the most common advantage of aggregation, but some snakes also bask (e.g., Northern Watersnakes [*Nerodia sipedon*] on a log), breed (e.g., Common Gartersnakes at a hibernaculum), or feed in groups (e.g., Cottonmouths [*Agkistrodon piscivorus*] at a drying pond). Gravid Timber Rattlesnakes (*Crotalus horridus*) and Prairie Rattlesnakes (*C. v. viridis*) bask in communal rookeries. Beyond the immediate advantage of sharing a scarce resource (e.g., food, mates, or sunlight), clumped individuals might more readily avoid predators (through group vigilance and confusion by numbers) or lose heat and water more slowly. In any case, proximity to other snakes for any of those reasons could incidentally facilitate various intraspecific interactions.

SEXUAL DIMORPHISM AND SOCIAL BEHAVIOR

Male and female snakes typically resemble each other externally, although they also differ in subtle, often poorly understood ways. Males have longer tails (and thus more subcaudal scales), required for storage of their hemipenes; females have longer bodies (and more ventral scales), providing additional room for developing eggs or young. Some burrowing snakes have short, stout tails that work as a lever, and the sexes of those species differ little in relative tail length. Other examples of sexual dimorphism in shape include larger eyes in male Painted Bronzebacks (*Dendrelaphis pictus*), longer tongues in male Brown Blunt-headed Vinesnakes (*Imantodes cenchoa*), and larger heads in one sex or the other of many species. Female snakes are generally larger than males. Overall size dimorphism is more common in medium to large species than in small ones; greater length is associated with males of those species that fight and with females of those that produce large clutches or litters. For example, although Western Diamond-backed Rattlesnakes (*Crotalus atrox*) vary in size geographically, males everywhere exceed females by about 10 percent in average body length and sometimes compete for mates. By contrast, female Terciopelos (*Bothrops asper*) reach more than 2 m in total length and deliver dozens of young, whereas adult males of that species are rarely more than 1.3 m long and do not fight.

Sexual dimorphism in color patterns, common among visually oriented organisms (e.g., some lizards, many birds), usually is restricted in snakes to numbers of markings (e.g., body blotches and tail rings, themselves correlated with relative body and tail lengths). Among several exceptions, female European Adders (*Vipera berus*) have reddish ground colors, whereas males are gray, with a more distinctive, wavy mid-dorsal stripe; male and female Banded Rock Rattlesnakes (*Crotalus lepidus klauberi*) are, respectively, greenish versus lavender or gray in overall ground color. Female Boomslangs (*Dispholidus typus*) are generally

Madagascan Vinesnakes (*Langaha nasuta*), female (right) and male (opposite).

brown, and males in forest habitats resemble them, whereas those in savannas are bright green or blue.

Perhaps such differences are associated with divergent antipredator strategies and the increased vulnerability of male snakes as they search for females. Other species-typical male traits could provide tactile cues during courtship, including chin tubercles in Diamond-backed Watersnakes (*Nerodia rhombifer*), undivided anterior subcaudal scales in Desert Blacksnakes (*Walterinnesia aegyptia*), keeled subcaudals in Horned Adders (*Bitis caudalis*), tubercles above the cloaca in some New World coralsnakes, and a rostral horn on Turtle-headed Seasnakes (*Emydocephalus annulatus*). Male Madagascan Vinesnakes (*Langaha nasuta*) have a spike-shaped snout, whereas the female's is leaflike, and the two sexes differ in overall color as well; such bizarre dimorphism, without parallel elsewhere in serpentdom, might reflect microhabitat differences.

Four vignettes illustrate diverse social interactions in nature. A male Striped Whipsnake (*Masticophis taeniatus*) crawls among Great Basin shrubs, then pauses and scans with head raised. He approaches a female under the edge of a rock, crawls over her back while precisely matching its curvature, and passes waves of S-shaped loops of his foreparts under her chin; periodically he presses her head to the ground and curls his tail around her cloacal region. Another male arrives; both suitors tumble down a hillside with their bodies tightly entwined, and the intruder leaves. The first male rejoins the female, and after more courtship they copulate.

A female Northern Watersnake (*Nerodia sipedon*) and two courting males lie beside a Pennsylvania stream. A third male flicks its tongue rapidly as it gains the shore and joins them. One male inserts a hemipenis, while the two others continue rubbing the female with their

chins; eventually the writhing mass rolls into the water, and they swim to shore as two more snakes briefly join the group. The female reenters the stream and remains motionless for two minutes, after which the male unwraps his tail, withdraws his hemipenis, and rejoins the others while she swims away.

Dozens of male Yellow-lipped Seakraits (*Laticauda colubrina*) lie among boulders on a tiny island off the coast of Borneo, intercepting females that arrive at night in the surf. As many as five or six males pile on each female, rubbing her with their chins and their undulating bodies, and even riding her back as she crawls. One male achieves intromission, and the losers turn their attention to other females.

In lowland Peruvian rain forest, a female Common Tree Boa (*Corallus hortulanus*) is draped in symmetrical coils over a large liana, as a male flicks its tongue over her skin and vibrates his spurs against her back and sides. As he coils nearby, they lower their hindparts and he wraps his tail around her gaping cloacal region. Later several Saddle-backed Tamarins (*Saguinus fuscicollis*) mob the copulating snakes, who slowly withdraw into a dense vine tangle, their entwined tails still hanging downward.

For reproduction to occur, snakes must encounter, persuade, discriminate among, and synchronize behavior with potential mates. Although chemical and tactile cues predominate in the social behavior of all snakes, visual cues (e.g., movement) are important for location and identification of females by male Fox Snakes (*Elaphe vulpina*), King Cobras (*Ophiophagus hannah*), and perhaps many other diurnal taxa. Common Gartersnakes (*Thamnophis sirtalis*) and some other species that mate soon after hibernating in aggregations do not have to search for mates; European Adders (*Vipera berus*), however, travel well-defined routes to mating sites shortly after they emerge in the spring. Viperine Watersnakes (*Natrix maura*) mate around

Taipans (*Oxyuranus scutellatus*), Australia, mating. (Photograph by John Weigel)

noon on cool spring days and earlier in the morning as warmer summer temperatures prevail. Rattlesnakes usually mate weeks or months after emerging from hibernation, and thus must travel hundreds of meters in search of the chemical trails of females to court; males of species that do not aggregate for hibernation must also search for receptive females. Male Plains Gartersnakes (*T. radix*) use pheromones released from the female's dorsal skin to recognize conspecifics, and they determine her direction by sensing different concentrations of odor on opposite sides of blades of grass. Probably all snakes are capable of similarly impressive chemical discriminations.

Courtship and mating entail three phases, all of them overwhelmingly dependent on tactile and chemical cues. Initial approach is followed by bouts of chasing and mounting, and during the latter the male may undulate his body in head-to-tail or tail-to-head waves, jerkily rub his chin and flick his tongue on the female's back, bite her, and press spurs against her skin. The second phase begins with attempts by the male to juxtapose his cloaca with the female's and insert a hemipenis. Repeated bouts of these tail-search copulatory attempts are accompanied by frequent tactile contact (as in the initial phase) and interspersed with inactivity on the part of both snakes. If the female is receptive to that male, she eventually gapes her cloaca and permits intromission.

Mating aggregations of normally terrestrial Common Gartersnakes (*Thamnophis sirtalis*) occasionally move up into shrubs, but serpentine coupling usually occurs in species-typical habitats. Boomslangs (*Dispholidus typus*) and Sri Lankan Tree Pitvipers (*Trimeresurus trigon-*

Male Black Mambas (*Dendroaspis polylepis*), eastern Africa, in combat. (Photograph by Adrian Warren)

ocephalus) mate while stretched across limbs and foliage, Mexican Earthsnakes (*Conopsis biserialis*) copulate in burrows under rocks, and Olive Seasnakes (*Aipysurus laevis*) do everything in the ocean. Differences in courtship behavior among higher taxa include use of the spurs by male pythons and boas, mating bites by some colubrids, and an emphasis on jerky tapping of the female's back by male vipers. The duration of copulation varies from a few minutes in many colubrids (e.g., Fox Snakes [*Elaphe vulpina*]) and about an hour in King Cobras (*Ophiophagus hannah*) to more than twenty-five hours in some viperids (e.g., Western Diamond-backed Rattlesnakes [*Crotalus atrox*]). Prolonged copulation might reduce the likelihood of the female mating with other males; special defensive abilities perhaps facilitate lengthy mating, whereas rapid copulation reduces the exposure of both sexes to predators.

In some mating aggregations several suitors struggle for intromission with a single larger female (e.g., Green Anacondas [*Eunectes murinus*], Grass Snakes [*Natrix natrix*]), with the most vigorous male perhaps favored to win. Outright combat between males occurs in more than 100 species and 50 genera of pythons, boas, colubrids, elapids, and vipers, ranging in size from Prairie Groundsnakes (*Sonora semiannulata*) and Shovel-nosed Snakes (*Chionactis occipitalis*) less than 30 cm long to Asian Rock Pythons (*Python molurus*), Black Mambas (*Dendroaspis polylepis*), and other large species. Combat typically occurs during the mating season, often in the immediate vicinity of a smaller female, and is controlled by chemical, tactile, and visual cues. Boas and pythons generally fight with their entire bodies on the ground, although male Madagascan Tree Boas (*Boa mandrita*) spar and spur each other with their suspended lower bodies while lying in adjacent coils on tree limbs. Some colubrids (e.g., Common Kingsnakes [*Lampropeltis getula*]) and elapids (Red-bellied Blacksnakes [*Pseudechis porphyriacus*]) also fight parallel to the ground, entwined throughout much of their bodies; other colubrids (e.g., Banded Ratsnakes [*Ptyas mucosus*]), some elapids (e.g., Black Mambas [*Dendroaspis polylepis*], and many viperids (especially pitvipers) elevate foreparts during their struggles. Combat is usually a wrestling match, in which opponents attempt to pin each other's head repeatedly to the ground, but Mole Snakes (*Pseudaspis cana*) slash their rivals with powerful bites. Fights in nature often last only a few minutes (e.g., Taipans [*Oxyuranus scutellatus*]), but sometimes they extend over an hour or more (e.g., Gopher Snakes [*Pituophis catenifer*], some rattlesnakes).

Prolonged defense of an area, common in lizards, does not occur in serpents. Instead, the goal of most fighting among male snakes is securing access to receptive females and, in some species, preventing further matings; females might delay mating and thus encourage combat, then copulate with the superior male. Less obvious but more complicated forms of sexual interference and intrasexual communication are probably widespread among serpents. Defeat stifles male Copperheads (*Agkistrodon contortrix*), and they refuse to mate afterward; when a male initiates courtship, female Copperheads sometimes mimic the first stages of combat and then terminate the encounter if he fails to respond forcefully. Male gartersnakes (*Thamnophis*) do not fight, but their cloacal secretions form postcopulatory plugs in females that discourage further matings for the rest of the season. Some male Common Gartersnakes (*T. sirtalis*) in breeding aggregations adopt female-like behavior, diverting the courtship of other males and thereby securing a mating advantage. Despite such tactics, many male snakes and females of at least some species mate with multiple partners during a breeding season.

OVIPARITY, VIVIPARITY, AND PARENTAL CARE

Most snakes lay shelled eggs, which vary in shape from broadly oval in pythons and Bushmasters (*Lachesis muta*), to more narrowly elliptical in many species (e.g., kingsnakes [*Lampropeltis*]), to extraordinarily slender in some blindsnakes and burrowing colubrids (e.g., Long-tailed Threadsnakes [*Leptotyphlops longicaudus*] lay two eggs, each 21 mm long and 4 mm wide). Snake eggs can be smooth or granular and often adhere to each other after deposition. The time from ovulation to deposition is usually two to five weeks, and embryonic development often is not far advanced when the eggs are laid. In a clutch of Bull Snake eggs (*Pituophis catenifer sayi*) that required 56 days for incubation, the embryos were clear, with pigmented eyes and visible organs, on day 8; some dorsal and ventral scales were present by day 16; and the embryos resembled adults in color and pattern by day 31 but, accompanied by considerable yolk, they were still smaller than hatchlings. Hemipenes of embryonic squamates initially develop externally and invert into the tail prior to hatching; this inversion occurred 44 days after the Bull Snake eggs were laid. Incubation periods for snake eggs generally average about 70 days; they are shorter at higher temperatures and range from as few as 4 days in some clutches of Smooth Greensnakes (*Liochlorophis vernalis*) to 160 days in Banded Seakraits (*Laticauda semifasciata*) and more than 300 days in Twin-spotted Snakes (*Liophis bimaculatus*).

Oviparity, or the deposition of shelled eggs, is a primitive characteristic of amniotes, retained in many snakes and perhaps even re-evolved from live-bearing ancestors within a few

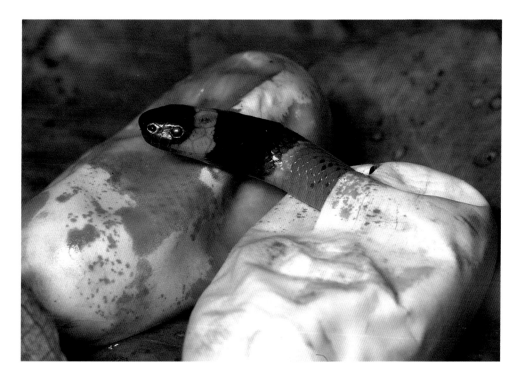

Central American Coralsnakes (*Micrurus nigrocinctus*), Costa Rica, hatching. An egg tooth is barely visible protruding from under the bottom edge of the snout.

groups (e.g., among vipers, Chapter 14). Live-bearing reptiles that supposedly keep their eggs in the oviducts and primarily nourish their embryos with yolk deposited before ovulation have been called "ovoviviparous," which implies that they simply provide an internal incubation environment; such species thus would differ from "true" viviparous ones, whose females actively nourish developing offspring through placental connections with their oviducts. In practice, ovoviviparity and viviparity have never been distinguished for most live-bearing snakes, because we know nothing about their fetal-maternal connections; exchange of oxygen and carbon dioxide must take place between all developing embryos and their mothers, as it does between incubating eggs and external environments. Common Gartersnakes (*Thamnophis sirtalis*) provide substantial amounts of nutrients to their developing offspring and receive wastes from embryonic metabolism; a variety of placental arrangements perhaps exists among other serpents.

Defined simply as giving birth to young rather than laying shelled eggs, viviparity has evolved more than thirty times among snakes, including at least fourteen times in colubrids and eight times in vipers. Some of the earliest offshoots within Alethinophidia are live-bearers (e.g., Asian pipesnakes [*Cylindrophis*]), and one ancient snake fossil contains developing young (see p. 274). These alternative modes of reproduction often characterize entire higher taxa, suggesting that they change in concert with new ways of life and thereafter are conservative in their evolution. Pythons, Asian sunbeam snakes (*Xenopeltis*), Neotropical Sunbeam Snakes (*Loxocemus bicolor*), and seakraits (*Laticauda*) all retain oviparity; Asian pipesnakes, Red Pipesnakes (*Anilius scytale*), shield-tailed snakes, boas, dwarf boas, and seasnakes are exclusively viviparous. Reproductive mode varies only occasionally in some higher taxa: most

Female Timber Rattlesnake (*Crotalus horridus*), Pennsylvania, with her young. (Photograph by Howard K. Reinert)

[*Agkistrodon piscivorus*], Timber Rattlesnakes [*Crotalus horridus*]). Some boids (e.g., Green Anacondas [*Eunectes murinus*]) consume fetal membranes and infertile eggs, reducing the vulnerability of their newborn to ants and predators. Juvenile Prairie Rattlesnakes (*C. v. viridis*) and Timber Rattlesnakes trail adults to dens for their first winter hibernation (Chapter 14), and perhaps the parent snakes alter their own behavior in ways that help the naive young find traditional hibernacula.

REPRODUCTIVE CYCLES AND FECUNDITY

Snakes are amazingly diverse in the timing, frequency, and magnitude of their reproductive efforts. All temperate zone snakes have well-defined breeding cycles that are strongly associated with warm weather. Females ovulate in spring and lay their eggs in early summer, with young hatching in late summer or early fall; viviparous temperate species extend gestation for roughly two more months, giving birth in late summer or early fall. Males of most temperate colubrids (e.g., gartersnakes [*Thamnophis*]) emerge early in the spring and court females as they come out of hibernation; however, many North American pitvipers (see the Special Topic in this chapter), Eurasian adders (*Vipera*), and some other species (e.g., Lined Snakes [*Tropidoclonion lineatum*]) mate in late summer or early fall. Females of species in the latter category store sperm during winter in special oviductal receptacles, ovulate and fertilize their eggs in early spring, and produce young in late summer or fall. In most snakes peak sperm

Bushmaster (*Lachesis muta*), Costa Rica, hatching.

production occurs several months before mating; in the extreme version of this dissociated pattern, a temperate pitviper's ova are fertilized in spring from a mating the previous summer, using sperm that in turn were produced almost a year earlier. Conversely, spermatogenesis coincides with or immediately precedes mating in a few Moroccan colubrids, several Australian elapids, and the Horned Viper (*Cerastes cerastes*).

Tropical snakes are less easily characterized as a group in terms of breeding seasonality, probably in large part because of the complexity of tropical climates. Although annual temperature cycles are less extreme than in temperate zones, equatorial regions usually have one or more wet-dry cycles; in some areas there also are pronounced topographic effects on climate over short distances. Some species might reproduce aseasonally (e.g., Catesby's Snail-eater [*Dipsas catesbyi*], Puff-faced Watersnake [*Homalopsis buccata*], Yellow-lipped Seakrait [*Laticauda colubrina*], several species of *Liophis*), in that sperm production and hatching or births extend over several months. Most such cases probably involve asynchronous reproduction by individual snakes, although some females might produce multiple clutches. Many tropical snakes have restricted breeding seasons, timed so that their young are born during a rainy season (e.g., Brown Vinesnake [*Oxybelis aeneus*], Jararaca [*Bothrops jararaca*]). Cycles can even vary within one species over a relatively small area: Jumping Pitvipers (*Atropoides nummifer*) give birth between August and November on the Atlantic lowlands of Costa Rica, where rainfall is sporadic to heavy throughout the year; young of the same species on the Pacific Coast are born between March and June, near the end of a well-defined dry season.

Snakes have relatively larger body cavities than most lizards, thanks to their tubular shape, and average almost twice as many offspring per clutch or litter as do limbed squamates of the

RADIOTELEMETRY AND THE ANNUAL CYCLE OF BLACK-TAILED RATTLESNAKES (*CROTALUS MOLOSSUS*)

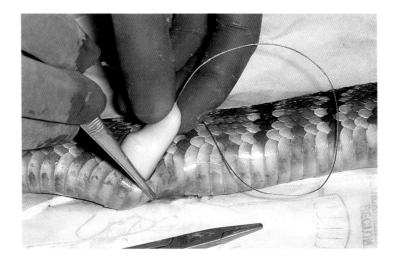

Surgical implantation of radiotelemetry transmitter in the body cavity of a Black-tailed Rattlesnake (*Crotalus molossus*), Arizona.

As an aspiring field biologist, I envied George B. Schaller's studies of Gorillas (*Gorilla gorilla*) and African Lions (*Panthera leo*), because he focused on the daily lives of individuals in nature. I wondered how a snake negotiated complex environments and located prey; above all, I was curious about what it was like to *be* a snake. Finding these profoundly cryptic animals was difficult, however, and repeatedly locating the same ones seemed impossible. In the ensuing decades, technology partially solved that problem. Telemetry of snakes dates to the 1960s and was revolutionized by Howard K. Reinert and David Cundall in the early 1980s. They implanted transmitters in the body cavities of Timber Rattlesnakes (*Crotalus horridus*), just above the belly scales and behind the stomach. An antenna ran from the radio under the skin, and the small surgical incisions were sutured shut. Their rattlers soon resumed normal activity, and the transmitters broadcast for several kilometers over a period of many months.

Using the Reinert-Cundall procedure, David L. Hardy Sr. and I have studied Black-tailed Rattlesnakes (*Crotalus molossus*) since 1988 in the Chiricahua Mountains of southeastern Arizona. We unobtrusively watch radio-tagged snakes from several meters away, supplementing field notes with photos and video recordings. As our observations accumulate, general patterns of activity emerge as well as individual idiosyncrasies. Finding a radio-tagged rattlesnake is never boring. In theory, one simply walks with a hand-held antenna in the direction of increasingly loud "beeps" from the telemetry receiver, which lead to the snake. In practice, silence can result from transmitter failure, an error in setting the reception frequency, or a snake that has moved out of range. Confusing directional information can mean the signal is bouncing from under a boulder or around a rock outcrop, or that the snake is in some unexpected place. Blistering temperatures, loose talus slopes, and cactus thickets also can make radio-tracking frustrating and difficult. Once Dave and I searched about 25 m² of scattered boulders and open ground, rechecking every crevice and worried that the "beep-ing" female was underfoot. Finally I lay flat and spotted a familiar color pattern, the edge of her coils. Among the dozens I'd seen in twenty years of searching their habitats, no other Blacktail had been tucked under the base of a yucca plant—but of course, without transmitters, I'd never have looked there.

Adult Black-tailed Rattlesnakes eat mainly woodrats (*Neotoma*), rabbits (*Sylvilagus*), and Rock Squirrels (*Spermophilus variegatus*). They hunt primarily by following the odor trails of favored prey, then position themselves for ambush near a runway or nest entrance. These life-and-death contests undoubtedly are complicated, since some mammals use multiple nests, and each nest might have more than one entrance. Some snakes adopt unusual hunting tactics too: among twenty-one Blacktails we have watched, only No. 12 sometimes positions

Black-tailed Rattlesnake (*Crotalus molossus*), Arizona.

herself on large tree roots and even vertically against trunks, presumably waiting for an arboreal rodent. Surprisingly, the rattlesnakes continue feeding into the fall months: On a chilly day in early October, two radio-tagged males were coiled perpendicular to the runways into woodrat nests, obviously hunting, and another was basking on a nest with a suspiciously rat-shaped lump at midbody. A fourth Blacktail lay in a small arroyo, distended with recently eaten prey (probably a rabbit) and unable to crawl; more than a week later she moved to a winter retreat, having consumed in one meal perhaps a third or more of her annual energy needs.

Blacktails begin feeding with the onset of warm spring temperatures, but their social activity coincides with monsoon storms in July and August. Males crawl hundreds of meters, following the chemical trails of receptive females and then at-

tempting to mate. Sexual encounters typically take place among rocks and low vegetation, although No. 8 unsuccessfully courted a female in a small tree. During courtship the male taps his chin jerkily up and down the female's spine, flicking his tongue on her skin and probing for the opening of her cloaca with his tail. Initially the female shakes him off with tail slaps, causing an audible clanking of rattle segments, but eventually she may open her cloaca for hemipenial insertion. During copulation the female controls the couple's overall movements: When sunlight shifted onto one pair, the male crawled back along his own body to their locked vents before following as she moved slowly to shade; another male crawled in reverse rectilinear style as the female moved away. Courting pairs remain together for days, and fights sometimes occur when a second suitor arrives; one

male even successfully wrestled a smaller rival while copulating with a female.

Chiricahua Blacktails hibernate in crevices among the rim rocks and do not usually aggregate. Individuals emerge on warm days to bask, and they sometimes change sites in midwinter. Occasionally they feed as early as March, although spring freezes occur sporadically. Each snake returns to the same area during successive active seasons, using some of the exact hunting and resting places it had used months earlier. Females bask near their hibernacula during gestation, give birth in late summer of the year following mating, and remain with their newborn for a week or so—until the young first shed their skins. Blacktails rely on concealment to avoid hawks, mammalian carnivores, and other potential enemies. Unusually mild-mannered as rattlesnakes go, their responses to us depend heavily on the immediate surroundings: in more than a hundred encounters, for example, No. 3 rattled only the few times I approached him suddenly in open situations.

Radiotelemetry allows us to wonder more accurately what it's like to be a snake. Transmitter size is a serious limiting factor, since a package exceeding 5–10 percent of the snake's weight might alter normal activities. Nevertheless, with smaller transmitter batteries, night-vision devices, and other technological innovations, biologists will someday study even the behavior of blindsnakes and other miniature creatures in nature.

same mass. Most serpents produce fewer than 50 eggs or young at a time, with numbers usually ranging from 2 to 16, 7 being the most frequent count; lizards often have 2 to 6 offspring, with 6 the mode. Across species, a clutch or litter amounts to roughly 10–45 percent of a female snake's mass before oviposition or birth, averaging about 30 percent. Among individuals and among species, female size best predicts variation in the number and size of offspring; larger snakes generally produce larger clutches or litters and larger offspring. A 3-m Reticulated Python (*Python reticulatus*) lays about 15 eggs, whereas a giant 6-m female could produce a clutch of about 100. Typical counts for some well-known colubrids, listed by increasing average female size, include 3 for Western Wormsnakes (*Carphophis vermis*), 7 for Rough Greensnakes (*Opheodrys aestivus*), 22 for Eastern Hog-nosed Snakes (*Heterodon platirhinos*), and 47 for Diamond-backed Watersnakes (*Nerodia rhombifer*). Species that occasionally produce about 100 offspring include North American Mudsnakes (*Farancia abacura*), Northern Watersnakes (*N. sipedon*), and Terciopelos (*Bothrops asper*); the records are 109 young born to a Tiger Snake (*Notechis scutatus*) and 156 in a litter of Puff Adders (*Bitis arietans*).

Mode of reproduction, lifestyle, and geography also influence reproductive output, so that generalizations based only on female size have many exceptions. For example, the overall clutch mass of oviparous snakes averages 20 percent higher than the overall litter mass of viviparous species, presumably because the latter endure their developing offspring longer than do egg-layers; clutches also tend to be smaller than litters, respectively averaging 7 and 11 young. Aquatic species often have fewer offspring than terrestrial serpents, and their embryos are positioned more narrowly, at midbody, reflecting the demands of locomotion in water. Irrespective of these other factors, however, clutch or litter size is usually higher in more seasonal environments. Thus, temperate snakes often lay larger clutches or litters than do tropical species, and within North America about 60 percent of snake species increase their clutch or litter size at more northern latitudes. Copperhead litters (*Agkistrodon contortrix*) average 6.2 in the northeastern United States, 5.3 in Kansas, and 3.0 in west Texas; Ring-necked Snake clutches (*Diadophis punctatus*), however, average 3.5 in Michigan, 4.2 in Kansas, and 5.2 in Florida. Snakes also produce fewer offspring at lower elevations than at higher ones, respectively averaging 10.5 and 12.2 in Common Gartersnakes (*Thamnophis sirtalis*), 3.9 and 6.0 in Rock Rattlesnakes (*Crotalus lepidus*), and 5.3 and 6.3 in Twin-spotted Rattlesnakes (*C. pricei*).

Beyond numbers of eggs or young, fecundity also depends on frequency of reproduction. Multiple clutches are common in small oviparous lizards but rare in serpents, because the latter require long periods to store sufficient fat for reproduction. Probably the majority of snakes in nature, both oviparous and viviparous, reproduce once each year or less frequently; a few tropical species (e.g., Brazilian Greensnakes [*Liophis viridis*]) breed at least twice each season. Many temperate vipers produce young in alternate years or even less frequently, as do North American hog-nosed snakes (*Heterodon*) and some northern populations of Common Gartersnakes (*Thamnophis sirtalis*). Among tropical serpents, Yucatan Pitvipers (*Porthidium yucatanicum*) produce litters annually; Pacific boas (*Candoia*), Orange-bellied Racers (*Mastigodryas melanolomus*), and Jararacas (*Bothrops jararaca*) reproduce less frequently. All things considered, food intake clearly is a major factor controlling fecundity in snakes. Captive snakes with unnaturally high food intake can produce larger and more frequent clutches, and Speckled Brownsnakes (*Pseudonaja guttata*) manage two clutches during years when unusu-

Javan Filesnake (*Acrochordus javanicus*). Females of some filesnakes require many years to reach maturity and reproduce infrequently.

ally high floods concentrate lizards and mice as prey. Conversely, Arafura Filesnakes (*Acrochordus arafurae*) have perhaps the lowest feeding rates of any snake, and females deliver a litter as infrequently as once every decade.

POPULATION BIOLOGY AND SYNTHESIS

Growth in snakes during the first few weeks of life depends largely on the supply of abdominal yolk (e.g., about 15–30 percent of their mass in newborn Copperheads [*Agkistrodon contortrix*]); success in feeding, effects of temperature on metabolism, and sexual differences in growth rates strongly influence later size increases. Even allowing for considerable individual variation, most female colubrids and elapids grow more rapidly than males, attain greater length than males, mature early, produce numerous offspring, and suffer high adult mortality. Many large viperids as well as some colubrids and elapids exhibit the opposite trends: females grow slowly, attain shorter lengths than males, mature late, produce fewer offspring, and have high adult survivorship. Age at maturity ranges from less than two years in some male colubrids (e.g., Western Wormsnakes [*Carphophis vermis*], Yucatan Slug-eaters [*Sibon sanniola*]) and Water Pythons (*Liasis fuscus*) to more than four in females of the Arafura Filesnake (*Acrochordus arafurae*) and several species of vipers (e.g., Timber Rattlesnakes [*Crotalus horridus*]). Moderate to large snakes often survive fifteen to twenty-five years in captivity, comparable to average life spans for those reaching adulthood in nature (e.g., sixteen years for Gopher Snakes [*Pituophis catenifer*], twenty-one for Western Rattlesnakes [*Crotalus viridis*]).

Ectotherms often have denser populations than endotherms with similar diets, since they require far less food, and some snakes are surprisingly common. Estimates of individuals per hectare (an area 100 m on each side, roughly the size of two football fields) include 16–97 Arafura Filesnakes (*Acrochordus arafurae*), 719–1,849 Ring-necked Snakes (*Diadophis punctatus*), 11–17 European Smoothsnakes (*Coronella austriaca*), 4–46 Four-striped Ratsnakes (*Elaphe quadrivirgata*), <1 Milksnake (*Lampropeltis triangulum*), 7 Grass Snakes (*Natrix natrix*), <1 Gopher Snake (*Pituophis catenifer*), 1,289 Striped Swampsnakes (*Regina alleni*), 6–9 Copperheads (*Agkistrodon contortrix*), and <1 Sidewinder (*Crotalus cerastes*). These numbers change seasonally because of births and deaths, of course, and are much higher in some habitats than others. When numbers of individuals are converted to biomass, the highest values for snakes (e.g., 7.1 kg/ha in Rough Greensnakes [*Opheodrys aestivus*], 30.8 kg/ha in Striped Swampsnakes) exceed the maximum values for species of birds or carnivores in similar habitats.

Snakes die from disease, extreme temperatures, starvation, predation, and even accidents with potential prey, but little is known about how those factors affect overall natural mortality. Tropical species that forage in bird and mammal nests (e.g., Tiger Ratsnakes [*Spilotes pullatus*]) frequently have ticks, and some pythons and vipers harbor huge grotesque tapeworms and other internal parasites. Survivorship through the first year is typically less than 50 percent, often much less; many young snakes probably are eaten, starve, or freeze during their first hibernation. Size and lifestyle influence survivorship profiles, as illustrated by four species of moderately large snakes in the Great Basin that suffer similar overall mortalities (18–25 percent each year). In Racers (*Coluber constrictor*) and Striped Whipsnakes (*Masticophis taeniatus*), both fast-moving diurnal species, 67–80 percent of deaths occur during spring and summer. Mortality is divided roughly equally between winter and the active seasons for Gopher Snakes (*Pituophis catenifer*), whereas more than 95 percent of deaths for Western Rattlesnakes (*Crotalus viridis*) take place during the latter period.

Combined patterns of survivorship and fecundity in turn yield diverse population structures. Among early-maturing colubrids with low survivorship, young adults with large clutches contribute disproportionately to population growth (e.g., Rough Greensnakes [*Opheodrys aestivus*], Yamakagashis [*Rhabdophis tigrinus*]). Late-maturing colubrids experience high adult survivorship and lower fecundity, and reproduction is spread more evenly over mature age groups (e.g., Western Wormsnakes [*Carphophis vermis*], Racers [*Coluber constrictor*]). The pattern for late-maturing viperids is similar to the latter, except that many of them breed less frequently than once a year (e.g., Western Rattlesnakes [*Crotalus viridis*], European Adders [*Vipera berus*]).

Differences in fecundity between close relatives accompany different mortality rates. Among likely examples, Louisiana Pinesnakes (*Pituophis ruthveni*) are secretive burrowers, evidently have few enemies, and lay a few huge eggs; Gopher Snakes (*P. catenifer*) are commonly encountered on the surface, frequently taken by hawks and other predators, and deposit larger clutches of smaller eggs. Boa Constrictors (*Boa constrictor*) in predator-rich neotropical forests deliver up to sixty young, but Madagascan Ground Boas (*B. madagascariensis*) on a predator-poor island produce only about six relatively large babies.

Two examples illustrate how reproductive strategies and population biology also are influenced by the interacting demands of particular feeding and climatic regimes. Mexican Gartersnakes (*Thamnophis eques*) are surface hunters that rely heavily on adult frogs as prey, whereas Black-bellied Gartersnakes (*T. melanogaster*) are aquatic foraging specialists. During a drought year when frogs disappeared at one site, Mexican Gartersnakes fed poorly and were unable to reproduce, whereas Black-bellied Gartersnakes caught tadpoles underwater and bred normally. Gopher Snakes (*Pituophis catenifer*) concentrate on deer mice (*Peromyscus*) in Idaho but also take other small mammals and reptiles; at the same site, adult Western Rattlesnakes (*Crotalus viridis*) feed largely on Townsend's Ground Squirrels (*Spermophilus townsendi*). During a year when the squirrels failed to produce litters, Western Rattlesnakes fed very little and did not breed, whereas Gopher Snakes emphasized several other prey species and reproduced normally.

Most serpents reproduce less frequently and suffer lower adult mortality rates than do many lizards, evidently a consequence of their radically different lifestyles. Compared to limbed squamates, snakes generally feed more infrequently on large items, practice less risky social behavior, and avoid detection by predators if at all possible. Those trends reach their extremes in Timber Rattlesnakes (*Crotalus horridus*) and other vipers, discussed further in Chapter 14.

PART TWO: DIVERSITY

It has never occurred to me to speak with elegant animals: I am not

curious about the opinion of wasps or of racing mares. Let them settle

matters while flying, let them win decorations by running! I want to

speak with flies, with the bitch that has recently littered, and to converse

with snakes. . . . I want to speak with many things and I will not leave

this planet without knowing what I came to seek, without investigating

this matter, and people do not suffice for me, I have to go much further

and I have to go much closer.

Pablo Neruda, Bestiary

Yellow Blunt-headed Vinesnake (*Imanto-des inornatus*), Costa Rica, in defensive posture. This species crosses gaps in vegetation by extending its slender fore-parts from branch to branch.

BLINDSNAKES

7

Walter Rose's *Reptiles and Amphibians of Southern Africa* introduced blindsnakes as follows: "From one point of view all snakes may be regarded as degenerate quadrupeds, but some have gone even further in having the appearance if not the actuality of degenerated degenerates. . . . No reptile approaches so nearly the stage of being 'sans teeth, sans eyes, sans taste, sans everything' as the members of the *Typhlops* and *Leptotyphlops* genera." These creatures are so routinely small that an African Western Threadsnake (*Leptotyphlops occidentalis*) about twice the length of an ordinary pencil was proclaimed "immense" at 322 mm, "fully 20 percent larger than the previously published record maximum" for that species! Blindsnakes are resistant to the stares of biologists: we ignore these shiny little creatures, periodically look again, and are frustrated by lack of resolution—we literally cannot see them very well.

Systematic monographs usually make for dry reading, with no evidence of the joys and frustrations of scholarly creativity, so it is surprising how frequently this one group of snakes evokes written outbursts. Frank Wall described the difficulties of studying Asian blindsnakes in 1918:

To count the rows [of scales] round the body is almost impossible unless both hands are free, and one has to accustom oneself to . . . a powerful watchmaker's lens. With this in the eye (and it must be remembered that the eye not in use must be kept open in order to retain the lens) the snake is grasped in both hands, and gradually rolled round as one rolls a cigarette, the eye never straying or blinking while the count is being made, and it is necessary to make a pinprick or some such mark on one of the scales from which the count is commenced, or transfix the specimen with a fine needle. . . . I have also used a special make of watchmaker's lens with double glasses, employed in the trade to examine the holes in watches in which gems are set. . . . With the best lens available, however, it is impossible to see the true outlines of the scales unless the light is allowed to strike obliquely across them, a trick which takes a little time to acquire dexterously.

Seventy years later, things aren't much better with expensive dissecting microscopes. After counting roughly 140,000 scales on 239 specimens, James R. Dixon and Christopher P. Kofron named a new species of South American blindsnake with the Greek word *argaleos,* meaning "troublesome, vexatious," and concluded their report with this curse: "We castigate the ancient lineage that begat *Liotyphlops,* for it is obviously the worst designed snake from which to obtain systematic data."

Studying the internal anatomy of a vertebrate less than 2 mm wide is even more difficult than investigating its exterior; some blindsnakes cram three chambers, valves, and all the necessary blood vessels into a heart the size of a caraway seed. In 1944 two respected herpetologists reported finding the pectoral girdle in *Liotyphlops,* a surprising discovery since that structure is otherwise completely absent in snakes. A year later Rosemary Warner repeated their dissections, painstakingly tracing the origins and insertions of various minute muscles. She decisively demonstrated that Emmett Reid Dunn and Joseph A. Tihen had mistaken a laryngeal cartilage for the hyobranchial apparatus and thus incorrectly identified the latter as a vestigial shoulder girdle. As Warner noted with dignified understatement, "The extreme thinness of the apparatus probably accounts for the difficulty of demonstration by gross dissection."

Some blindsnakes indeed are the smallest living serpents, whether judged as adults or neonates. The difficulties of studying such tiny animals, however, reflect a shortage of technology and imagination rather than a lack of interesting biology. The skulls of blindsnakes are spectacularly diverse and bizarre—a few species possess teeth on both jaws (as in most snakes), some have teeth only on the mandibles, others have teeth only on the oddly transverse maxillary bones—and no one has a clue as to the functional significance, if any, of these differences. Blindsnakes aggregate with others of their kind, and they have complex interactions with their insect prey and potential avian predators. One species with an extraordinarily wide distribution reproduces in the complete absence of males. As a group, these "degenerated degenerates" might hold major keys to understanding the origin of snakes. Moreover, the fact that they have so few externally varying characters compared to other snakes raises interesting questions about adaptive radiation and how food and living space are partitioned among co-occurring species.

Oddly enough, there do exist a couple of species of gigantic blindsnakes, reaching almost a meter in length and 5 cm in diameter. Having never seen one of these creatures alive, I imagine that finding one must be like suddenly meeting an amoeba as big as a turtle or a sperm the size of a banana slug, crawling over the ground and actually large enough to see and pick up. Some of the largest species are evidently not too rare, but as yet no one has capitalized on their size for studies of blindsnake biology.

The Scolecophidia includes three families of very odd creatures, collectively called blindsnakes. All have polished, round, equal-sized scales throughout a cylindrical body; all have highly modified skulls. The lower jaws are attached to each other anteriorly, as in most vertebrates and unlike all other snakes. Teeth are lacking on the palatine and pterygoid bones and are variably present on the maxillae and dentaries. The skin on their heads is often richly supplied with small glands of unknown function. Although the Seven-striped Blindsnake (*Leptotyphlops septemstriatus*) has a distinct pupil and colored iris, in some species the eyes are scarcely visible externally, and in others they appear as vague dark spots under enlarged head scales; some scolecophidians even lack eyes entirely. As befits an organism with little or no vision, blindsnakes rely heavily on chemical modalities for feeding, breeding, and defense.

Our scanty knowledge, based mainly on studies in the United States and Australia, suggests that scolecophidians are similar among themselves and distinctive from other snakes in many aspects of morphology, behavior, and ecology. Among the more than 300 species of blindsnakes, however, there is substantial diversity in size, diet, reproduction, and internal anatomy. A detailed understanding of their biology, as yet unavailable, will be crucial for interpreting the overall evolution of squamates, because living scolecophidians represent one half of the initial, ancient diversification within snakes (see Chapter 1, fig. 2). Whether we interpret various features as unique to snakes overall or only to Alethinophidia depends in part on the presence or absence of those attributes in blindsnakes.

STRUCTURE, TAXONOMY, AND DISTRIBUTION

Pelvic girdles are variably present in each of the three scolecophidian families, and some African leptotyphlopids also have hind limbs in the form of external spurs. Leptotyphlopids, typhlopids, and some anomalepidids have a pouchlike cecum between the small and large intestines. The cloacal opening is covered with several scales in anomalepidids and most typhlopids and by a single scale in leptotyphlopids and the Madagascan *Xenotyphlops grandidieri*. The tail of blindsnakes is always short (1–18 percent of total length), its blunt tip capped with a tiny, sharp spine. Some species have a pair of retrocloacal glands, unique among snakes, and hemipenes that are odd in several respects. Unlike all other serpents, some blindsnakes shed the highly polished skin in compacted, detached rings.

The family Leptotyphlopidae includes about 80 species of small snakes (maximum total length about 460 mm), found throughout warmer parts of the world; a few species extend well into temperate regions. Except for the monotypic *Rhinoleptus* of western Africa, all are placed in *Leptotyphlops*. English names for this family—slender blindsnakes, threadsnakes, and wormsnakes—reflect their attenuate build and relatively few rows of scales around the body. Leptotyphlopid skulls are unusual among snakes in lacking all teeth on the upper jaws. Because of an intramandibular hinge on each side, teeth on the dentary bones are arranged almost transversely.

The family Typhlopidae includes about 215 species in 6 genera, confined to warmer parts of the world. Reaching almost a meter in length, Schlegel's Blindsnake (*Rhinotyphlops schlegelii*) of Africa is by far the largest scolecophidian. At least among Australian *Ramphotyphlops*, males have significantly longer tails than females, but the latter reach much greater adult size. In contrast to leptotyphlopids, typhlopids have more rows of scales around the body, their lower jaws are toothless, and the upper teeth are confined to movable, horizontally arranged and paddlelike maxillary bones. Unlike most other scolecophidians, species of *Acutotyphlops* have numerous small head scales. Whereas most blindsnakes have typical hemipenes, those of *Ramphotyphlops* are modified for storage in the comparatively short tail: each organ is only partially inverted when not in use, and the retractor muscle is coiled or kinked. *Ramphotyphlops angusticeps* has more than six hundred vertebrae, perhaps the greatest number of any

Western Threadsnake (*Leptotyphlops occidentalis*), southern Africa. The eyes in this species are readily visible, although still covered by head scales.

snake. The skull of typhlopids is so peculiar that in 1954 a serious proposal was made, now firmly discounted, that they are not even snakes.

The Anomalepididae is even less known than the previous two families and includes only about 15 species in 4 neotropical genera. Snakes of this family are less derived in some respects than other scolecophidians. They have teeth on the upper and lower jawbones. Females of *Liotyphlops* and *Typhlophis* possess two oviducts, the primitive condition in snakes, whereas the left oviduct is vestigial or absent in *Anomalepis* and *Helminthophis.* We know almost nothing about the behavior and ecology of anomalepidids, beyond their restriction to tropical habitats.

Evolutionary relationships among and within higher taxa of blindsnakes are not well understood. Some features are uniformly primitive among anomalepidid genera, whereas others are derived in leptotyphlopids or typhlopids. The poorly known Madagascan *Xenotyphlops grandidieri* is especially intriguing for its anatomical resemblance to leptotyphlopids.

HABITATS AND GENERAL BIOLOGY

Blindsnakes occur in diverse temperate and tropical habitats, ranging from deserts and prairies to rain forests. These snakes are usually considered fossorial, and most species are typically found crawling on the ground at night or under rocks and other cover objects. They some-

Pink-headed Blindsnake (*Helminthophis frontalis*), Costa Rica. The snake's eye is a vague dark spot beneath a light-colored scale on the side of its head.

times wash to the surface during floods and have been uncovered during digging operations at depths of up to several meters. Several species commonly inhabit termite mounds. Blind-snakes travel above ground by lateral undulation, using their spine-tipped tails as levers for pushing off, and burrow readily. Most blindsnakes have rounded heads, with tiny, underslung lower jaws; several species of *Leptotyphlops, Ramphotyphlops,* and *Rhinotyphlops* have enlarged, beaklike or shovel-like snouts. Blindsnakes have an enlarged rostral scale, sometimes covering much of the front of the head, and in *Xenotyphlops grandidieri* that scale is adorned with flexible papillae. The relationships among external morphological variation (especially snout shape and scale structure), habitat preferences, and digging behavior in scolecophidians remain to be explored in detail.

Fossorial and arboreal lifestyles seem disparate, but some South American, North American, Madagascan, and Australasian leptotyphlopids and typhlopids are occasionally found several meters up in trees. Some Philippine Island typhlopids live among the roots of tree ferns, which implies that they simply inhabit a subterranean microhabitat above ground.

Schlegel's Blindsnake (*Rhinotyphlops schlegelii*), southern Africa. The enlarged, sharp-edged rostral scale of this species presumably facilitates digging.

However, Sulu Blindsnakes (*Ramphotyphlops suluensis*) regularly crawl over vegetation at night, evidently to forage, and they use their comparatively long, prehensile tails during locomotion on stems.

The significance of color pattern variation among blindsnakes is unknown. Scolecophidians are often unicolored and somewhat translucent, with hues of black, gray, brown, lilac, or pink. Exceptions include some Schlegel's Blindsnakes (*Rhinotyphlops schlegelii*), with prominent dark checkers on a light gray background, and strikingly bicolored Malayan Blindsnakes (*Typhlops muelleri*), with a dark dorsum and shiny white or yellow venter. A few species have the head and tail tip distinctively offset from the otherwise uniform body color; *Ramphotyphlops grypus* of Australia is light tan with a black tail cap, black nape and head, and a light spot on the snout. In Costa Rica, the Pink-headed Blindsnake (*Helminthophis frontalis*) has a pink tail tip and head; on close inspection, a pattern of dark red lines on the head proves to be blood vessels, visible through the pigmentless skin. Some Amazonian *Leptotyphlops* are almost gaudy, with prominent dorsal stripes of black and red or yellow. Judging from trends among other limbless vertebrates, darkly pigmented and contrastingly colored blindsnakes might be more active above ground than pale, unicolored taxa.

Neotropical Blindsnake (*Leptotyphlops goudotii*), Costa Rica.

FEEDING BIOLOGY

All scolecophidians—even the largest African and Asian typhlopids—feed exclusively on invertebrates, primarily ants and termites. In Arizona, those insects make up 64 percent and 54 percent of the diets of Texas Blindsnakes (*Leptotyphlops dulcis*) and Western Blindsnakes (*L. humilis*), respectively. Both species also consume the larvae and soft-bodied adults of various other arthropods, many of them known commensals of ant and termite nests. Four species of Australian *Ramphotyphlops* feed almost entirely on ant and termite larvae, with larger individuals and species consuming larger prey. North American and Australian scolecophidians often contain twenty or more items in their stomachs, usually of one species, which suggests frequent feeding on clumped, abundant prey. One Australian *R. nigrescens* had eaten 1,431 larvae and pupae of a single species of ant! The stomachs of a few Brazilian anomalepidids (*Liotyphlops* and *Typhlophis*) contained ant larvae and eggs. *Acutotyphlops subocularis* of New Guinea is exceptional in that the stomachs of individuals typically contain a single earthworm.

Hunting and the handling of prey have been described for only a few species of scolecophidians. Texas Blindsnakes (*Leptotyphlops dulcis*) follow the pheromone trails of army ants, thereby finding and eating the broods of those insects. Secretions of the blindsnakes' cloacal scent glands aid in repelling the attacking ants. One Puerto Rican Blindsnake (*Typhlops richardi*) was crawling toward a termite nest on a tree trunk, waving its head from side to side and constantly flicking its tongue. Australian *Ramphotyphlops nigrescens* use chemical cues to locate small communal ants and large solitary ants; they perhaps locate small solitary ants by chance. Several species of African *Typhlops* crawl in the trails of safari ants, but whether the snakes are feeding there has not been studied.

Unfortunately, we lack the details on ingestion behavior that might explain differences in skull structure among the three families of blindsnakes. A Texas Blindsnake (*Leptotyphlops dulcis*) first grasps a termite's abdomen, then breaks off its head by pressing it against the substrate, and finally swallows the soft hindparts. In contrast, that species arches its anterior over ant larvae and pupae, then forces its mouth down over the intact prey. A Neotropical Blindsnake (*Leptotyphlops goudotii*) collapses the abdomens of termites, swallows their contents, and brushes the collapsed remains against the substrate and out of its mouth. At least some typhlopids swallow prey intact, judging from the condition of items in their stomachs. Perhaps those differences in ingestion behavior are related to the size of the prey, since Cape Threadsnakes (*L. conjunctus*) swallow small termites whole but leave larger termites dead with their abdomens mangled. *Acutotyphlops subocularis* swallows earthworms intact.

All blindsnakes are similar in that one pair of elements in either the upper or lower jaws is movable and armed with a more or less transverse row of teeth (the maxillae in anomalepidids and typhlopids, the dentaries in leptotyphlopids). All possess as well a pair of geniomucosalis muscles, found elsewhere among serpents only in certain seasnakes, that originate on the mandibles and insert on the floor of the mouth. Perhaps either combination of musculoskeletal modifications can drain the abdomens of large insect prey or pull the mouth over entire smaller items.

SOCIAL BEHAVIOR AND REPRODUCTIVE BIOLOGY

Fossorial snakes are notoriously difficult to observe, and we have only scattered glimpses of their social lives. Sixty-four African *Typhlops nigricans* found under a slate slab and nineteen Puerto Rican *T. platycephalus* beneath a fallen coconut tree might simply have congregated in response to a favorable microhabitat during the dry season and winter, respectively. Spring aggregations of Texas Blindsnakes (*Leptotyphlops dulcis*) include both sexes and perhaps are associated with mating. The male of most lizards and some colubroid snakes grasps the female with his mouth during copulation; many other snakes lack that behavior, however, and it probably would be impossible for such slippery creatures as blindsnakes. A male European Blindsnake (*T. vermicularis*) coils tightly around the female's posterior during copulation, with his head facing away from her.

Some female anomalepidids and all female leptotyphlopids and typhlopids have only one oviduct—perhaps a consequence of their unusually slender build. Most scolecophidians probably are oviparous, the eggs of some species so tiny they resemble grains of rice in size and shape. Clutch size ranges from a single egg (2.5 × 25 mm) in Asian *Leptotyphlops blanfordii* to as many as sixty eggs (each 10 × 20 mm) in the giant Schlegel's Blindsnake (*Rhinotyphlops schlegelii*). Eggs of Schlegel's Blindsnakes have well-developed embryos when laid and hatch in five to six weeks, and those of another African species, Bibron's Blindsnake (*Typhlops*

Costa Rican Blindsnake (*Typhlops costaricensis*).

bibronii), hatch in as few as five days after deposition. Diard's Blindsnake (*T. diardii*) may be viviparous in some parts of Asia and oviparous elsewhere, but evidence for this variation is scanty. Australian typhlopids have well-defined reproductive cycles, with vitellogenesis in the spring and oviposition in the summer, and other scolecophidians perhaps are similar. At 53 mm in total length, newly hatched Flowerpot Blindsnakes (*Ramphotyphlops braminus*) must be among the smallest serpents.

Some blindsnakes remain with their eggs during incubation. A female *Rhinotyphlops caecus*, three newly hatched young, and three unhatched eggs were discovered together in a termite mound in central Africa. As many as six female Texas Blindsnakes (*Leptotyphlops dulcis*), each coiled with eggs, have been found near each other under blocks of buried sandstone in Kansas. Forty-two eggs were at one site, and the presence of older, empty eggshells indicated use during more than one breeding season.

The Flowerpot Blindsnake (*Ramphotyphlops braminus*), named for its worldwide transport in garden soil, is the only known parthenogenetic snake. As is frequently also the case in unisexual species of lizards, it has an unusually high number of chromosomes. Flowerpot Blindsnakes exhibit low variability in the number of scales, perhaps an indication of the reduced genetic variation typical of clonal organisms. Large "colonies" of this all-female species sometimes live in rotting wood, but the significance of those aggregations is unknown.

Peters' Threadsnake (*Leptotyphlops scutifrons*), southern Africa.

PREDATION, DEFENSE, AND COMMENSALISM

A variety of amphibians, reptiles, birds, mammals, and some invertebrates eat blindsnakes. Several *Leptotyphlops* were even found in the stomachs of Amazonian fish. Most blindsnakes are probably taken incidentally by generalist predators, such as toads and some birds (e.g., Caribbean blindsnakes [*Typhlops*] are occasionally eaten by American Kestrels [*Falco sparverius*] and Snowy Egrets [*Egretta thula*]). Some North American desert scorpions and the Sonoran Coralsnake (*Micruroides euryxanthus*) frequently eat *Leptotyphlops*.

Blindsnakes initially respond to discovery and contact by rapid burrowing, and Long-tailed Threadsnakes (*Leptotyphlops longicaudus*) readily take to water when pursued. If grasped by ants or humans, blindsnakes violently writhe and smear themselves with excreta and foul-smelling cloacal scent gland secretions. The Texas Blindsnake (*L. dulcis*) and Peters' Threadsnake (*L. scutifrons*) sometimes feign death when seized; both species assume a silvery coloration under duress, apparently as a result of fluids smeared about while they writhe. The distinctive, brightly colored heads and tails of some scolecophidians perhaps confuse a predator as the snakes rapidly coil and uncoil. Like those of skinks and other slippery lizards, the polished scales of blindsnakes are probably difficult for some predators to grasp. Blindsnakes routinely push their spiny tails against a captor. Although that behavior is often interpreted as defensive and has led to a widespread belief among rural people that these little snakes can sting, they might simply be trying to crawl away.

Eastern Screech-Owls (*Otus asio*) feed mainly on insects but also occasionally bring small vertebrates, including Texas Blindsnakes (*Leptotyphlops dulcis*), to their nestlings. Presumably the snakes are caught at night, as they wander above ground in search of the ant colonies upon whose larvae they feed. Among seventy-seven screech-owl nests in central Texas, 18 percent contained live *Leptotyphlops dulcis*. Four dead blindsnakes were found in owl nests, and many other live ones brought in by the owls showed scars and other evidence of capture and transport. Apparently most blindsnakes escape from parent birds after arrival at the nest and survive by eating the larvae of scavenging and parasitic insects in the nest debris. Arthropods in the nests torment the young owls, and about 30 percent of broods fail for various reasons. Young owls in nests that contain blindsnakes grow faster and suffer lower mortality than those lacking the snakes, so evidently the owls benefit from the reptiles' presence.

As yet nothing suggests that Texas Blindsnakes (*Leptotyphlops dulcis*) benefit from their association with screech-owls. The snakes run the risk of death or at least decreased vigor in the nests, but some do survive, feed, and persist there after the young birds have fledged. One female blindsnake even deposited a clutch of eggs shortly after removal from a screech-owl nest. Thrashers, motmots, and other owls eat *Leptotyphlops* elsewhere in North America, and other birds eat *Typhlops* in Africa and the Middle East. It remains to be seen if complex interactions among blindsnakes, arthropods, and nesting vertebrates are widespread.

PIPESNAKES, BOAS, AND OTHER BASAL GROUPS

8

Often collectively labeled "primitive" snakes, almost a dozen small lineages are anatomically and behaviorally intermediate between blindsnakes and all others. These basal groups—so-called because they arose early in the overall diversification of serpents—are thus especially important for understanding snake evolution. Among the most familiar of all vertebrates, by virtue of large size and legendary hunting prowess, a few of them even have name recognition way out of proportion to our knowledge of their biology. Boa Constrictors (*Boa constrictor*) and Asian Rock Pythons (*Python molurus*), for example, are prominent in the pet trade and popular lore yet little studied in the wild. In fact, among the roughly 160 species of basal snakes worldwide, the natural histories of only two Australian pythons are well known. Fascinating in their own right, some of these serpents also interact with wild primates and thereby might illustrate early stages through which our own relationships with snakes have passed.

My experience has been typical of our general dependence on captive observations and anecdotes for insights into the biology of basal snakes. I wrote a Ph.D. dissertation about them but relied almost entirely on animals in zoos to study feeding tactics and explore their implications for behavioral evolution. I watched constricting coil application by Red-tailed Pipesnakes (*Cylindrophis ruffus*), Caicos Dwarf Boas (*Tropidophis greenwayi*), and Common Tree Boas (*Corallus hortulanus*), among a total sample of 48 species in 26 genera; René Honegger at the Zurich Zoo kindly supplemented my findings with photographs of a Javan Filesnake (*Acrochrodus javanicus*) constricting fish. My friend Jonathan A. Campbell was just then embarking on landmark studies of Mexican herpetology, and one day his cryptic phone message about precious cargo sent me to the local airport. Soon I was watching a Oaxacan Dwarf Boa (*Exiliboa placata*), then known only from less than a dozen specimens, constrict a frog with movements like those used by Green Anacondas (*Eunectes murinus*) to kill crocodilians and large waterfowl. Because all basal snakes apply coils in the same way, I concluded that similar constricting behavior was used by their most recent common ancestor, more than 65 million years ago.

I can probably recall every individual basal snake among thousands of other serpents I've seen in the field. There was the Neotropical Sunbeam Snake (*Loxocemus bicolor*) on a dark road in Mexico and then the first Boa Constrictor, stretched across the eaves of a porch in Panama, to herald my arrival in the tropics. I found little Caicos Dwarf Boas under limestone rocks in low Caribbean scrub, and a silver-and-black Turks Island Boa (*Epicrates chrysogaster*) sheltered by palmetto leaves on the glistening white sand of Big Ambergris Cay. In Costa Rica I spotted a young Annulated Tree Boa (*Corallus annulatus*) resting by day on an understory shrub, like just one among thousands of dead, orange-brown leaves that tumbled from the canopy; at night my headlamp caught the eye-shine from an adult of the same species, beyond my reach in a rain forest tree. A supple, blue-gray Rosy Boa (*Charina trivirgata*) crawled between granite boulders on a cool, sunny day in southern California, and one freshly caught Rubber Boa (*C. bottae*) in Mendocino County regurgitated three young Western Moles (*Scapanus latimanus*). The most frustrating miss in forty years of snake hunting was when Marcio Martins exclaimed that a Red Pipesnake (*Anilius scytale*) was escaping in aquatic vegetation between us, and I never even glimpsed the brightly banded Amazonian prize! Perhaps because encounters with basal snakes in nature have been rare events for me, they seem all the more like snapshots of ancient serpent lifestyles.

For Boa Constrictors, as with other giant snakes, we lack convincing studies of geographic variation within the species because of the practical problems of preserving, curating, and studying large specimens. We have only fragments of the natural history of this widely famous reptile. Boas range from northern Mexico to Argentina, in habitats as different as rocky, semidesert scrub and lowland rain forest; mainly terrestrial, they also are slow, steady climbers. Boa Constrictors feed entirely on vertebrates, including Green Iguanas (*Iguana iguana*) and other lizards, occasional birds, and mammals as different as spiny rats (*Proechymys*) and Tamanduas (*Tamandua mexicana*); one large Boa in Trinidad ate an adult Ocelot (*Felis pardalis*). These heavy-bodied snakes hunt by coiling for days at burrows of Agoutis (*Dasyprocta punctata*), fruiting trees frequented by Blue-gray Tanagers (*Thraupis episcopus*), and other strategic ambush sites.

Some tantalizing chance observations suggest that the best is yet to come, that there is much more to being a huge serpent than we realize. An ornithologist in Suriname watched a 1.75-m Boa Constrictor successively snatch three territorial male Cocks-of-the-Rock (*Rupicola rupicola*) off a traditional courtship site over a period of several weeks; after each meal the snake retreated to a nearby tree cavity, where the birds' numbered leg bands later appeared in its droppings. At La Selva, Manuel Santana removed a pregnant 3-m Boa Constrictor from within a tree buttress, and a few days later we returned her with several dozen newborn young to the same site. The babies had been essentially immobile throughout their stay at the field station, even when handled, but all slithered into the hollow trunk immediately after release; the extremely agitated mother backed against the entrance hole, hissing and lunging at us while covering their retreat.

Modern humans in several cultures eat large snakes, other primates eat smaller serpents, and—since big snakes are geologically at least as old as primates—the dinner tables undoubtedly have been turned on many occasions. A Costa Rican Boa Constrictor killed an immature White-faced Monkey (*Cebus capucinus*) while other troop members screamed and threw sticks; then the snake hissed and struck at the live monkeys but didn't abandon its dead prey. Moustached Tamarins (*Saguinus mystax*) often crossed the edge of a Peruvian lake on fallen tree trunks, until a Green Anaconda (*Eunectes murinus*), hidden by aquatic vegetation, struck from below and ate one. The surviving tamarins avoided that route for days, and perhaps their descendants still chatter nervously when approaching water. African Rock Pythons (*Python sebae*) eat Vervets (*Cercopithecus aethiops*), and those monkeys adopt vigilant bipedal postures and give specific alarm cries when a snake appears. Surprisingly, they react to predator tracks with curiosity rather than fear, then jump back when drag marks lead to a big python in the grass. Several, usually vague, older accounts mention giant snakes taking humans, and I've seen photos of a Reticulated Python (*P. reticulatus*) with an Indonesian adult in its stomach and of a Green Anaconda with an obvious stomach bulge said to be a person. One well-documented case involved a South African teenager seized on the leg by a 4.5-meter-long African Rock Python as he ran on a path. Neighbors soon scared off the snake by pelting it with stones, but by then the boy was dead and partly consumed; the python was later located and transplanted to a distant park (in the United States it would have been destroyed, judging from reactions to recent predation on humans by Mountain Lions [*Felis concolor*]). Several people have been killed by pet pythons in the past few years, reflecting a careless lack of respect for wild animals by urban humans.

Boas and other basal forms represent only 6 percent of the world's snake fauna, and they lack the elaborate defensive repertoires and dietary versatility of vipers and other more advanced groups—no python rattles its tail or eats centipedes. Nevertheless, these serpents hint at evolutionary limits and opportunities during the early stages of a highly successful adaptive radiation. Among them are modern representatives of the first lineages that swallowed relatively enormous prey, like Red-tailed Pipesnakes that consume eels nearly equaling their own mass. Other basal groups include the only truly giant serpents, those massive predators on antelope, wildcats, and even primates. We know that human responses to snakes result from complex, still poorly understood inborn biases, cultural traditions, and individual experience. Basal snakes teach us that ambush predation on large vertebrates arose prior to the origin of venomous species, and perhaps so did our mingled curiosity and fear of these creatures.

The widely known boas and pythons have traditionally been lumped with several small, more obscure lineages as the Henophidia and informally called "primitive" snakes. Recent biochemical and anatomical studies, however, show that these snakes were grouped on the basis of retained primitive features rather than shared derived traits. Henophidia thus is paraphyletic, defined by default when more advanced serpents are removed, and I collectively refer to "henophidians" plus Acrochordidae as basal snakes. Most species in thirteen groups of basal snakes have vestiges of a pelvic girdle, hind limbs, and two well-developed lungs; with the exception of acrochordids, basal snakes lack the highly flexible jaw mechanisms that characterize more advanced serpents, and their adaptive radiation has been modest in terms of antipredator mechanisms and feeding biology. Nevertheless, the fossil record and relict distributions of living members of these groups indicate more widespread geographic ranges in the past (see the Special Topic in this chapter, and also p. 278). They represent some major stages in the evolutionary history of snakes and several minor but interesting experiments (e.g., divided upper jaws in Round Island boas, physiological heat production in some brooding

Isolated populations of species or higher taxa are called relicts. For example, Old World sand boas (*Eryx*), the Calabar Burrowing Boa (*Charina reinhardtii*), and their New World relatives (e.g., Rubber Boas [*C. bottae*]) are related to extinct erycines from throughout the Northern Hemisphere that have left an extensive fossil record (see also p. 274). Such disjunctions are common, and two phenomena are primarily responsible for them. Dispersal occurs when individuals colonize new regions, as adjacent areas become more favorable or previously separated habitats connect. Conversely, vicariance is the fragmentation of widespread distributions that results from mountain building, rising sea levels, and other disruptive events, usually on regional or even global scales.

We know that dispersal occurs because organisms occupy extensive areas formerly devoid of habitat (e.g., Massasaugas [*Sistrurus catenatus*] in previously glaciated regions of the northeastern United States; see p. 288), and chance observations tell us that dispersing organisms surmount inhospitable habitats (e.g., a Boa Constrictor [*Boa constrictor*] at sea, floating on vegetation). Vicariance is implied when the order of successive divergences among organisms parallels the timing of separation among their respective landmasses; conversely, dispersal is implied when the relative timing of

those divergences is not congruent. Pythons, boas, and other basal snakes provide many tantalizing hints of both vicariance and dispersal, although in most cases we still lack the phylogenetic and geological information necessary for firmly interpreting their histories.

Biogeographers have long explained disjunct distributions by dispersal from postulated centers of origin, but by the early 1970s geologists revolutionized this outlook. The now prevalent theory that present-day continents were fused into supercontinents that subsequently drifted apart, actually proposed by Alfred Wegener in 1915, provides a mechanism for vicariance and an alternative to dispersalist explanations. In the late Paleozoic (roughly 350 million years ago [mya]), most of the earth's dry land formed an enormous mass called Pangaea; by late Jurassic (about 150 mya) that supercontinent had split into northern and southern parts, known as Laurasia and Gondwanaland, respectively. Fragmentation of Gondwanaland in the Cretaceous (145–65 mya) yielded South America, Africa, Antarctica, India, and Australia, although some land connections persisted among them into the Paleocene (65–55 mya) and perhaps later (former relationships among the southern continents are evident by matching their current shorelines, especially between eastern South America and western Af-

rica). A huge ocean between Gondwanaland and Laurasia, the Tethys Sea, gradually shrunk as the southern continents drifted northward, and the Himalayas were uplifted as India collided with Laurasia in the Eocene (45 mya) or earlier. Pangaea's breakup drastically and repeatedly changed the size and shape of oceans, in turn affecting climates, shorelines, sea levels, and the distribution of organisms.

Now recall the distinction between dispersal and vicariance. Initial divergences among early snake lineages perhaps occurred on Gondwanaland and its descendant southern continents, because blindsnakes and basal alethinophidians (e.g., shield-tailed snakes, Red Pipesnakes [*Anilius scytale*]) are largely or entirely restricted to those regions today; the modern presence on northern continents of a few species (e.g., Texas Blindsnakes [*Leptotyphlops dulcis*]) is presumably due to dispersal. If the first divergence in pythons had been between an ancestral Australian stock and others, ancient vicariance between Australia and Africa, with subsequent dispersal of some *Python* to Asia, could plausibly account for their modern distributions. Instead, the closest relative of *Python* is *Morelia,* not the entire Australian and Asian radiation, which implies that *Python* arose within the latter and then dispersed north and west as far as Africa. Likewise, the initial divergence

Woma Python (*Aspidites ramsayi*), Australia. A remnant of the early radiation of pythons, this species often eats reptiles.

of Pacific boas (*Candoia*) from all other Boidae could indicate trans-oceanic migration of the former from the New World, where most members of the family now occur. Origin of Pythonidae in the same general region as Pacific boas, however, raises the interesting alternative that Boidae arose there as well; if that were the case, boids dispersed from the Old to the New World rather than vice versa. And as a final example, if all Old and New World erycines (including fossils) were each others' closest relatives, vicariance of the two continents by seaways would provide a simple explanation for their divergence and current distributions. The presence of *Charina* in North America and Africa, however, suggests that their common ancestor with sand boas (*Eryx*) was in the Old World.

Greater Antillean boas (*Epicrates*) and dwarf boas (*Tropidophis*) offer exciting possibilities for untangling vicariance and dispersal as underly-ing causes of distribution and speciation on islands. The Greater Antilles formed between North and South America in the early Cretaceous period, 130–110 mya, and drifted as a block into what is now the Caribbean about 80 mya. By the late Oligocene or early Miocene epoch, about 30–20 mya, the landmass that was Cuba plus Hispaniola divided; about that same time, as Hispaniola moved eastward, Puerto Rico broke off as a separate island. Accordingly, snakes might have diverged as their homelands divided and drifted, independently reached various islands by rafting, or participated in both processes. Phylogenetic studies of morphology show that Antillean boas are each others' closest relatives and share an ancestor with the mainland Rainbow Boa (*E. cenchria*); their divergence is consistent with proposals that ancestral boas drifted northeastward into the Caribbean on land that became the Greater Antilles and that the Cuban Boa (*E. angulifer*) split from a common ancestor of all others when that island broke free from the precursor of Hispaniola and Puerto Rico.

The ancestor of Jamaican Boas (*Epicrates subflavus*) must have arrived by water, however, since that island was inundated until long after the other large islands separated; judging from their phylogenetic relationships, some other species of *Epicrates* on Hispaniola or Puerto Rico also must have dispersed to their present distributions. Likewise, evidence from rates of molecular genetic differentiation implies that all Caribbean dwarf boas diverged from their common ancestor with mainland *Tropidophis* much later than predicted by geological splitting of the major Antillean landmasses, and thus they too must have come by dispersal over water.

pythons); moreover, although basal alethinophidians are relatively small serpents, some gigantic boas and pythons are the longest and heaviest living squamates.

In this chapter I first characterize each group of basal snakes in terms of general appearance, distribution, anatomy, and habits. Focusing on locomotion, feeding, reproduction, and defense in the better known boas and pythons, I then discuss the implications of their natural history for understanding early evolution in snakes.

PIPESNAKES AND OTHER BASAL ALETHINOPHIDIANS

Several superficially similar groups of snakes have often been combined as the Anilioidea; they might represent as many as six independent lineages of basal alethinophidians, each more closely related to successively divergent serpents than to earlier offshoots. Basal alethinophidians have stout skulls and only a few, large, teeth, reflecting both retention of lizardlike features and specializations for burrowing; their lower jaws are rather tightly linked in front by connective tissue. Most of these snakes have very short tails, undifferentiated or slightly widened ventral scales, and iridescent dorsal scales; most, too, have remnants of pelvic girdles and vestigial hind legs, the latter sometimes externally visible as tiny spurs.

Two species of dwarf pipesnakes (*Anomochilus*) are found on the Malayan Peninsula, Sumatra, and Borneo; they are sole survivors of an early divergence in snake evolution, anatomically intermediate between all other alethinophidians and blindsnakes. Dwarf pipesnakes outwardly resemble Asian pipesnakes (*Cylindrophis*); patterned with bold yellow or white spots and a red tail band against a dark ground color, they reach a length of almost 40 cm. Dwarf pipesnakes lack a chin groove and have no teeth on their pterygoid or palatine bones; some have no spectacles over their minute eyes or have only one spectacle. These odd little serpents are represented by fewer than ten museum specimens, and we know almost nothing about their natural history. One female was found to contain four eggs with leathery shells and early-stage embryos.

Although often classified together, the next five lineages perhaps arose independently and early in alethinophidian evolution. The first group, made up of 46 species in 9 genera of shield-tailed snakes (Uropeltidae), is restricted to peninsular India and Sri Lanka; although most species were originally confined to montane rain forests, today some shield-tailed snakes are extremely common in agricultural regions. Radically modified for fossorial locomotion, their heads are elongate and their tails pluglike; the skull is especially sturdy, and usually only the maxillaries and mandibles have teeth. The anterior trunk muscles of shield-tailed snakes are enlarged and specialized for shoving the pointed head through soil, so that the posterior is more or less dragged along—a "freight train" approach to burrowing. Ranging from 20 to 80 cm in length, shield-tailed snakes are viviparous and feed mainly on earthworms.

In the second group, 9 species of Asian pipesnakes (*Cylindrophis*) are shiny, stout, blunt-headed and very short-tailed burrowers. The Red-tailed Pipesnake (*C. ruffus*) reaches 70 cm, whereas a large Sri Lankan Pipesnake (*C. maculatus*) is only 40 cm long. The former is dull brown dorsally, and the latter is bright copper colored with black cross bars. All Asian pipesnakes have black-and-white checkered bellies; the underside of their tails is often mottled with orange or red. Asian pipesnakes are viviparous and feed on snakes, caecilians, and eels; their prey sometimes nearly equals the predator's own mass. The third group consists only of the strikingly beautiful, 70–90-cm-long Red Pipesnake (*Anilius scytale*), native to equatorial South America, east of the Andes. Unlike most other alethinophidians, this semiaquatic and

Red-tailed Pipesnake (*Cylindrophis ruffus*), southeastern Asia. Representative of an early stage in snake evolution, pipesnakes eat elongate, heavy prey.

Neotropical Sunbeam Snake (*Loxocemus bicolor*), Costa Rica. Note the pointed snout, used to dig up the reptile eggs and lizards on which it feeds.

burrowing species lacks spectacles over the eyes. The Red Pipesnake is viviparous and eats other snakes, amphisbaenians, caecilians, and eels.

Two groups of sunbeam snakes are independent lineages, both of meter-long serpents with somewhat more flexible jaw mechanisms than other basal alethinophidians. The Neotropical Sunbeam Snake (*Loxocemus bicolor*) lives in tropical dry forest and open habitats on the west

Boelen's Python (*Python boeleni*),
New Guinea.

Bredl's Carpet Python (*Morelia bredli*),
Australia.

coast of Mexico and Central America. This species is uniformly brown or has a white belly sharply set off from the dark dorsum; in addition to that pattern dimorphism, many individuals have small, highly irregular white blotches. Neotropical Sunbeam Snakes dig with upturned snouts, eating whip-tailed lizards (*Cnemidophorus*) and rodents, as well as the eggs of spiny-tailed iguanas (*Ctenosaura*) and sea turtles. Two species of Asian sunbeam snakes (*Xenopeltis*), found in southern Asia and nearby archipelagoes, have strangely flattened heads and strongly iridescent scales. Like some colubrids, Asian sunbeam snakes have folding teeth and eat skinks; they also feed on frogs and rodents. Neotropical and Asian sunbeam snakes lay eggs.

BOAS, PYTHONS, AND SAND BOAS

Boas, pythons, and sand boas may represent two or three independent offshoots near the origin of Macrostomata, although they have often been classified as a single family, Boidae. Except for a few species (e.g., Rough-scaled Sand Boa [*Eryx conicus*], Viper Boa [*Candoia aspera*]), these snakes have smooth scales. Carpet Pythons (*Morelia spilota*), Emerald Tree Boas (*Corallus caninus*), and some of their relatives have strongly pitted labial scales, associated with temperature-sensing. Although Boa Constrictors (*Boa constrictor*) lack those labial pits, their facial nerve endings are exquisitely sensitive to temperature variation, and perhaps other basal macrostomatans also have thermoreceptive abilities.

Most of the roughly 27 species in 8 genera of pythons (Pythonidae) have teeth on their premaxillary bones (directly beneath the snout), and all are oviparous; they are restricted to equatorial Africa, Asia, and Australasia. The Australian genus *Aspidites* is the sister taxon of all other living pythons; these are slender terrestrial snakes, and unlike other pythons, they often prey on squamates. Once regarded as dietary specialists in a land especially rich with lizards, the Black-headed Python (*Aspidites melanocephalus*) and Woma (*A. ramsayi*) instead might simply retain a diet typical of more basal alethinophidians (e.g., Asian sunbeam snakes [*Xenopeltis*]).

Several other genera of Australasian pythons, however, span diverse sizes, color patterns, and natural histories. They include the Pigmy Python (*Antaresia perthensis*) of arid central Australia, less than 1 m long, as well as much larger Water Pythons (*Liasis fuscus*) and Scrub Pythons (*Morelia amethystina*)—the latter perhaps reaching 8 m in length. Diamond Pythons (*M. s. spilota*) and their relatives are generally more slender than other Australasian pythons and are at least occasionally arboreal; the Green Tree Python (*M. viridis*) is behaviorally and anatomically specialized for life in the tropical rain forest canopy.

Snakes of the genus *Python* arose from within the Australasian radiation (their closest relative is *Morelia*). They are generally terrestrial, and their common ancestor dispersed northward onto mainland Asia and westward as far as Africa (see the Special Topic in this chapter). Today this group encompasses fairly small but stoutly built Ball Pythons (*P. regius*) and Angolan Pythons (*P. anchietae*), as well as truly enormous Asian Rock Pythons (*P. molurus*) and African Rock Pythons (*P. sebae*); longer and more slender than the others, Reticulated Pythons (*P. reticulatus*) might even exceed 10 m.

The 35 species in 5 genera of Old and New World boas (Boidae) exceed pythons in overall ecological and morphological diversity. They lack premaxillary teeth, and all species give birth to live young. Three species of Pacific boas (*Candoia*) are restricted to New Guinea, the Solo-

Pacific Viper Boa (*Candoia aspera*), New Guinea. Despite the allusion to vipers in its common name, this species perhaps mimics death adders (*Acanthophis*).

Common Tree Boa (*Corallus hortulanus*), Costa Rica. As is typical of arboreal snakes, tree boas are more slender than their terrestrial relatives.

mons, and several other island groups; they parallel the more species-rich radiation of South American and West Indian boas in some respects. Pacific Ground Boas (*C. carinata*) and Viper Boas (*C. aspera*) are usually less than 1 m in length; the latter is stout-bodied and perhaps mimics a venomous elapid, the Common Death Adder (*Acanthophis antarcticus*). Pacific Tree Boas (*C. bibronii*) are more elongate, reaching 2 m. Pacific Ground Boas and Viper Boas are sometimes diurnally active, whereas the Tree Boa is nocturnal. All three Pacific boas feed mainly on lizards, especially skinks, as juveniles; as adults they take various rodents and other mammals as large as bandicoots.

Three species of Madagascan boas formerly were assigned to the endemic genera *Acrantophis* and *Sanzinia,* but recent studies suggest they are so closely related to the New World Boa Constrictor (*Boa constrictor*) as to be placed in the same genus. Madagascan Ground Boas (*B. madagascariensis*) and Dumeril's Boas (*B. dumerili*) are generally placid snakes that reach almost 2 m in length and frequent the vicinity of water in tropical forests. Madagascan Tree Boas (*B. mandrita*) are slightly shorter and more slender; they are widespread in diverse habitats.

At under 2 m, the usually terrestrial Rainbow Boa (*Epicrates cenchria*) of Central and South America probably resembles in length and build the ancestor of a modest radiation of Caribbean boas; the nine island species range from stout, 3–4-m-long Cuban Boas (*E. angulifer*) to slender Vine Boas (*E. gracilis*) less than a third that long. Four species of South American anacondas (*Eunectes*) are specialized for aquatic living, with dorsal eyes and nostrils; some female Green Anacondas (*E. murinus*) exceed 6 m, although they perhaps rarely reach 10 m, and are certainly the heaviest living snakes.

Four species of neotropical arboreal boas (*Corallus*) include the Common Tree Boa (*C. hortulanus*) and Emerald Tree Boa (*C. caninus*). Like other tree-climbing boas and pythons, species of *Corallus* have relatively slender bodies, long prehensile tails, and large heads. Their large front teeth often have been regarded as adaptations for gripping through the feathers of birds, but in fact these snakes mainly eat lizards, mouse opossums (*Marmosa*), and other mammals.

Scattered across the Northern Hemisphere, several small terrestrial Old World sand boas and their relatives (collectively the Erycinae) often show specializations for burrowing and perhaps are not closely related to other boas. Eleven species of *Eryx* occur over much of western Asia and parts of northern Africa. Brown Sand Boas (*Eryx johnii*) reach almost 1 m, whereas some other species are barely half that length as adults. The Arabian Sand Boa (*Eryx jayakari*) is particularly well suited for life in sandy habitats by virtue of its concave ventral surface, countersunk lower jaw, and dorsally placed eyes and nostrils. Usually nocturnal, sand boas feed on lizards, rodents, and occasionally birds; the especially stout Rough-scaled Sand Boa (*E. conicus*) even eats Indian Palm Squirrels (*Funambulis palmarum*).

Among other erycines, Rubber Boas (*Charina bottae*) inhabit moist forests and cold desert shrub land in northwestern North America; ranging in color from uniform gray and lavender through shades of brown, they climbs trees to rob birds' nests and they explore tunnels in search of nestling mammals. The somewhat larger Rosy Boa (*C. trivirgata*) is found in boulder-strewn desert uplands of the southwestern United States and northwestern Mexico; it feeds on nestling woodrats (*Neotoma*) and other mammals. The Calabar Burrowing Boa (*C. reinhardtii*) of western African rain forest searches for nestling mice in leaf litter and low shrubs. Sometimes classified as a python, this species has oddly irregular, sometimes brightly

Rough-scaled Sand Boa (*Eryx conicus*), India. Sand boas everywhere are short, stocky, secretive snakes.

hued color patterns. Species of *Charina* are active by day and night, probably depending on temperature; all perhaps use their clublike tails to shield their heads from protective female rodents during hunting as well as to deflect attacks from other predators. Most erycines are viviparous, but the Calabar Burrowing Boa lays eggs.

DWARF BOAS AND ROUND ISLAND BOAS

Dwarf boas (Tropidophiidae) and Round Island boas (Bolyeriidae) might be close relatives; in any case, they diverged more recently than other basal macrostomatans and prior to the origins of acrochordids and other more advanced snakes. Dwarf boas are nocturnal and viviparous. Among 4 genera, *Tropidophis* includes 3 species in South America and 13 species on Caribbean islands. Like the best-known Caicos Dwarf Boa (*T. greenwayi*), most tropidophiids are drab-colored, mainly terrestrial predators on frogs and lizards; most species reach maximum lengths of 34 to 72 cm. Nine Cuban species, however, vary in color pattern, diet, and perhaps habitat preferences. Black-tailed Dwarf Boas (*T. melanurus*) reach 1.06 m and eat birds and rodents; one was found 9 m above ground in a bromeliad. The brightly colored Banded Dwarf Boa (*T. semicinctus*) and other smaller Cuban species eat only frogs and lizards.

Two species of eyelash dwarf boas (*Trachyboa*) are perhaps the strangest looking of all basal snakes, by virtue of heavily keeled, protuberant scales and their habit of freezing in seemingly random postures when handled. Two species of Central American dwarf boas (*Ungaliophis*) are small, gaudily patterned arboreal or fossorial snakes that eat lizards. Oaxacan Dwarf Boas (*Exiliboa placata*) are small, shiny black snakes with a white spot over the cloacal region; they

Central American Dwarf Boa (*Ungali-ophis panamensis*), Costa Rica.

live in cool cloud forest and eat rain frogs (*Eleutherodactylus*), salamanders, and amphibian eggs. *Exiliboa* and *Ungaliophis* may not be closely related to other dwarf boas.

Confined to a small landmass in the western Indian Ocean, two species of Round Island boas are unique among terrestrial vertebrates in having each maxilla divided into movable anterior and posterior parts. The Round Island Burrowing Boa (*Bolyeria multocarinata*), which went extinct probably during the past few decades, reached 95 cm; its natural history is unknown. Round Island Ground Boas (*Casarea dussumieri*) are found under palm fronds and in burrows of Wedge-tailed Shearwaters (*Puffinus pacificus*) as well as low in trees; up to 128 cm in length, they are oviparous and eat skinks and day geckos (*Phelsuma*). These snakes can pull the anterior segment of each maxilla downward, trapping slippery prey behind enlarged teeth at the front of the mouth; the overall jaw shape of Round Island boas thus recalls that of some Asian colubrids that also eat slippery skinks and geckos.

ACROCHORDIDAE

Three species of Australasian filesnakes (*Acrochordus*) are stunningly homely, strictly aquatic inhabitants of the Indo-Australian region; fossil acrochordids show that their relatives once were widespread in the Old World. Although resembling advanced snakes in some aspects of internal anatomy, acrochordids have a puzzling mixture of primitive and highly specialized traits. Among the latter are its exceptionally flabby, tubercular skin; its marked vertical flattening of the body when in water; and its method of shedding old skin by forming a body knot that it moves posteriorly.

Javan Filesnake (*Acrochordus javanicus*), with its extraordinarily long tongue partly extended.

The Little Filesnake (*Acrochordus granulatus*) commonly lives in mangrove swamps and other shallow saltwater habitats, but it is also found at depths of 4–20 m and as far as 10 km out to sea. This species is banded or blotched in black and white, thus vaguely resembling seakraits (*Laticauda*) and some seasnakes (e.g., *Hydrophis*). Javan Filesnakes (*A. javanicus*) and Arafura Filesnakes (*A. arafurae*) inhabit freshwaters and estuaries, and the former aggregate at times in burrows beneath tree roots in riverbanks; both species are dull brown or gray in color. Acrochordids are primarily nocturnal, although Arafura Filesnakes sometimes follow changing shadows beneath vegetation, perhaps to conceal themselves from avian predators. Little Filesnakes feed primarily on gobies and occasionally marine crustaceans, Javan Filesnakes eat eels and catfish, and Arafura Filesnakes take certain large freshwater fish. These bizarre aquatic serpents have an extraordinarily slow lifestyle, with very low rates of feeding and reproduction—perhaps giving birth once every eight to ten years. Arafura Filesnakes achieve high population densities, approaching one hundred snakes and an overall biomass of more than 50 kg per hectare.

NATURAL HISTORY AND THE DIVERSIFICATION OF BASAL SNAKES

The largest living scolecophidians (Schlegel's Blindsnake [*Rhinotyphlops schlegelii*]) and basal alethinophidians (e.g., Red Pipesnakes [*Anilius scytale*]) reach only about 1 m; most species in these groups are <50 cm in length, which suggests that the earliest serpents were also relatively small. All blindsnakes and basal alethinophidians are mainly burrowers as well, although a few species in each group are sometimes aquatic or arboreal. By contrast, all basal macrosto-

Juvenile Ringed Python (*Bothrochilus boa*), Bismarck Archipelago; adults of this species are a uniform olive-brown or brown ringed with black.

matans are active above ground, and some species are anatomically and behaviorally modified for diverse habitats. Pacific Viper Boas (*Candoia aspera*) sidewind on mud flats, adopting a locomotor mode otherwise characteristic of various advanced snakes. Some dwarf boas and erycines climb occasionally, and arboreal specializations have evolved independently in Green Pythons (*Morelia viridis*), Pacific Tree Boas (*C. bibronii*), neotropical tree boas (*Corallus*), and Vine Boas (*Epicrates gracilis*). Red Pipesnakes frequent wet places, anacondas (*Eunectes*) hunt and even mate in water, and Australasian filesnakes (*Acrochordus*) are highly adapted for aquatic life.

Some basal serpents probably mirror early stages in the evolution of snake feeding biology. Living remnants of the oldest alethinophidian lineages specialize on elongate prey, either heavy vertebrates (eaten, e.g., by Red-tailed Pipesnakes [*Cylindrophis ruffus*]) or earthworms (eaten by shield-tailed snakes); the latter diet is shared with at least one species of blindsnake. More recently divergent, Neotropical Sunbeam Snakes (*Loxocemus bicolor*) and Asian sunbeam snakes (*Xenopeltis*) take a wider range of shapes and kinds of vertebrate prey and thereby foreshadow macrostomatans. Broad diets and shifts in prey preferences with age and size— usually from frogs and lizards to birds and mammals—occur within individual pythons, boas, erycines, and Black-tailed Dwarf Boas (*Tropidophis melanurus*). Those snakes and other basal macrostomatans have increased gapes relative to earlier groups, and this capacity for swallowing heavy, bulky prey reaches a zenith in boas and pythons feeding on large mammals. One 47-kg African Rock Python (*Python sebae*) ate a 26-kg Uganda Kob (*Adenota kob*); a Papuan Python (*Liasis papuanus*) consumed a 23-kg Agile Wallaby (*Macropus agilis*); and Boa Constrictors (*Boa constrictor*) occasionally consume White-tailed Deer (*Odocoileus virginianus*)

Stimson's Python (*Antaresia stimsoni*),
Australia.

weighing more than they do. A 4.6-m African Rock Python swallowed a Thomson's Gazelle (*Gazella thomsoni*) in about forty-five minutes without difficulty, although one spikelike horn briefly pierced the big snake's body wall. Among other prey species, Asian Rock Pythons (*P. molurus*) take Purple Swamp-hens (*Porphyrio porphyrio*), Rufous-tailed Hares (*Lepus nigricollis*), Spotted Deer (*Axis axis*), Barking Deer (*Muntiacus muntjac*), and even adult Leopards (*Panthera pardus*). Prey of a size approaching a snake's maximum gape are potentially dangerous, however; one 4.9-m Asian Rock Python choked on a Hog Deer (*A. porcinus*) with 34-cm-long antlers.

Early alethinophidians were able to swallow prey so large that it could not be subdued by brute force; it is not surprising, then, that innovations for minimizing danger to limbless predators accompanied evolutionary increases in gape. Constriction clearly arose prior to predation on mammals, in serpents anatomically and ecologically similar to Asian pipesnakes (*Cylindrophis*), and was retained in subsequent transitional groups. At least as practiced by boas and pythons, constricting coils are somewhat analogous to a "choke hold," interfering with breathing and blood circulation so that the victim is immobilized within a minute or so. Blindsnakes and basal alethinophidians search for their prey, as do erycines and dwarf boas, but boas and pythons use ambush tactics under a wide variety of circumstances. An Angolan Python (*Python anchietae*) catches Namaqua Doves (*Oena capensis*) that drink at a spring; a Diamond Python (*Morelia s. spilota*) waits for days, 6 m up in the crotch of a tree, to seize a

Brush-tailed Possum (*Trichosurus vulpecula*). Submerged Green Anacondas (*Eunectes murinus*) snatch Wattled Jacanas (*Jacana jacana*) as the birds walk on vegetation in a shallow lagoon, and as many as fourteen Puerto Rican Boas (*Epicrates inornatus*) congregate around a cave entrance to capture emerging bats. Juveniles of the Green Tree Python (*M. viridis*) and several other basal macrostomatans lie in ambush and lure prey with their tails, and many pythons and boas use infrared-sensitive lip scales to sense potential prey.

Beyond reliance on chemical cues and seasonal aggregations by a few species, too little is known about blindsnakes and basal alethinophidians for us to speculate about sociality in their common ancestor. Pythons and boas have rich behavioral repertoires, however, encompassing male-male combat and seasonal assemblages (e.g., a dozen Asian Rock Pythons [*Python molurus*] in the burrow of an Indian Porcupine [*Hystrix indica*], four male Boa Constrictors [*Boa constrictor*] on a ledge with a female). Although the first serpents almost certainly laid eggs, parental care and viviparity probably evolved repeatedly in the early history of alethinophidians. African Rock Pythons (*P. sebae*) and Reticulated Pythons (*P. reticulatus*), both restricted to lowland equatorial regions, remain with their developing clutches throughout incubation and perhaps protect the eggs from monitor lizards (*Varanus*) and other predators. Young African Rock Pythons bask at the nest site for about ten days after hatching, where they stay until they have shed their skin for the first time.

In addition to guarding their nests, Asian Rock Pythons (*Python molurus*) and several Australasian pythons in relatively cooler climates produce excess metabolic heat by shivering and thus incubate their eggs. A female Diamond Python (*Morelia s. spilota*) builds her nest by burrowing slightly into sandy soil at the base of a tree or bush, then pushing leaf litter into a mound 25–30 cm in diameter and 5–10 cm deep. She emerges for twenty to ninety minutes on sunny days to bask but does not feed during the eight to ten weeks of incubation. Thanks to a well-insulated nest and heat produced by shivering, her eggs are on average 9°C warmer than surrounding air, even on cloudy days, and as much as 18°C warmer late at night. Incubation is costly for female pythons, who may lose as much as 15 percent of their body mass from the increased metabolic rates while shivering.

To judge from blindsnakes and basal alethinophidians, the first serpents used inaccessibility, cryptic coloration, and noxious chemical discharges to avoid predation. Unlike Gila Monsters (*Heloderma suspectum*) and monitors (*Varanus*), early snakes could not rely on strong jaws and claws for defense, and they also lacked the wide array of antipredator mechanisms later found in vipers and other advanced species. Many basal snakes struggle, void their cloacal contents, and make their heads less obvious when captured; some large pythons and boas also bite readily, using their sturdy teeth as weapons. With few exceptions (e.g., some Turks Island Boas [*Epicrates chrysogaster*]), striped color patterns, usually seen in snakes that emphasize locomotor defense, are rare among basal snakes; this implies that locomotor escape is not among their primary defensive tactics.

Savanna Twigsnake (*Thelotornis capen-sis*), southern Africa, with its neck

OLD WORLD COLUBRIDS

9

Roughly two hundred thousand years after the origin of modern humans on the famous Dark Continent, we crouched around a small cooking fire and watched night fall on the Mubwindi Swamp in southwestern Uganda. With Bob Drewes, Jens Vindum, and Jim O'Brien from the California Academy of Sciences, I was in the Bwindi-Impenetrable Forest Reserve to survey amphibians and reptiles for conservation purposes. Jan Kalina, an avian ecologist who ran the Bwindi project with her husband Tom Butynski, had organized the expedition, and several Ugandan park personnel and students rounded out the group. Initially known as "fat boys in the forest," we four herpetologists rapidly lost weight because of the rugged terrain and because Jan, a slim vegetarian, had planned the meals.

More than 2,100 m above sea level, our camp on a terrace beside the swamp was often cold and drenched with rain. At dusk pairs of ibis flew silently and low over the water, as if stealthy jet fighters in tight formation, then squawked like New Year's Eve horns when they landed. Cicadas screeched incessantly, and a frog chorus filled the gathering blackness with boinks, chirps, and trills. Huddled under a tarp shelter, I ached all over from walking up paths almost vertical in front of my face, struggling to keep up with the Ugandans. Later, after a hot meal of potatoes and peanut sauce, my thoughts bounced from noisy turacos and other brightly colored birds to shockingly large elephant dung on the tunnel-like forest trails. I was especially puzzled by the strong body odor, somehow both strange and familiar, of the game guard I'd followed all day. Neither our ethnic and dietary differences nor his lifetime in the forest explained Vincent Bashekura's provocative aroma, yet I couldn't place it.

My hiking companion was the most outgoing of those who guided us in Bwindi. Vincent wore khaki pants and a green Bwindi Reserve T-shirt, had a ready sense of humor, and was enthusiastic about everything. He believed I took "special medicine" to handle snakes, and I was in awe of his wilderness skills. Vincent fashioned rope out of sapling bark, made toast from moldy bread skewered over hot coals, scoured cooking pots with grass, and carried an old bolt-action 30.06 rifle "to intimidate poachers." Within hours of our meeting, this cheerful, generous man was teasing me about not keeping up; minutes later he'd doubled back from hundreds of meters up a hill to show me the trail of a Gorilla (*Gorilla gorilla*). Next morning, as a chilly dawn crept over our sleeping bags, I had a strange dream of someone walking around and methodically tossing small stones against the tent. Then, awakened fully by the screams of nearby Chimpanzees (*Pan troglodytes*), I discovered the earlier sounds were made by frogs we'd collected the night before, hopping and croaking softly in a plastic bag next to my head. A few moments later, still lying there and listening to the low murmuring of the Ugandans while they made tea, I finally realized that Vincent smelled like thirty years of campfire smoke.

Some modern snake genera differentiated at least twenty million years ago, whereas hominids diverged from our common ancestor with Gorillas and Chimps four to six million years before the present, in the early Pliocene epoch. Stone tools go back maybe two million years, fire use by *Homo erectus* a mere four hundred thousand years, and the earliest written records are about five thousand years old. Thus, throughout the vagaries of history, our ancestors have faced roughly the same snake faunas as we encounter now. Like people everywhere, and even other primates, early humans probably regarded limbless reptiles with a mixture of fear and curiosity, of accurate lore and wild legends. Doubtless some hairy, bipedal nest raider marveled at an egg-eating snake (*Dasypeltis*) caught in the act, and surely ancestral, savanna-ranging hominids respected the threat display of a cornered twigsnake (*Thelotornis*). Although our kin undoubtedly killed them all along, snakes probably were common in the Pliocene, just as they have been in central Africa until recently.

Today, even as its wilderness dwindles in the wake of burgeoning human impact, Africa is perhaps the least studied large land mass on earth relative to its biodiversity. This tragic irony was obvious even during our brief reconnaissance. In a month we four experienced collectors found only six snakes where Charles Pitman, author of *A Guide to the Snakes of Uganda,* once described them as "extraordinarily common." Nevertheless, there was a surprise among the few we caught. Nocturnal colubrids are typically drab brown or gray and stay hidden during the day, but we discovered an elegant, emerald *Dipsadoboa* with catlike pupils, possibly a new species, that hunts frogs at night and sleeps exposed on foliage by day. Perhaps at such elevations the green Bwindi snake must bask even as it sleeps to ensure high temperatures for digestion, and thus requires a camouflage unsuited to its relatives in the warmer lowlands.

It is difficult to rationalize the disparity between conservation as conducted in comfortable Berkeley coffee shops or Washington boardrooms and the difficult, sometimes dangerous task of saving tropical forests. Jan and Tom are respected scientists, approaching midlife with hopes for a family, yet they were living on yearly contracts and had to travel hours over rough roads for basic medical care. Vincent's salary covered only a fraction of his family's subsistence needs, so I was humbled that he didn't poach or steal. Throughout rural Uganda, the threat of death was omnipresent in an almost casual way. Breaking camp one morning on the edge of Bwindi, we were suddenly accosted by four Forestry Department officials. While an aide with an assault rifle hovered nearby, their leader, a thin, ominous man with a sparse goatee, angrily berated us for affronts to his jurisdiction. Three weeks earlier some trespassing soldiers had disarmed our game guards, forced them to disrobe and lie on the ground, and fired bursts from automatic weapons around them. Thinking the others dead, Vincent then made a dash for the forest and arrived at reserve headquarters three days later, almost incoherent from fatigue and fear. Now, fingers on their rifle triggers, these same game guards regarded the forestry people with smoldering glares. One whispered to me, "They are all corrupt and we hate them. They hate us and sometimes we shoot at each other." For several minutes we smiled at the forestry officer and silently planned where to dive for cover if things escalated, while Jan calmly but firmly dealt with his complaints and defused the confrontation.

Bwindi means "darkest of dark places," and our journey was like a descent through successively older layers of human history coupled with apocalyptic visions of the future. It began with twenty hours of jet flight and ended with a jolting twelve-hour truck ride across an almost totally converted landscape. As we traveled through neighboring Kenya, Nairobi National Park evoked a surprisingly powerful nostalgia for North American prairies. Surrounded by thousands of Zebras (*Equus burchellii*), Wildebeest (*Connochaetes taurinus*), and Cape Buffalo (*Synceros caffer*), I imagined the richness of other equally grand and now extinct savanna faunas: a couple of centuries back the Great Plains held seventy million Bison (*Bison bison*), and only twelve thousand years ago mammoths and giant ground sloths inhabited the southwestern United States. Looking south at the Virunga Volcanoes from a forested ridge near the Rwanda-Uganda border, I never had felt more remote from Western culture, yet we saw no undisturbed habitat on most of the drive there from Kampala—only people and livestock everywhere.

On that first night in the Mubwindi Swamp, my eyes burned from drifting campfire smoke, and I remembered a whimsical phrase from teenage hiking trips. "Vincent, in the United States we say, 'Smoke follows beauty.'" After a pause, he grinned and replied in precisely enunciated English, "Here we say, 'Smoke follows one who defecates too close to the trail.'" Looking for snakes in Africa, cradle of our primal quest for fire, I sometimes glimpsed a darker irony in those lighthearted comments. With so many trails crowded and foul, where will the mingled destinies of snakes and people lead next?

The early evolution of colubrids followed fragmentation of a worldwide landmass at the end of the Mesozoic era. Europe and Asia, collectively the largest remaining parcel of a northern supercontinent called Laurasia, include habitats ranging from cold tundra and steppes to deserts and tropical rain forests. The influences of glaciation and epicontinental seas on modern Eurasian snake distributions have been pervasive, in that advancing ice sheets and enormous bodies of water have episodically compressed and splintered entire biotas. Among extant Old World colubrid lineages, only colubrines and natricines are shared with North America, the other large remnant of Laurasia.

The Indian subcontinent and Australia, both chunks of the former southern supercontinent called Gondwanaland, faunistically resemble Eurasia to degrees that reflect their different geographical relationships with the ancient northern landmass. India collided with Eurasia about fifty million years ago, its impact uplifting the Himalayan mountains, and has a colubrid fauna generally similar to that of adjacent areas. By contrast, Australia, which has been surrounded by seawater throughout much of its history, has mainly blindsnakes, pythons, and elapids; the few colubrids found there are related to groups centered elsewhere in the Australasian region and have probably arrived much more recently. The colubrid snakes of Sumatra, Java, and Borneo mostly are related to those of southeastern Asia; those large islands lie on the Sunda Shelf, an immense extension of the Asian mainland that was periodically exposed and flooded during Pleistocene climatic fluctuations. A continental shelf also surrounds northern Australia and New Guinea, and emergent landbridges across what is now the Torres Straits sometimes allowed faunal exchange between those two regions.

Africa is the largest existing piece of Gondwanaland, the former southern supercontinent. Biogeographically, its northern coastal region resembles Europe in many aspects of flora and fauna (some primarily African groups also penetrate southern Europe and western Asia; e.g., sandsnakes [Psammophiini], discussed below). In terms of area, the largest African biome is the vast Sahara Desert, followed by the equatorial tropical savannas and forests; southern Africa is unusually diverse climatically and is characterized by several unique, relict habitats. African colubrid assemblages reflect a combination of several ancient lineages, some perhaps survivors of Gondwanian groups and others surely the remnants of more recent dispersals from the north. The former Tethys Sea, which periodically covered parts of central Eurasia and what is now the Arabian Peninsula, had especially profound effects on colonization by groups of snakes originating elsewhere. As yet no evolutionary links have been found between the colubrids of Africa and those of other major Gondwanian derivatives—South America, India, and Australia.

Colubrids are so morphologically, behaviorally, and ecologically diverse that generalizations about the group as a whole are difficult. Although many Eurasian and African colubrids are often discussed in terms of formal groups, their assignment is frequently ambiguous, and taxonomic recognition would be misleading. In this chapter I discuss common or otherwise interesting Old World colubrids in terms of convenient assemblages, while emphasizing uncertainties about their relationships. African stiletto snakes (*Atractaspis*) and their associates, at times placed in the Colubridae, are discussed in Chapter 11.

COLUBRINAE

Eurasian colubrines include many well-known, endemic Old World genera. The Smooth Snake (*Coronella austriaca*) and its relatives are moderately small constrictors that feed primarily on lizards. Thirteen species of Eurasian dwarfsnakes (*Eirenis*) range from small arthropod eaters to moderate-sized predators on lizards. The Asian Banded Ratsnake (*Ptyas mucosus*) reaches the unusual length of at least 2.3 m, and females guard their clutches. Other prominent Eurasian colubrids include several dozen species of whipsnakes (*Coluber*) and Old World racers (*Elaphe*); those genera include North American species as well, although in each case the Old and New World congeners are probably not closely related. The whipsnakes are relatively slender, whereas Old World racers are heavier-bodied constrictors. Some large Asian colubrines (e.g., Common Malayan Racer, *E. flavolineata*) have impressive threat displays in which the neck region is elevated and inflated. Most colubrines are oviparous, but Smooth Snakes, Asian vinesnakes (*Ahaetulla*), and mock vipers (*Psammodynastes*) give birth to live young.

The large and diverse Asian genus *Oligodon* includes more than 70 species, all <1 m long and collectively called "kukrisnakes," because their enlarged rear fangs are shaped like the famed kukri knives of Ghurka soldiers. These diurnal, terrestrial serpents range from unicolored to striped and blotched, usually in shades of brown and gray. Some species display brightly patterned venters when threatened. Although some kukrisnakes are inoffensive, the

Common Malayan Racer (*Elaphe flavo-lineata*), Borneo, in a gaping threat display.

Banded Kukrisnake (*Oligodon arnensis*), India, in defensive posture, with its head and neck in an **S**-coil and the back of its head enlarged.

White-spotted Catsnake (*Boiga drapezii*), southeastern Asia, in defensive posture.

defensive repertoires of Common Kukrisnakes (*O. cyclurus*) and Taiwan Kukrisnakes (*O. formosanus*) include slashing strikes, butting of the head, and repeated eversion of the hemipenes. Most kukrisnakes take prey as diverse as frogs, lizards, birds, and mammals but seem to have a predilection for eggs of the first three. The Chinese Kukrisnake (*O. cinereus*) feeds exclusively on arthropods, especially spiders and grasshoppers.

Several nocturnal, rear-fanged colubrines are sometimes grouped in a separate tribe or subfamily, the boigines, but its monophyly is unlikely. Usually called "catsnakes" or "treesnakes," all have smooth scales and vertical pupils. The Afro-Asian genus *Boiga* includes more than 20 species of moderate-to-large, slender arboreal snakes (see the Special Topic in this chapter). Blanding's Treesnake (*B. blandingi*) of central Africa is polymorphic in color, velvety black or blotched black and brown, and raids colonies of birds and bats. The strikingly black-and-yellow-banded Mangrove Treesnake (*B. dendrophila*) of Asia is a habitat specialist, usually found along waterways. Its diet is especially broad compared to that of other members of the genus and includes frogs, lizards, birds, snakes, mammals—rarely even a mouse deer! Some species of *Telescopus* are drably patterned, but the African Tigersnake (*T. semiannulatus*) has stunning black saddles on a yellow-orange background. The African Herald Snake (*Crotaphopeltis hotamboeia*) and treesnakes of the genus *Dipsadoboa* eat frogs and their eggs, but the former is terrestrial, whereas the latter climb well.

Several genera of Asian colubrines of uncertain relationships occupy diverse habitats but share cranial modifications that facilitate eating skinks. In fact, wolfsnakes (*Lycodon,* meaning

THE BROWN TREESNAKE (*BOIGA IRREGULARIS*): A DISASTROUS VAGABOND

Brown Treesnake (*Boiga irregularis*), Australia.

In the 1970s biologists on the Pacific island of Guam became alarmed at rapid declines in populations of native birds. Disease, parasites, habitat alteration, and other "ordinary" causes of extinction were investigated before Julie A. Savidge demonstrated that an introduced snake was responsible for decimating the avifauna. Other ornithologists initially were skeptical of her findings and, as Joe T. Marshall noted, "Seemingly in a state of shock, some . . . asked questions or volunteered statements that were unscientific, unchivalrous, and embarrassing to the rest of us. Few could believe that a mere snake was such an efficient predator and could build up the numbers commensurate with such devastation."

Brown Treesnakes (*Boiga irregularis*) occur naturally in New Guinea, the Solomons and other offshore islands, and northern Australia. They feed on lizards as juveniles, whereas adults—up to almost 3 m in total length—eat mainly rodents on New Guinea and birds on smaller islands. The Guam immigrants were "doing what comes naturally," and a better snake invader could scarcely be designed: both terrestrial and arboreal, large, and well prepared with venom and constricting behavior to handle a variety of prey. Moreover, Guam has no native snakes, so the newly arrived Brown Treesnakes faced no competitors or skilled predators on serpents. Without evolved defenses, such as specialized nesting behavior in the face of snake predation, the island's avifauna was extremely vulnerable.

Brown Treesnakes were accidentally introduced to Guam on cargo shipments shortly after the end of World War II. By the mid-1980s populations of *Boiga irregularis* were extraordinarily dense on the island, and most of the ten species of native forest birds were extirpated or extremely rare. Native fruit bats ("flying foxes") and lizards also declined drastically from snake predation. Brown Treesnakes on Guam have taken an economic toll as well, making inroads on subsistence and commercial poultry farms and causing numerous power outages by climbing on transformers. Some *Boiga* have powerful neurotoxic venom, and several young children on Guam, bitten while sleeping, have suffered serious symptoms.

Massive efforts to understand and control Brown Treesnakes have been under way for more than a decade, led by the United States Fish and Wildlife Service and the Mariana Islands Department of Natural Resources. A multidisciplinary team of scientists from several institutions has studied the snake on Guam, in its native range, and in captivity. Their investigations have ranged from stomach analyses of museum specimens and radiotelemetry of free-living snakes to scanning electron microscopy of venom glands. Initial results confirmed the cause of this tragedy, and now Brown Treesnakes are being trapped to reduce their populations on Guam. Careful inspection of cargo should help prevent their spread to other Pacific archipelagoes. Several zoos are maintaining captive populations of Guam birds, in hopes of eventually reconstituting that island's avifauna. First the snake must be stopped.

Asian Green Vinesnake (*Ahaetulla prasina*), Borneo; note the enlarged front teeth on the lower jaws, used to grip the skink.

"wolf tooth") are named for the enlarged front teeth with which they catch those slippery lizards. Asian vinesnakes (*Ahaetulla*) have slender brown, gray, or green bodies, up to 2 m in total length, and elongate snouts. Mock vipers (*Psammodynastes*) have a pronounced brow, or canthal ridge, and otherwise resemble some sympatric vipers in appearance and behavior. The most unusual among species specialized for feeding on skinks are the 5 species of Asian flying-snakes (*Chrysopelea*). These big-eyed, meter-long, often brightly marked snakes climb readily. Their common name comes from the ability to launch themselves from a coil, straightening the body like a released spring, and travel through the air in a controlled glide. During "flight" the body is flattened and concave ventrally, a position facilitated by hinges on either side of each ventral scale. Although other colubrines sometimes dive out of trees (e.g., African bushsnakes [*Philothamnus*], neotropical parrotsnakes [*Leptophis*]), only the flyingsnakes are structurally modified for this task.

Most other Asian colubrines are slender diurnal snakes. Like flyingsnakes (*Chrysopelea*), the 10 species of bronzebacks (*Dendrelaphis*) are excellent climbers and feed mainly on lizards. Named for the iridescence of their dorsal scales, bronzebacks inflate their necks when threatened, exposing bright blue, yellow, or red interscalar skin. Other African colubrines include some of that continent's most famous snakes. Boomslangs (*Dispholidus typus*) and twigsnakes (*Thelotornis*) are predators on lizards and birds, and among the few colubrids capable of fatally envenoming humans. African egg-eaters (*Dasypeltis*) are probably the most highly modified of all serpents for ingesting bulky objects. African colubrines also include a few specialized burrowing genera. Twelve species of shovel-snouted snakes (*Prosymna*) use their upturned rostral scale to dig for lizard eggs. Leaf-nosed snakes (*Lytorhynchus*) of northern Africa and

Striped Shovel-snouted Snake (*Prosymna bivitatta*), southern Africa.

the Middle East have snout modifications similar to those of some New World colubrids (patch-nosed snakes [*Salvadora*], leaf-nosed snakes [*Phyllorhynchus*]) and likewise feed on lizards.

NATRICINAE

Widespread in the Old World and North America, this group of 230 species includes the familiar Grass Snake (*Natrix natrix*) and its relatives, a few African species, and a large radiation of Asian watersnakes. Natricines usually have broad heads and keeled scales, are of moderate size (1 m or less in total length), and are foul-smelling. Terrestrial members of this group tend to be striped or unicolored, whereas their aquatic relatives are spotted or banded. Most Old World natricines are oviparous, and female Grass Snakes nest communally in trash piles. Females of some species of *Amphiesma, Natrix, Rhabdophis,* and *Xenochrophis* remain with their eggs during incubation, and some *Amphiesma* are viviparous. Natricines vary greatly in temperament, and the bites of two species of *Rhabdophis* are potentially fatal to humans.

Several Asian natricine genera, collectively called "keelbacks," are ecologically and behaviorally diverse. The Checkered Keelback (*Xenochrophis piscator*) is a common snake in flooded rice paddies; it feeds on fish and lays up to a hundred eggs in a clutch. Most species of *Rhabdophis* are terrestrial or aquatic, feeding mainly on fish and frogs, but the Speckle-bellied Keelback (*R. chrysargus*) climbs in bushes along streams and eats frogs, lizards, rodents, and even birds. Some montane streamsnakes (*Opisthotropis*) are found under stones near streams and feed on earthworms, while the Bicolored Streamsnake (*O. lateralis*) resembles North

American natricines of the genus *Regina* in its striped color pattern and diet of crustaceans. Many Asian natricines elevate and spread their necks when threatened, and in *Macropisthodon* and some species of *Rhabdophis*, a white, bitter-tasting substance oozes from nuchal glands during that display.

African natricines include the Viperine Watersnake (*Natrix maura*) north of the Sahara Desert and various watersnakes (*Afronatrix, Hydraethiops*) and marshsnakes (*Natriciteres*) in southern parts of that continent. Marshsnakes eat frogs and aquatic spiders, and they frequently escape predators by breaking their tails.

HOMALOPSINAE

Homalopsines include about 32 species in 10 genera of watersnakes in southeastern Asia and Australia; all are viviparous. *Enhydris* contains 20 species, whereas most of the other genera are monotypic. Homalopsines inhabit freshwater or marine habitats, mainly the former, and their slitlike, valvular nostrils and tight-fitting rostral and labial scales keep water from entering when they are submerged. Homalopsines have grooved rear fangs and use venom to immobilize their prey. Most species are inoffensive; although some Bocourt's Watersnakes (*E. bocourti*) and Puff-faced Watersnakes (*Homalopsis buccata*) bite fiercely when handled, their venoms evidently have no serious effects on humans.

Adult Puff-faced Watersnakes (*Homalopsis buccata*) have dark body blotches on a lighter ground color, a pattern similar to that of some North American watersnakes (*Nerodia*); juveniles are brightly banded with black and pink or yellow. Bocourt's Watersnake (*Enhydris bocourti*) is the longest and stoutest homalopsine, reaching >1 m in length, and has a color pattern strongly reminiscent of North American Diamond-backed Watersnakes (*N. rhombifer*) and Brown Watersnakes (*N. taxispilota*). The Dog-faced Watersnake (*Cerberus rynchops*) sidewinds on tropical mud flats and catches fish underwater. Like some other homalopsines, this estuarine species is polymorphic for dark brown and red colors; it thereby resembles another North American natricine, the Salt Marsh Watersnake (*N. clarkii*).

Two homalopsines have no close parallels among unrelated watersnakes elsewhere. The White-bellied Mangrove Snake (*Fordonia leucobalia*) prowls for crustaceans at night on tidal mud flats; prey are first pinned by the snake's forebody, then constricted and envenomed. This species tears the legs off larger crabs before swallowing their bodies and perhaps partially crushes prey with its unusually blunt teeth. Tentacled Snakes (*Erpeton tentaculatus*) are noteworthy for their long, paired rostral protuberances of uncertain, perhaps cryptic or tactile function. Moreover, Tentacled Snakes have heavily keeled dorsal scales, extremely narrow ventral scales, and no distinct subcaudal scales. Hanging by their tails in slow-moving streams, these strange snakes seize passing fish with a lightning-fast strike and swallow them in a few seconds. Tentacled Snakes become rigid when touched, thus perhaps resembling a twig, and as a result this species is called "boardlike snake" in Thailand.

Although all homalopsines exhibit adaptations to aquatic and marine environments, the Keel-bellied Watersnake (*Bitia hydroides*) superficially resembles a true seasnake more than does any other colubrid. This species has a small head and neck tapering to a larger posterior body, reduced ventral scales, and a short, feebly compressed tail. Keel-bellied Watersnakes are banded with yellow or olive and dark gray and are thus similar to several true seasnakes in coloration. Their dorsal scales are triangular and separated by exposed skin, unlike those of any other snake.

Tentacled Snake (*Erpeton tentaculatus*), southeastern Asia.

XENODERMINAE

This group includes about 15 species in 4 genera, all restricted to southeastern Asia and the Indonesian Archipelago. Xenodermines are dark-colored snakes, 35 to 65 cm in total length; peculiar in one aspect or another, they remain poorly known. Almost the entire chin of *Fimbrios klossi* is covered by a pair of large scales, and its scalloped, protuberant anterior labial scales give the mouth a slightly fringed appearance. *Stoliczkaia* has small, keeled dorsal scales that are juxtaposed or separated by bare skin, and the Javan Mudsnake (*Xenodermis javanicus*) has heavily keeled scales separated by areas of bare skin and three dorsal rows of large, knobby tubercles. Javan Mudsnakes are oviparous. These nocturnal, slow-moving, fossorial predators on frogs differ from other burrowing snakes in their strange scalation, distinctively enlarged heads, and long tails. Rather, their overall appearance is vaguely similar to that of certain aquatic and burrowing lizards (e.g., the Bornean Earless Monitor [*Lanthanotus borneensis*]).

CALAMARINAE

Calamarines include 9 genera of small, shiny-scaled Asian reedsnakes. With about 50 species, *Calamaria* is the largest calamarine genus and reaches its greatest diversity in Borneo and Sumatra; other genera include mountain reedsnakes (*Macrocalamus*) and dwarf reedsnakes (*Pseudorhabdion*). All calamarines share structural modifications associated with burrowing, including a reduced number of dorsal scale rows, fused head scales, and consolidated skull bones. Most reedsnakes evidently eat earthworms and insect larvae, but some eat skinks, which perhaps accounts for puzzling variation in their dentition. All calamarines are oviparous.

The color patterns and defensive tail displays of some calamarines are strikingly similar to those of sympatric venomous Asian elapids. The Variable Reedsnake (*Calamaria lumbricoidea*) is brown dorsally with a yellow-and-black-checkered belly, as are the Spotted Coralsnake (*Calliophis gracilis*) and the Brown Long-glanded Coralsnake (*Maticora intestinalis*). In some parts of its range the Pink-headed Reedsnake (*Calamaria schlegeli*) is iridescent blue-black with an immaculate orange-red head, a color pattern like that of the Blue Long-glanded Coralsnake (*M. bivirgata*) and the Red-headed Krait (*Bungarus flaviceps*); elsewhere *C. schlegeli* has a dark head. The possibility that reedsnakes mimic dangerous elapids deserves further study.

PAREATINAE

This small group of drab-colored, inoffensive Asian snakes includes 19 species of slugsnakes (*Pareas*) and the Blunt-headed Treesnake (*Aplopeltura boa*). Pareatines have no mental groove; they reach a maximum length of about 75 cm and are oviparous. *Pareas* includes some species that are mainly terrestrial and others that are arboreal. Most slugsnakes are slender creatures, with large heads and eyes, and the arboreal species have laterally compressed bodies, thin necks, and enlarged vertebral scales. Slugsnakes feed on mollusks, pulling snails from their shells with alternate movements of their elongate mandibles and needlelike teeth; the Blunt-headed Treesnake eats lizards. Pareatines parallel certain dipsadine colubrids of like body form that have similar diets, but their New World counterparts independently evolved unusually slender foreparts as arboreal predators on frogs and lizards (blunt-headed vinesnakes [*Imantodes*]) or on mollusks (snail-eaters [*Dipsas*]).

PSAMMOPHIINI

This well-defined group includes about 35 species in 8 genera of mostly slender, diurnal serpents, most of them Afro-Asian sandsnakes and grass snakes (*Psammophis*). An exception is the Montpellier Snake (*Malpolon monspessulanus*), found in southwestern Europe as well as northern Africa. Psammophiines are rear-fanged, and most are fast-moving predators on terrestrial lizards. Species of *Psammophis* are evidently graced with good vision; these snakes often travel with their heads held off the ground, and they climb readily. Barksnakes (*Hemirhagerrhis*) are secretive arboreal creatures, however, and beaked snakes (*Rhamphiophis*) use their sharply angled snouts to dig up Naked Mole Rats (*Heterocephalus glaber*) and other prey. Several psammophiines show maxillary modifications associated with eating skinks, but they are evidently not closely related to other such snakes. Some psammophiines bite fiercely when captured, and *M. moilensis* of western Asia spreads a hood when threatened. Psammophiines have small, almost vestigial hemipenes and peculiar, "detached" copulatory behavior. During copulation, Stripe-bellied Sandsnakes (*P. subtaeniatus*) lie still with their cloacas about

Namibian Sandsnake (*Psammophis leightoni*). Like many slender, diurnal serpents with large eyes, these snakes chase lizards as prey.

1.5 cm apart and one of the male's wormlike hemipenes visibly connecting them. Spotted Skaapstekers (*Psammophylax rhombeatus*) coil around their eggs, under a rock, throughout the 35–45 day incubation period.

Some psammophiines use stereotyped polishing movements to anoint their skins with a colorless, fast-drying fluid. This liquid is secreted by glands on either side of the snout and emerges from a pore inside each nostril. The Montpellier Snake (*Malpolon monspessulanus*) applies the fluid in about ninety seconds by lifting its belly and wiping its head on the ventral and caudal scales. A Short-snouted Grass Snake (*Psammophis sibilans*) uses vigorous circular swipes on alternate sides to cover its entire body, while the behavior of two other species in that genus is distinctive but more like that of *Malpolon*. The snakes polish themselves fre-

Striped Skaapsteker (*Psammophylax tritaeniata*), southern Africa.

quently, especially after ecdysis and feeding. Montpellier Snakes are unusually resistant to desiccation, so perhaps their nasal gland secretion and remarkable polishing behavior aid in preventing water loss.

BOODONTINI

This group might contain as many as 45 species in 15 genera, but the allocation of many of them is uncertain. At the very least, boodontines encompass several moderate-sized, smooth-scaled, oviparous, nocturnal, constricting snakes. Among them are a dozen or more species of

Brown Housesnake (*Lamprophis fuliginosus*), southern Africa. This widespread species constricts its rodent prey.

African housesnakes (*Lamprophis*), terrestrial serpents that feed largely on rodents. Six species of African watersnakes (*Lycodonomorphus*) hunt fish underwater, constrict their prey, and even swallow small items while submerged. Ten species of African filesnakes (*Mehelya*, unrelated to Australasian filesnakes) eat mainly other snakes. One 1.6-m Cape Filesnake (*M. capensis*) contained a 1-m Olive Grass Snake (*Psammophis phillipii*), an 82-cm African Rock Python (*Python sebae*), a 53-cm Brown Watersnake (*Lycodonomorphus rufulus*), and a 48-cm Spitting Cobra (*Naja mossambica*)! The triangular cross-sectional shape, the basis for the common name of these snake-eaters, might identify them as non-prey if seized by other filesnakes.

MISCELLANEOUS ASIAN, AFRICAN, AND MADAGASCAN COLUBRIDS

There is considerable morphological and ecological diversity among the many Old World colubrids whose relationships are still uncertain. *Thermophis baileyi* is a particularly intriguing and poorly known Asian colubrid snake. Found at elevations above 4,000 m and active year-round, this species is known only from the vicinity of a few hot springs in Tibet. Originally assigned to *Natrix*, *T. baileyi* is structurally intermediate between colubrines and natricines, and its relationships remain unknown.

Mole Snake (*Pseudaspis cana*), southern
Africa; juveniles of this species are
brightly blotched.

Cape Wolfsnake (*Lycophidion capense*),
southern Africa.

African slug-eaters (*Duberria*) bear an uncanny resemblance in color pattern to North American Sharp-tailed Snakes (*Contia tenuis*), the latter also specializing on slugs. Adult Mole Snakes (*Pseudaspis cana*) can exceed 2 m in total length, and males injure each other with deep slashing bites during combat over females. Female Mole Snakes have been known to give birth to as many as eighty-four young, which, unlike their unicolored parents, resemble sympatric venomous adders (*Bitis*) in color pattern. African wolfsnakes (*Cryptolycus, Lycophidion*), so-named for their long anterior teeth, feed mainly on skinks; they also have flat heads and peculiarly shiny skin, perhaps associated with their squeezing through crevices.

The poorly studied colubrid fauna of Madagascar has possible affinities with that of Africa; *Mimophis* is a psammophiine, and *Geodipsas* occurs on both landmasses. Among 14 other Madagascan colubrid genera, vinesnakes (*Langaha*) have bizarre sexual dimorphisms but otherwise superficially resemble Asian vinesnakes (*Ahaetulla*) and New World vinesnakes (*Oxybelis, Xenoxybelis*). Other endemic taxa include hog-nosed snakes (*Leioheterodon*), blunt-headed vinesnakes (*Lycodryas*), and skink-eaters with folding teeth (*Liophidium*).

Bird Snake (*Pseustes poecilonotus*),
Costa Rica.

10

My first exposure to the philosophical problems posed by snakes and birds had a pragmatic, rural slant: Mother's family in Texas did not deal kindly with "chicken snakes" in the hen-house. Snakes versus eggs and pullets was a straightforward issue for Grandpa Gibson, quickly resolved with whatever was handy for killing the marauding serpent. My later, more esoteric pursuits have led me to overturn rocks and logs in search of lizards and snakes, whereas those who watch wild birds usually keep their eyes glued skyward. Our interests converge because some snakes are especially proficient nest robbers, and because notions of good and evil are sometimes misused in that context.

For more than forty years Alexander F. Skutch has studied the bird life of tropical forests in Costa Rica, and he is especially well known for his careful, descriptive accounts of nesting biology. Skutch often mentions snakes in his popular writings, but never favorably and sometimes inaccurately (like many rural Costa Ricans, he confuses three dissimilar species under the local name "Zopilota"). I can't quarrel with liking birds, of course, and some neotropical snakes surely are expert bird hunters; they even crawl along branches and down into the hanging, woven grass nests of oropendolas, approaching them the way African Boomslangs (*Dispholidus typus*) do colonies of weaverbirds (*Ploceus*). Skutch dislikes predators in general, however, and condemns snakes with a special vengeance. Among raptors he praises only the Laughing Falcon (*Herpetotheres cachinnans*), which eats mainly snakes.

Skutch grudgingly allows for predators in distant wilderness areas, but prefers a "principle of harmonious association" for his own surroundings, one in which "every member is compatible with every other, and there is mutual exchange of benefits." To that end he kills Bird Snakes (*Pseustes poecilonotus*), Tiger Ratsnakes (*Spilotes pullatus*), and other harmless species and singles them out to explicate a strange worldview. After dismissing danger to humans as a justification for killing these predators, Skutch wrongly claims that snakes are never "really social" and notes that with few exceptions, they are "devoid of parental solicitude." Then he really gets steamed up: "The serpent . . . crams itself with animal life that is often warm and vibrant, to prolong an existence in which we detect no joy and no emotion. It reveals the depth to which evolution can sink when it takes the downward path and strips animals to the irreducible minimum able to perpetuate a predatory life in its naked horror. The contemplation of such an existence has a horrid fascination for the human mind and distresses a sensitive spirit."

There are bird-watchers whose tolerance for snakes at least borders on admiration, as I learned from Marcia Bonta's fine article in *Bird Watcher's Digest* about a Black Ratsnake's (*Elaphe o. obsoleta*) successful raid on the nest of a House Wren (*Troglodytes aedon*). (Despite their common name, those serpents climb well and often eat birds; Henry S. Fitch used the cries of mobbing Blue Jays [*Cyanocitta cristata*] and other birds to locate ratsnakes for his Kansas field study.) This Black Ratsnake scaled a porch support of the Bontas' Pennsylvania farmhouse, negotiated almost insurmountable eaves and gutters, and cleaned out the wrens' nest. Bonta's meticulous description of the hour-and-forty-minute-long event is valuable natural history, unparalleled in the literature on snakes. Family and friends thought she should have killed the snake or at least prevented its attack, and *Bird Watcher's Digest* received a flood of protest mail. Twenty letters criticized the article as "asinine" and "disgusting," and one terrified reader stapled the pages together, "since each time I picked up the magazine [it fell] open to that awful picture of the snake." Along with six other correspondents, I preferred Bonta's conclusion, that "in the case of the nestlings and the snake, respecting the intricate web of life forced us to applaud the winner even though we had been rooting for the losers."

Soon I heard from Bethune Gibson, of Sedona, Arizona, who'd read my letter to *Bird Watcher's Digest*. A vivacious seventy-eight-year-old with a keen interest in wildlife, Beth now offered me her own detailed account of hunting behavior by a free-living snake. She has a large picture window with a rolled-up sunscreen along its top, rarely used since the mulberry trees out front grew large enough to provide shade. Every year, House Finches (*Carpodachus mexicanus*) nest in each end of the roll. One day in May, Beth looked up from her reading by the window and saw "the tail half of a snake wriggling like crazy, the front half already into the rolled screen, where finches had recently hatched." The snake was slender, about a meter long, and "a beautiful and exquisite rosy brown." Judging from size, shape, and color pattern, it was surely a Coachwhip (*Masticophis flagellum*), a fast-moving, nonvenomous species that eats everything from grasshoppers to small rattlers and rabbits. Coachwhips occasionally take young birds, and by comparison with the pendulous constructions of some tropical species, an exposed nest in a desert shrub or rolled up in a sunshade would be an easy bowl of snacks.

This Coachwhip had evidently scaled the vine-covered cement wall adjacent to Beth's window. It then crossed on vegetation beside the house and, still undetected, entered the shade roll. Beth "hastily checked the end of its tail before going outside to discourage its lunch of baby finches and noted that there were no rattlesnake rings!" At this point, daughter Linda and her Benedictine monk friend arrived, and at Beth's request—she does not approve of needlessly killing snakes—Abbot Leonard calmly tapped on the rolled-up shade with a broom handle. Linda thought the snake "looked a bit 'nose out of joint' " as it backed out with a baby bird in its mouth. Beth told me, "You won't believe this, but that snake looked peevish." On two other occasions Coachwhips have peered in on her, perhaps attracted by the finch nests, and one of them stared at the window for a full two minutes. Incidental to all this, in one of life's pleasant ironies, Beth and I might be related through

Speckled Racer (*Drymobius margaritiferus*), Costa Rica.

Green-headed Racer (*Leptodrymus pulcherrimus*), Costa Rica. As is true of many striped species, these are fast-moving snakes.

Scorpion-eater (*Stenorrhina freminvillei*),
Costa Rica.

and feed mainly on frogs. The neotropical Neck-banded Snake (*Scaphiodontophis annulatus*) specializes on skinks, with the aid of folding teeth that prevent the slippery lizards from escaping.

Of the New World colubrines, Indigo Snakes (*Drymarchon corais*) are especially impressive for their great size, wide distribution, and broad diet. Found from the southeastern United States to Argentina, often near water, these shiny serpents are commonly known as "Sabaneras" (savanna creatures) or "Cribos" in Latin America. This species (or complex of species) grows to more than 3 m long and eats an extraordinarily wide range of vertebrates. One individual contained three mice, two small Common Snapping Turtles (*Chelydra serpentina*), and two Mexican Burrowing Frogs (*Rhinophrynus dorsalis*); another regurgitated a toad (*Bufo*), a hatchling Gopher Tortoise (*Gopherus polyphemus*), a Southern Hog-nosed Snake (*Heterodon simus*), and a Pigmy Rattlesnake (*Sistrurus miliarius*). Other serpents as well, including large and venomous species, are often its prey: a 2.95-m Guatemalan Indigo Snake that was swallowing a 1.68-m Boa Constrictor (*Boa constrictor*) already had in its stomach a Jumping Pitviper (*Atropoides nummifer*) almost .95 m long!

Many temperate North American colubrines are large and usually diurnal; they are among the most popular snakes as pets. One clade includes Eastern Pinesnakes (*Pituophis melanoleucus*) and their relatives, as well as various ratsnakes (*Bogertophis* and *Elaphe*), all powerful constrictors that feed mainly on mammals and occasionally on birds. Most species of ratsnakes and some Gopher Snakes (*P. catenifer*) are semiarboreal, whereas Eastern Pinesnakes are decidedly fossorial. All species of *Pituophis* have a unique cartilaginous keel in front of the glottis; this structure amplifies hissing and enhances their resemblance to venomous rattlesnakes. Kingsnakes (*Lampropeltis*) and Milksnakes (*L. triangulum*) are rather heterogeneous in behavior and ecology (see the Special Topic in this chapter). They may not be monophyletic and are traditionally grouped apart from ratsnakes because of their single anal scale and smooth dorsal scales (their generic name means "bright skin"). The widespread Common Kingsnake (*L. getula*) is a dietary generalist that occasionally eats pitvipers and is immune to their venoms. The Sonoran Mountain Kingsnake (*L. pyromelana*) eats lizards, mice, and nestling birds. Oddly enough, this beautiful, long-tailed species is both secretive, often sheltering in rock piles, and sometimes semiarboreal in its hunting and escape behavior. Related taxa include scarletsnakes (*Cemophora*) and long-nosed snakes (*Rhinocheilus*)—two groups of burrowing predators on squamates and their eggs, both of them mimics of New World coralsnakes—and the Short-tailed Snake (*Stilosoma extenuatum*), which feeds only on blackheaded snakes (*Tantilla*).

A few genera of temperate colubrines are rear-fanged, such as the two species of lyresnakes (*Trimorphodon*). Lyresnakes superficially resemble some Old World catsnakes (e.g., *Boiga*, *Telescopus*) in shape and habits, but they independently evolved rear fangs within the larger clade of nonvenomous New World ratsnakes (*Elaphe*) and their relatives. Both species of *Trimorphodon* have broad, flattened heads and vertical pupils. Both are nocturnal, inhabit arid regions, and feed largely on diurnal lizards, especially iguanas (*Ctenosaura, Iguana*) and spiny lizards (*Sceloporus*); adult lyresnakes occasionally also eat birds, bats, and rodents. Their main prey typically shelter in crevices at night, and the venom of lyresnakes probably facilitates extraction of the otherwise tightly wedged lizards.

About a dozen genera of small North American colubrine snakes are placed in a separate tribe, Sonorini. Many are restricted to arid areas and all specialize on arthropods as prey. Sonorinines include shovel-nosed snakes (*Chionactis*) that eat primarily roaches and scorpions; hook-nosed snakes (*Ficimia, Gyalopion,* and *Pseudoficimia*) that feed mainly on spiders; black-headed snakes (*Tantilla,* about 50 species), most of which specialize on centipedes; and Mexican earthsnakes (*Conopsis*) that eat insect larvae. Sandsnakes (*Chilomeniscus*) and shovel-nosed snakes have countersunk lower jaws, concave bellies, and valvular nostrils that facilitate locomotion under sand. Prairie Groundsnakes (*Sonora semiannulata*) are often stunningly polymorphic in color patterns, with individuals unicolored, striped, or brightly banded. The File-tailed Groundsnake (*Sonora aemula*) of northern Mexico is a mimic of New World coralsnakes and has a peculiar spinose tail, perhaps used in burrowing. Scorpion-eaters (*Stenorrhina*) are tropical members of this group that eat scorpions and tarantulas. Most sonorinines are oviparous, but some montane Mexican earthsnakes give birth to live young.

NATRICINAE

The 52 species of New World watersnakes (*Nerodia*), gartersnakes (*Thamnophis*), and their relatives are restricted to the Northern Hemisphere. General resemblances among North American watersnakes perhaps approximate the immigrant stock from which gartersnakes and other, more terrestrial, New World natricines evolved. All North American natricines are viviparous, in contrast to most of their Old World relatives.

TEMPERATE AND TROPICAL MILKSNAKES (*LAMPROPELTIS TRIANGULUM*): A SINGLE SPECIES?

Scarlet Kingsnake (*Lampropeltis triangulum elapsoides*), Florida.

The Milksnake (*Lampropeltis triangulum*) is among the world's most widely distributed terrestrial snakes, rivaled only by the European Adder (*Vipera berus*), the Indigo Snake (*Drymarchon corais*), and the Brown Vinesnake (*Oxybelis aeneus*) for total area occupied. (Flowerpot Blindsnakes [*Ramphotyphlops braminus*] and Yellow-bellied Seasnakes [*Pelamis platurus*] are more widespread globally, but their distributions are far less continuous than that of *L. triangulum*.) None of these may in fact be single species, but among them *L. triangulum* is surely the most variable in external morphology, ecology, and behavior.

In the northeastern United States, *Lampropeltis t. triangulum* inspired the common name Eastern Milksnake, because its propensity for frequenting barns in search of mice resulted in the mistaken belief that snakes drain cows dry. This gray snake with brown blotches has a rounded snout and reaches a maxi-

mum length of about 1 m. Elsewhere Milksnakes generally resemble venomous New World coralsnakes, but they run the gamut in size and ecology. The sharp-snouted Scarlet Kingsnake (*L. t. elapsoides*) of the southeastern United States is less than .5 m long and eats mainly skinks, whereas round-snouted Tropical Milksnakes (*L. t. hondurensis*) on the Atlantic versant of Costa Rica reach almost 2 m and prey mainly on rodents.

Geographic variation of color patterns in *Lampropeltis triangulum* is correlated with the absence or presence of particular sympatric elapids. Examples include increasing dissimilarity in the northeastern United States to any venomous coralsnake; broad red bands in sympatry of the Sinaloan Milksnake (*L. t. sinaloae*) and West Mexican Coralsnake (*Micrurus distans*); orange-and-black Tropical Milksnakes and similarly bicolored Central American Coralsnakes (*M. nigrocinctus*) in Hondu-

ras; and a tendency toward secondary black rings in Milksnakes of southern Mexico, where the complexly banded Elegant Coralsnake (*M. elegans*) is found. Like venomous coralsnakes, Tropical Milksnakes in lowland Central America are mainly diurnal and bite fiercely when seized.

Adult Black Milksnakes (*Lampropeltis triangulum gaigeae*) from the highlands of Costa Rica and Panama are an immaculate, shiny black, whereas their juveniles are brightly colored and resemble other tropical *L. triangulum*. Possible ecological explanations for this unique and derived color pattern are that in some areas the venomous Allen's Coralsnake (*Micrurus alleni*) undergoes a similar pattern shift, and thus color change in the Milksnakes maintains mimicry, or that black coloration aids thermoregulation in cool, montane habitats.

Milksnakes draw considerable attention and high prices in the pet

Tropical Milksnake (*Lampropeltis triangulum hondurensis*), Costa Rica; this individual is about 1.4 m long and is eating a Spiny Pocket Mouse (*Heteromys desmarestianus*).

trade, owing to their shiny, bright color patterns. Nevertheless, for all the attraction of this species, several interesting facets of its biology remain unstudied. The significance of geographic variation for problems of speciation and definition of species remains to be explored. Many of the 25 subspecies of *Lampropeltis triangulum* probably do not warrant taxonomic recognition, while others might qualify as separate species (see the Appendix). The ways that changing biotas and climates have shaped Milksnake distributions in the western United States is worthy of attention, and the relationships of *L. triangulum* to other tricolored species (e.g., the Sonoran Mountain Kingsnake [*L. pyromelana*] and California Mountain Kingsnake [*L. zonata*]) are not well understood. Only for the Eastern Milksnake and Red Milksnake (*L. t. syspila*) in the United States do we have even barely adequate natural history information. Other under-studied problems with Milksnakes include species status relative to population overlap of the small Scarlet Kingsnake and large Eastern Milksnake in parts of the United States, geographic variation in body size and tooth morphology relative to diets of rodents versus skinks, and color pattern in relation to coralsnake mimicry.

Western Ribbonsnake (*Thamnophis proximus*), Costa Rica. This neotropical representative lacks the bright colors typical of many species of *Thamnophis*.

New World watersnakes are typically stout-bodied, dull-colored, and ill-tempered animals; some species resemble Cottonmouths (*Agkistrodon piscivorus*) in color pattern and perhaps are mimics of those pitvipers. They inhabit streams and ponds and feed primarily on fish and frogs. The Salt Marsh Watersnake (*Nerodia clarkii*), found in estuaries on the Gulf Coast of the United States, tolerates frequent immersion in saltwater simply by not drinking it. Most species of *Nerodia* are distributed over large parts of eastern North America, but the Brazos River Watersnake (*N. harteri*) in Texas has a restricted distribution and is threatened by the construction of dams.

Gartersnakes (*Thamnophis*) occur throughout most of North and Central America, in habitats ranging from lowland swamps to deserts and montane meadows. Often brightly colored and attractively striped, they are profoundly bad-smelling animals! These usually semiaquatic snakes typically feed on slimy invertebrates, fish, and amphibians. Some populations of the Western Terrestrial Gartersnake (*T. elegans*) eat rodents, subduing them by constriction. Although most species of *Thamnophis* are diurnal, Checkered Gartersnakes (*T. marcianus*) sometimes forage nocturnally in breeding aggregations of desert toads.

Other aquatic North American natricines include crayfish snakes (4 species of *Regina*) and Black Swampsnakes (*Seminatrix pygaea*). Adult crayfish snakes feed entirely on crustaceans, but juvenile Striped Swampsnakes (*R. alleni*) also eat dragonfly larvae. Queen Snakes (*R. septemvittata*) carry narrow dietary preferences to the extreme by eating only crayfish that have freshly shed their shells. Fully terrestrial New World natricines include Kirtland's Snakes (*Clonophis kirtlandi*), Brown Snakes (*Storeria dekayi*) and their relatives, Lined Snakes (*Tropidoclonion lineatum*), and North American earthsnakes (*Virginia*)—all small predators on earthworms and other soft-bodied invertebrates. These genera resemble in habits and overall

Brown Blunt-headed Vinesnake (*Iman-todes cenchoa*), Ecuador.

appearance certain Central American "goo-eaters," described below, as well as some Asiatic natricines.

DIPSADINAE

For decades many neotropical snake genera were lumped together as Xenodontinae or simply referred to as "xenodontines," largely on the basis of primitive hemipenial similarities. Recent studies reveal two well-defined lineages with uncertain relationships to each other and to other colubrids. Those two groups have been known informally as Central American Xeno-dontines and South American Xenodontines, or formally as Dipsadinae and Xenodontinae, respectively.

Several genera of rather unspecialized, terrestrial dipsadines include black-striped snakes (*Coniophanes*) and littersnakes (*Rhadinaea*). These usually dark little serpents glide over and within forest floor litter, where they feed on frogs, lizards, and other ectothermic vertebrates. Snakes of those genera are usually cryptically colored and have relatively normal tails, whereas two related species of halloween snakes (*Pliocercus,* evidently a derivative of *Rhadinaea*) com-bine mimicry of venomous New World coralsnakes with an exceptionally fragile tail. Snakes of all three genera have enlarged rear teeth and immobilize prey with venom.

Another dipsadine clade of terrestrial and arboreal predators on ectotherms includes some of the most frequently seen nocturnal snakes in neotropical forests. Cat-eyed snakes (*Lepto-deira*) usually eat frogs and their eggs, encountering prey by simply crawling slowly along leaves and vines until they contact food. Blunt-headed vinesnakes (*Imantodes*) are specialized relatives of cat-eyed snakes, characterized by more attenuate bodies and large heads with bulging eyes. Some blunt-headed vinesnakes eat frogs and lizards, hunting in a manner similar to cat-eyed snakes. Other *Imantodes,* with exceptionally slender necks and enlarged mid-dorsal scales, extend their foreparts from limbs and pluck sleeping lizards (*Anolis*) from leaves. Although most dipsadines are tropical forest snakes, desert nightsnakes (*Hypsiglena*) are small, terrestrial predators on lizards in drier parts of Mexico and the western United States.

A large subgroup of dipsadines is informally called the "goo-eaters," because their diet consists entirely of soft-bodied invertebrates. Some goo-eaters are small, relatively generalized inhabitants of leaf litter, such as the Red Coffeesnake (*Ninia sebae*), and are reminiscent in appearance and behavior of certain North American natricines (e.g., the Brown Snake [*Storeria dekayi*]). Several other genera of neotropical burrowing snakes (*Adelphicos, Atractus, Geophis*) exhibit the shiny bodies, fused head scales, and tiny eyes of unrelated fossorial taxa elsewhere, including North American wormsnakes (*Carphophis*) and several Old World genera.

Three genera of terrestrial and semiarboreal goo-eaters (*Dipsas, Sibon, Sibynomorphus*) parallel the Old World pareatine colubrids as specialized predators on slugs and snails, and their feeding behavior is better known than that of those Asian counterparts. An Amazonian Snail-eater (*D. indica*) follows the mucus trail of a slug for several meters, frequently flicking its unusually long tongue, and then seizes the mollusk with long, needlelike teeth in a short, accurate strike. It holds the prey in its coils while extracting the body from the shell, pulling with its elongate mandibles from side to side. The teeth of goo-eaters are ill-suited for defensive biting, and most species of *Dipsas* and *Sibon* are probably viper mimics by virtue of similar color patterns and threat behavior.

XENODONTINAE

This diverse radiation of mostly South American snakes is composed chiefly of moderate to large predators on ectothermic vertebrates. Some phylogenetic resolution within this subfamily is indicated by finer, tribal groupings.

Among several genera of Xenodontini are coralsnake mimics (*Erythrolamprus,* which feeds on snakes) and brightly colored neotropical hog-nosed snakes (*Lystrophis*). The large genus *Liophis* is ecologically reminiscent of North American natricines, with 40 species, ranging from terrestrial to fairly aquatic. Like some New World gartersnakes, certain *Liophis* eat amphibians with extremely toxic skin (e.g., dart-poison frogs). Seven species of false pitvipers (*Waglerophis, Xenodon*) feed mainly on toads (*Bufo*) and often are remarkably similar to sympatric venomous pitvipers in color pattern and behavior (photos, pp. 114–15). In addition to several internal anatomical features, these xenodontinine genera share the behavioral trait of flattening and elevating the foreparts of their bodies, thus creating a hoodlike effect (photo, p. 112).

The Pseudoboini includes several genera of powerful, rear-fanged constrictors on vipers, lizards, and rodents. Adult *Phimophis, Pseudoboa, Tripanurgos,* and some calico snakes (*Oxyrhopus*) are brown or red, their black heads interrupted or bordered with a yellow or white collar. Juvenile mussuranas (*Clelia*) retain that color pattern, whereas adults of some species are shiny dark brown or black. Some species of calico snakes are brightly banded mimics of venomous coralsnakes. Most pseudoboines are largely terrestrial, but *Siphlophis* and *Tripanurgos* are more slender, arboreal snakes. Some species of *Phimophis* and *Pseudoboa* have upturned snouts and are perhaps burrowers.

Above: Calico Snake (*Oxyrhopus petola*), Peru.

Left: Banded Treesnake (*Tripanurgos compressus*), Costa Rica.

White-headed Snake (*Enulius sclateri*), Costa Rica. Snakes of this strange genus have no known close relatives and feed on reptile eggs.

Big-headed Coralsnake Mimic (*Rhinobothryum lentiginosum*), Ecuador.

South American xenodontines of uncertain affinities include more than a dozen species of neotropical watersnakes (*Helicops*), reminiscent ecologically and morphologically of some New World natricines and Old World homalopsine colubrids. False water cobras (*Hydrodynastes*) are large aquatic snakes that sometimes hunt by probing with their tails for fish among giant shoreline bromeliads. Seventeen species in the racerlike genus *Philodryas* span a broad range of color patterns and ecologies in temperate and tropical South America. Species of *Apostolepis* and *Elapomorphus* are slender, brightly colored predators on other burrowing snakes and amphisbaenians; some of them look remarkably like venomous coralsnakes in head shape and scalation, and their venom injection mechanisms are more proteroglyph-like than those of most other rear-fanged snakes. *Tropidodryas striaticeps*, a nocturnal Brazilian snake, is highly unusual in that its tail is prehensile, and juveniles use caudal luring to attract lizard prey; it thus resembles some boids and viperids in locomotor and hunting behavior. Among the few viviparous neotropical colubrids, two species of South American mock pit-vipers (*Tomodon*) have enormous rear fangs that probably aid in feeding on slugs as well as enhance their resemblance, during gaping threat displays, to venomous true pitvipers.

OTHER NEW WORLD COLUBRIDS

Several taxa are not yet confidently placed in subgroups within New World colubrids. In Mexico and Central America, road guarders (*Conophis*) are brightly striped, diurnal chasers of lizards in savanna habitats, whereas long-tailed snakes (*Enulius*) are predators on reptile eggs. Species of *Rhinobothryum* of Central and South America are poorly known, possibly arboreal, mimics of New World coralsnakes. Wormsnakes (*Carphophis*) are shiny, bright pink-and-black or pink-and-brown burrowers that eat only earthworms, whereas Sharp-tailed Snakes (*Contia tenuis*) use long, needlelike teeth to feed exclusively on slugs. Ring-necked Snakes (*Diadophis punctatus*) show amazing diversity across their transcontinental distribution, ranging from 30-cm predators on salamanders in the eastern United States that lack fangs and tail displays, to 90-cm giants in the Southwest that envenomate reptile prey with rear fangs and have spectacular defensive tail displays. North American Mudsnakes (*Farancia abacura*) and Rainbow Snakes (*F. erytrogramma*) are large, semiaquatic predators on elongate giant salamanders and eels, respectively. Three species of North American hog-nosed snakes (*Heterodon*) dig up toads with their spadelike snouts and have among the most complex defensive repertoires of any snake (such as death-feigning, shown on p. 31).

STILETTO SNAKES AND OTHER AFRICAN ENIGMAS

11

Stiletto snakes (*Atractaspis*) are misleading and confusing creatures that often attract the attention of laypeople and naturalists. Their skulls are compact, as in many burrowing reptiles, yet among the most derived and movable compared to the feeding mechanisms of other snakes. Stiletto snakes are uniquely adapted for harvesting the subterranean nestlings of rodents, but they are provisionally included in a larger, more diverse group that might hold important keys to understanding the evolution of all snake venoms.

The generic name for stiletto snakes comes from the Greek words *atraktos,* "a thin shaft or arrow," and *aspis,* "a viper." The first two species described, in the 1840s, were assigned with cobras to the family Elapidae, but soon thereafter and for another hundred years, stiletto snakes were mistakenly classified as vipers. No viper, however, remotely resembles these shiny, slender serpents. In 1906 Frank Wall misidentified one as a new genus and species of "elapine," once again implying relationships with cobras and their relatives. For decades, herpetologists wrongly believed that the "overdeveloped," movable front fangs of *Atractaspis* exceeded the length of the head and were incapable of simultaneous erection. As early as 1898, they also noted that these peculiar black, gray, or brown snakes strikingly resembled certain rear-fanged but harmless burrowers, the purple-glossed snakes (*Amblyodipsas*); some stiletto snakes have cream-colored bellies, which simplify the distinction when their undersides can be glimpsed safely. Because of a long-standing emphasis on fang type for classifying venomous snakes, however, the possibility that stiletto snakes and those rear-fanged species might actually be close relatives was not raised until 1965, by Monique Bourgeois's studies on the skulls of African snakes.

Stiletto snakes are well known and feared by rural people in Africa, with good reason. In the Sudan, more than two dozen vernacular names are applied to them, including "father of blackness," "father of burrowing," "father of ten minutes," "father of jabbing," "father of diving," "father of prodding," "father of lancing," "shroud bearer," "father of the club," "rope," "snake of seven steps," "snake bracelet," "glistening snake," and "bite dead." Stiletto snakes respond to threats with a puzzling, immobile headstand, then strike suddenly with a slashing, backward movement. These irascible little creatures often bite inquisitive children and amateur snake catchers, but the lesson that stiletto snakes—and thus other snakes resembling them—cannot be picked up safely took years to sink in as common wisdom. The late Robert Mertens, an extraordinarily accomplished German herpetologist, was famous for his encyclopedic ability to identify amphibians and reptiles from anywhere, often to subspecies. Friends commonly brought live animals from all over the world to his office in Frankfurt am Main's Senckenberg Museum, and Mertens delighted in challenging visiting colleagues to identify his captives. Once, however, he reached into a bag said to contain a purple-glossed snake, exclaimed, "Oh, a white belly . . . ," and promptly was bitten by a stiletto snake.

Scientists strive for a stable, formal nomenclature, and some people prefer a uniform system of common names as well. Stiletto snakes and their relatives have proved unusually recalcitrant in both respects. Neither vipers nor cobras, species of *Atractaspis* have been assigned to at least six different taxonomic groups, and the evidence at hand suggests that they are related to certain other African burrowers, including purple-glossed snakes and perhaps harlequin snakes (*Homoroselaps*). Neither "mole vipers" nor "burrowing adders," as these strange serpents have been called, are appropriate English common labels. "Side-stabbing snake" and "burrowing asp" are reasonable names, but taking natural history into account, I prefer Donald Broadley's suggestion: "The drab coloration and innocent appearance of these snakes, together with their habits, strongly suggest the stealthy assassin with concealed stiletto."

A genus of 15 species found only in Africa and the Middle East, *Atractaspis* was long placed in the family Viperidae because of its dentition, which includes a pair of anterior hollow, movable fangs. However, studies in the 1960s by Monique Bourgeois, Elazar Kochva, and others indicated that these dark little snakes represented a separate lineage of venomous snakes, one that perhaps also included several colubrid genera then known informally as "aparallactines" (based on the generic name for centipede-eaters [*Aparallactus*]). Almost simultaneously, Samuel B. McDowell suggested that two species of African harlequin snakes (*Homoroselaps*), traditionally assigned to the Elapidae, also belonged in this group. Although subsequent studies are contradictory with regard to its exact composition, Atractaspididae might include as many as 65 species in 14 genera.

Stiletto snakes (*Atractaspis*) and their taxonomic associates occupy habitats as diverse as rain forest, grassland, and semidesert. They usually are found underneath objects, in termite nests, or crawling above ground at night, especially after rains. Like many other burrowing colubrids and elapids, atractaspidids have shiny, smooth scales; heads that are not distinct from the neck region; small eyes with round pupils; and a short, rather stout tail. Atractaspidids typically feed on elongate reptiles and other vertebrates, although one species eats earthworms. These snakes respond to threat with erratic body movements, head-hiding, and cloacal discharge. Most atractaspidids are oviparous, but Jackson's Centipede-eater (*Aparallactus jacksoni*) and some Natal Purple-glossed Snakes (*Amblyodipsas concolor*) are viviparous.

The African atractaspidids encompass all major modes of dentition and venom-injection mechanisms found in snakes (see Chapter 4 for a review of these modes). One species lacks

Common Purple-glossed Snake (*Amblyodipsas polylepis*), southern Africa; snakes of this genus are often very difficult to distinguish from highly venomous stiletto snakes (*Atractaspis*).

grooved teeth and is perhaps nonvenomous (*Aparallactus modestus*). Most, however, are rear-fanged, like many colubrids; some have fixed front fangs (harlequin snakes [*Homoroselaps*]), like those of cobras and their relatives; and others have movable front fangs (stiletto snakes [*Atractaspis*]), somewhat like those of vipers, although used differently. Atractaspidids might be remnants of an old evolutionary divergence among advanced snakes, perhaps even encompassing the origin of elapids and one or more lineages of colubrids. Because of their potential importance for understanding the evolution of venoms, sorting out relationships among these snakes and other major lineages of colubroids is one of the most intriguing challenges now facing systematic herpetologists.

OPISTHOGLYPHOUS AND AGLYPHOUS ATRACTASPIDIDS

Most atractaspidids have grooved rear fangs that are preceded on the maxillary bones by three to six smaller, solid teeth. The Natal Blacksnake (*Macrelaps microlepidotus*), heavy-bodied and up to 1 m long, has the most generally primitive morphology among atractaspidids. It also has a broad diet of frogs, skinks, snakes, and rodents. Nine species of purple-glossed snakes (*Amblyodipsas*) are rather stout and reach maximum lengths of about 1 m. They feed primarily on blindsnakes but also eat amphisbaenians and Giant Legless Skinks (*Acontias plumbeus*). The anterior mandibular teeth of purple-glossed snakes are enlarged, which perhaps aids

Bicolored Quill-snouted Snake (*Xenocalamus bicolor*), southern Africa; its highly modified snout is used to dig for amphisbaenians.

Cape Centipede-eater (*Aparallactus capensis*), southern Africa.

Spotted Harlequin Snake (*Homoroselaps lacteus*), southern Africa, eating a Black Threadsnake (*Leptotyphlops nigricans*). (Photograph by William R. Branch)

them in grasping shiny-scaled prey. Two species of black-and-yellow burrowing snakes (*Chilorhinophis*) are slender, brightly striped predators on amphisbaenians and snakes. They display a coiled, upturned orange ventral surface of the tail when threatened. Five species of quill-snouted snakes (*Xenocalamus*) have bizarrely elongate and flattened heads, tiny eyes that are displaced laterally by an enormous frontal scale, and underslung lower jaws. Quill-snouted snakes inhabit sandy regions and feed solely on amphisbaenians, usually a single prey species at a particular locality.

Eleven species of centipede-eaters (*Aparallactus*) are remarkably similar to New World black-headed snakes (*Tantilla*) in external appearance and ecology. Most members of each genus are some shade of brown with a dark cap on the head, and most eat centipedes. Predation on such formidable creatures is made possible by the snake's injection of venom and its own immunity to the prey's venom. Several species of *Aparallactus* occur in savannas or semiarid terrain, but the Black Centipede-eater (*A. guentheri*), a dark snake with anterior yellow crossbands, is found in tropical rain forest; its color pattern approaches that of the brightly banded *T. semicincta* of northern South America. Lacking grooved rear fangs, *Aparallactus modestus* is thus perhaps nonvenomous and eats earthworms.

PROTEROGLYPHOUS ATRACTASPIDIDS: *HOMOROSELAPS*

Two species of southern African harlequin snakes (*Homoroselaps*) have enlarged, immovable front fangs and no solid teeth on their maxillary bones. Until just over two decades ago, they were placed in *Elaps,* the type genus for the medically important Elapidae. With evidence that harlequin snakes are not really elapids, the International Commission on Zoological Nomenclature formally changed their generic name to *Homoroselaps* and thereby insured stability for the family name of cobras and their relatives. Harlequin snakes are found under stones and in termitaria, reach a maximum length of about 60 cm, and feed on legless lizards and blindsnakes. These beautiful little snakes are variously banded, spotted, or striped in black, yellow, and orange.

SOLENOGLYPHOUS ATRACTASPIDIDS: *ATRACTASPIS*

Stiletto snakes (*Atractaspis*), deceptively similar to purple-glossed snakes (*Amblyodipsas*) in external appearance, reach a maximum total length of slightly more than 1 m. Most species are dark brown or black, without markings; some Southern Stiletto Snakes (*A. bibronii*) have cream-colored bellies, the West African Stiletto Snake (*A. corpulenta*) has a white tail, and the Ogaden Stiletto Snake (*A. leucomelas*) has a pale head and mid-dorsal stripe.

Several aspects of the anatomy of stiletto snakes (*Atractaspis*) are unusual or unique among snakes. Their maxillary bone pivots on a lateral, socketlike joint with the prefrontal bone. A single, elongate hollow fang has a cutting edge opposite its orifice. The toothless pterygoid bones form continuous, fused and curved struts with the ectopterygoids; as in other atractaspidids, the palatines are independent from the pterygoids and presumably immovable. The dentaries have only a few small anterior teeth. The venom glands in some species (e.g., Small-scaled Stiletto Snakes [*A. microlepidota*]) are exceptionally long, extending under the skin for about 15 percent of the body length behind the head, whereas those of others (e.g., Southern Stiletto Snakes [*A. bibronii*]) are of normal dimensions. The most anterior vertebrae of stiletto snakes are smaller than others, which probably facilitates acute downward movements of the head during feeding and defense.

Stiletto snakes (*Atractaspis*) typically eat nestling rodents, limbless lizards, other snakes, and amphisbaenians; occasionally they consume frogs and reptile eggs. The average number of prey in stomachs containing rodents is about two and a half, which suggests that the snakes usually consume an entire nest of mammals at one time. Reptilian prey are usually encountered singly, although one Southern Stiletto Snake (*A. bibronii*) from Tanzania contained an amphisbaenian and a black-and-yellow burrowing snake (*Chilorhinophis*) that in turn had eaten another amphisbaenian! Stiletto snakes bite live mice repeatedly until they become immobilized; during each strike, a single fang is extended ventrolaterally from the closed mouth and jerked backward from whichever side of the head is next to the prey. When more than one nestling rodent is available, a stiletto snake strikes all potential prey before it begins swallowing—unlike real vipers, which typically refuse to strike additional prey until the first has been eaten. Stiletto snakes swallow mainly by ventral flexions of the head and anterior trunk, rather than by unilateral jaw movements, as in most other snakes; the lower jaws of *Atractaspis* evidently serve only to stabilize prey in the mouth and counter downward movements of the head during ingestion.

When restrained above ground, stiletto snakes (*Atractaspis*) rapidly coil and uncoil their flattened bodies, elevate their coiled tails, discharge their cloacal contents, and periodically freeze in a contorted posture with their heads tucked down. If touched anywhere on the body, they jerk and strike with one fang from the unusual head-down posture. Because of their peculiar fang erection mechanism, stiletto snakes can bite even if held tightly at the back of the head. In all respects except the unique headstand and biting method, their defensive repertoire resembles that of unrelated burrowing colubrids and elapids in other areas. Raptors sometimes drop stiletto snakes in flight, perhaps because the prey effectively bite the birds.

ENVENOMIZATION OF HUMANS BY ATRACTASPIDIDS

Most atractaspidids are inoffensive or too small to envenom a person effectively. Bites from the Natal Blacksnake (*Macrelaps microlepidotus*) are rare, but in one case temporary loss of consciousness occurred within thirty minutes. A few bites from Spotted Harlequin Snakes (*Homoroselaps lacteus*) have produced pain and mild neurotoxic symptoms. Stiletto snakes (*Atractaspis*) bites are common, usually producing severe pain, some swelling and local damage, and mild neurotoxic effects. The venom contains unique cardiotoxic peptides called "sarafotoxins," after the Hebrew common name for *A. engaddensis* (Saraf 'En Gedi). Very few deaths have resulted from accidents with stiletto snakes, but large individuals of the Small-scaled Stiletto Snake (*A. microlepidota*) and other long-glanded species are certainly dangerous.

Cape Cobra (*Naja nivea*), southern Africa.

COBRAS, CORALSNAKES, AND THEIR RELATIVES

12

What traits unify dozens of species in a worldwide assemblage of serpents? Among its diverse members, are there common features that captivate us, recurrent themes in our perceptions despite differences among the organisms in appearance and behavior? Zoologists recognize a group called Elapidae for cobras, New World coralsnakes, and their relatives—all with relatively immovable, venom-conducting fangs at the front of the mouth. Beyond such common attributes, elapids encompass brightly colored creatures no larger than some earthworms, much longer serpents that accurately squirt venom at the eyes of their enemies, and 5-m giants that construct a nest out of vegetation, within which their eggs are guarded. As a nine-year-old crawling under huge tree roots on a central Texas stream bank, I was enthralled to find the frog I was chasing replaced by a prowling Harlequin Coralsnake (*Micrurus fulvius*). By then I had my first reptile book and knew that this one's colors fit "red and yellow, kill a fellow," rather than "red and black, venom lack," but I probably hadn't yet read anything about cobras. Twenty years after that first one, watching another Harlequin Coralsnake follow the trail of a Rough Earthsnake (*Virginia striatula*) over moss in its terrarium, I promptly changed master's thesis topics from lizard reproduction to snake feeding behavior. Since then I've encountered other coralsnakes in Arizona, Mexico, Costa Rica, and Panama, but I still rely entirely on other naturalists for their observations and impressions of Old World elapids.

With prominent eyes and heavy reliance on visually guided responses, cobras seem intelligent; their keepers say caged individuals readily anticipate cleaning and feeding routines. Often austere and confrontational, all speed and eyes, some cobras seem downright belligerent. When I consulted a zoo curator friend about how to catch large African elapids, he recalled a captive Gold's Tree Cobra (*Pseudohaje goldii*): "We were not in control of that animal." Big elapids are among the scariest of vertebrates, and surely no other venomous snakes can match Black Mambas (*Dendroaspis polylepis*), King Cobras (*Ophiophagus hannah*), and Taipans (*Oxyuranus scutellatus*) for shock value. Back off just a few meters from a Terciopelo (*Bothrops asper*) or a Western Diamond-backed Rattlesnake (*Crotalus atrox*) and you are out of striking range, able to respond safely to the next move of those vipers. Black

Mambas are so superior in close quarters, by dint of size and speed, that they have us on the run. Those famous serpents are called terrifying, menacing, defiant, brave, and dreadful; their heads are usually described as coffin-shaped. When I asked a South African biologist how we would capture Black Mambas if I came to visit, he said, "I don't catch them," as matter-of-factly as, "Yeah, I listen to Dire Straits." Still, Donald G. Broadley, Africa's most widely experienced herpetologist, said about these supposedly aggressive creatures: "If the 'intimidated' person has the presence of mind to remain quite still, the mamba will soon drop to the ground and slither away, sometimes crawling over the person's feet or through his legs."

By contrast with the larger elapids, New World coralsnakes have beady little eyes and seemingly poor vision; they are preoccupied foragers, rooting under surface cover for small prey snakes and protected by contrastingly colored warning patterns. Step within their tactile and chemical sensory realms, however, and things get very complicated, difficult to comprehend effectively. When one grasps a freshly disturbed coralsnake with tongs, the tail loops up like a head and flops back and forth; often the real head is concealed momentarily under a coil, only to emerge snapping at everything within reach. The entire body is flattened, its colors made brighter by exposed skin between the scales. A riled-up coralsnake's movements are peculiar, jerky, and unpredictable; sometimes its vent gapes and gurgles, and male Southern Coralsnakes (*Micrurus frontalis*) pump their sex organs in and out of sight during defensive displays. Within seconds of a coralsnake's discovery, all hell breaks loose, and then the snake suddenly disappears into the leaf litter.

Scanning a thesaurus with those creatures in mind, I match them with confusing, cylindrical, harlequin, kaleidoscopic, lacquerlike, nervous, ornamented, protean, shiny, supple, surreal, treacherous, and unpredictable. Nevertheless, maybe coralsnakes, usually regarded as secretive and earthbound, are not as different from cobras and mambas as they seem at first glance. The late Archie Carr, who probably had seen as many North and Central American coralsnakes in the field as any herpetologist, said these brightly banded serpents "have a venturesome streak—a talent for the unexpected." Attracted to a commotion more than 6 m up in a Cabbage Palm (*Sabal palmetto*), he found a Harlequin Coralsnake eating a good-sized Yellow Ratsnake (*Elaphe obsoleta quadrivittata*)!

Whether large or small, most elapids are wriggly, uncontrolled, wild, and excitable. These snakes usually seem nervous, rather like ophidian terrorists, as if they were ready and willing to engage us one step further than would a viper. Then there's that word "neurotoxic," chilling at some bone-deep level: they thrash or charge and bite, the toxins flow, and then we have muscle spasms and stop breathing. All these things a person remembers, even if only subconsciously, when a long tan snake rises with its hood spread or dark mouth gaping widely. Despite their deadly bites and attentive responses, however, cobras display in vertical postures from which they can't strike quickly or upward. It's probably no coincidence that snake charmers don't wave flutes in front of pythons or vipers, or that this profession is restricted to the Old World.

Perhaps elapids seem intelligent because we have obvious, visible interactions with them, although of course there's really no evidence that vipers or even blindsnakes aren't equally perceptive and equally cognizant. For now, I can only daydream about cobras and mambas, inspired by the writings of others, and wish that someone would study the natural history of those snakes in more detail. Some large African and Australian elapids look a lot like harmless serpents elsewhere, and I've thought about how much harder it would be to catch fast-moving Coachwhips (*Masticophis flagellum*) if they were deadly.

Highly venomous snakes with enlarged, canal-like front fangs and usually neurotoxic venoms are placed together in Elapidae. Unlike other front-fanged serpents (vipers and some atractaspidids), many elapids also have several small, solid teeth behind each maxillary fang. Most species have heads not greatly wider than their necks, long cylindrical bodies, and smooth scales. One visible characteristic of all elapids, lack of a loreal scale, otherwise characterizes only atractaspidids (e.g., quill-snouted snakes [*Xenocalamus*]) and a few burrowing colubrids (e.g., black-headed snakes [*Tantilla*]), which implies that the common ancestor of cobras, seasnakes, and their relatives also might have been fossorial.

Today elapids are cosmopolitan and diverse in morphology, ecology, and behavior. The 272 species in 62 genera range from modest representation in Afro-Asian snake assemblages (e.g., several kinds of cobras; mambas [*Dendroaspis*]), through the species-rich but generally similar-looking New World coralsnakes, to the exceedingly diversified Australian elapid fauna. Most elapids lay eggs, but the African Rinkhals (*Hemachatus haemachatus*) and several Australian species of higher altitudes and latitudes are viviparous. Many elapids feed primarily on frogs, lizards, and snakes, but more unusual diets occur in several lineages. Venomous seakraits (*Laticauda*) and seasnakes are sometimes placed in a separate family, Hydrophiidae—totaling about 60 species—but they actually were derived one or more times from within Australasian terrestrial elapids (Chapter 13). Here I survey terrestrial and freshwater elapids in terms of major taxonomic and geographic groupings.

COBRAS

Cobras of the genus *Naja* are found over much of southern Asia and Africa, and closely related fossil cobras are known from widespread localities in Europe. The 16 living species are moderate to large snakes; Egyptian Cobras (*N. haje*), Forest Cobras (*N. melanoleuca*), and Black-necked Spitting Cobras (*N. nigricollis*) sometimes approach 3 m in length. All cobras have an expandable hood, and usually the ventral surface of the neck is brightly marked; the body is often mottled or unicolored, but some species instead have conspicuous dorsal bands (e.g., some Egyptian Cobras). Several species in Africa and Asia have fangs modified for "spitting" venom at the eyes of an adversary. Cobras frequently hide in termite mounds and rodent tunnels; they prey on vertebrates ranging from toads (*Bufo*) to young chickens but sometimes eat other snakes (e.g., one 2.4-m Egyptian Cobra contained a 74-cm Puff Adder [*Bitis arietans*]), and Forest Cobras occasionally take slow-moving fish. Females of at least some species of cobras guard their clutches throughout incubation.

Other African elapids include 2 species of water cobras (*Boulengerina*), which reach more than 2.4 m; these dull brown, sometimes banded snakes inhabit rocky shorelines of large lakes and feed mainly on fish. Two species of tree cobras (*Pseudohaje*) are slender, satiny black snakes with large eyes; sometimes more than 2.4 m long, they are almost hoodless and perhaps feed mainly on treefrogs. The Rinkhals (*Hemachatus haemachatus*) averages 1 m in length

Coral Cobra (*Aspidelaps lubricus*),
southern Africa.

and, unlike most other elapids, has keeled scales; it gives birth to as many as sixty young. Restricted to southern Africa, the Rinkhals is an effective spitter; this species readily shams death when threatened, and like some other snakes with similar defensive behavior (e.g., North American hog-nosed snakes [*Heterodon*]), it eats mainly toads.

Stout-bodied cobras of the genus *Aspidelaps* are 50–75 cm long; restricted to southern Africa, they are variably specialized for fossorial locomotion. The Coral Cobra (*A. lubricus*) has only a slightly enlarged snout, whereas Shield-nosed Cobras (*A. scutatus*) have perhaps the most highly modified rostrum of any serpent. Shield-nosed Cobras use their snouts like rounded bulldozer blades, pushing into sandy alluvial soil and then moving it to the side with a bend in the neck; their tubercular tail scales recall those of the File-tailed Groundsnake (*Sonora aemula*) and are of equally unknown function. Although habitat specialists, Coral Cobras and Shield-nosed Cobras are dietary generalists; they eat frogs, lizards and their eggs, rodents, and even large termites, when those insects swarm after rain showers. Both species of *Aspidelaps* spread a narrow hood and hiss loudly when disturbed. Some female Shield-nosed Cobras remain coiled around their eggs throughout incubation and become more aggressive during that period.

Shield-nosed Cobra (*Aspidelaps scutatus*), southern Africa.

At about 5.5 m in maximum length, the King Cobra (*Ophiophagus hannah*) of Asia is the longest venomous snake; although more slender than most vipers, one well-fed 4.7-m King Cobra weighed about 12 kg! Juveniles are shiny black with narrow yellow bands, whereas geographic variation encompasses uniformly olive-colored and dark-banded adults. This species feeds mainly on other snakes, including even Reticulated Pythons (*Python reticulatus*), and occasionally takes monitor lizards (*Varanus*). King Cobras spread a narrow hood and growl loudly when threatened, but their legendary aggressiveness is exaggerated. Nonvenomous Banded Ratsnakes (*Ptyas mucosus*) resemble this species in coloration, size, and defensive behavior. Female King Cobras construct a nest by collecting dead vegetation and soil with a body loop, piling these materials into a compact mass about 70 cm in diameter; bamboo thickets are preferred nesting sites. One female will lay twenty to forty-three eggs within the nest and will remain coiled above or near them throughout incubation; however, the role of this behavior (whether for thermoregulation or to guard against nest predators, e.g.) is unknown.

OTHER AFRICAN AND ASIAN ELAPIDS

The infamous mambas of the genus *Dendroaspis* are slender, fast-moving diurnal creatures, widely distributed in Africa south of the Sahara Desert; their English name comes from a Zulu word meaning "big snake." Three species of green mambas are arboreal and mild man-

Green Mamba (*Dendroaspis angusticeps*), Africa.

nered; they grow to more than 2 m and are largely restricted to rain forest. The Black Mamba (*D. polylepis*; its English name refers to the dark lining of its mouth) is primarily terrestrial, favoring rock outcrops in open habitats and gallery forests; large adults exceed 3 m in length and are more prone to bite than other species of *Dendroaspis*. All mambas feed mainly on endothermic vertebrates, especially rodents (e.g., Naked-soled Gerbils [*Tatera leucogaster*]) and birds; Black Mambas even eat hares (*Lepus*), rock hyraxes (*Procavia*), and bush babies (*Galago*), and one was seen catching flying termites as they emerged from holes. Mambas have more movable maxillae than other elapids and sometimes bite predators and prey repeatedly; they may gape and spread a small hood when threatened, which perhaps reflects a common ancestry with cobras. Some Black Mambas repeatedly use the same lair, such as a hollow tree or a termitarium.

Other Afro-Asian elapids do not spread hoods, and their relationships are uncertain. With brightly banded color patterns, the 7 species of African gartersnakes (*Elapsoidea*) somewhat resemble New World coralsnakes, at least as juveniles; some species are almost uniformly dark as adults. African gartersnakes feed mainly on other snakes and lizards, including their eggs, but occasionally eat mammals; one species may specialize on caecilians. These small elapids, <1 m long, are less prone to bite than their New World counterparts. The shiny Middle Eastern Desert Blacksnake (*Walterinnesia aegyptia*) forages mainly after midnight, at air temperatures as low as 8–12°C. Hunting for skinks, agamid lizards, and Green Toads (*Bufo viridis*), this meter-long snake explores holes and refuges under rocks before returning to an abandoned rodent burrow for shelter during the day. Young Desert Blacksnakes have narrow pinkish-brown crossbands.

Among 3 genera of brightly marked, tropical Asian elapids, 13 species of kraits (*Bungarus*) are nocturnal snakes that sometimes attain lengths of more than 1.5 m. Most species have an enlarged mid-dorsal scale row and peculiar, protrusible vertebral spines that might play a role

Angolan Gartersnake (*Elapsoidea semi-annulata*); juveniles are brightly banded, like many venomous elapids elsewhere.

in defensive body thrashing. The Many-banded Krait (*B. multicinctus*) eats mainly fish; other species of *Bungarus* are primarily snake-eaters, although occasionally also taking frogs, Musk Shrews (*Suncus murinus*), and even Yellow Monitor Lizards (*Varanus flavescens*). An emaciated Banded Krait (*B. fasciatus*), found underground with eggshells and newborn young, evidently accompanied her clutch during incubation. Most Asian coralsnakes (*Calliophis*) and long-glanded coralsnakes (*Maticora*) are smaller and more slender than kraits. Asian coralsnakes are red or brown with dorsal spots, rings, or stripes, whereas both species of *Maticora* are striped. These brightly colored burrowers eat only snakes, and species of *Calliophis* perhaps specialize on blindsnakes and reedsnakes (*Calamaria*). Like their New World namesakes, Asian coralsnakes and long-glanded coralsnakes display elevated, coiled tails when disturbed.

AUSTRALASIAN ELAPIDS

The largest and most diverse radiation of elapids occupies Australia, New Guinea, and various offshore islands; encompassing more than 75 species, these snakes include more than a fourth of the world's elapid fauna. Australasian elapids are remarkably diverse, exhibiting morphologies and ecologies that run the gamut from those of harmless North American whipsnakes (*Masticophis*) to those of New World coralsnakes and even vipers. This impressive adaptive diversification is correlated with a near absence of other advanced snakes, except for a few colubrines and homalopsines that probably arrived recently by dispersal over water (a consequence of the long isolation of Australia).

Many Australian elapids look and to some extent act like harmless terrestrial colubrids elsewhere; widespread examples include the Red-bellied Blacksnake (*Pseudechis porphyriacus*) and the Western Brownsnake (*Pseudonaja nuchalis*). Only the Broad-headed Snake (*Hoplocephalus bungaroides*) and its relatives are habitual climbers, inhabiting vertical rock crevices.

Western Brownsnake (*Pseudonaja nuchalis*), Australia, in threat posture.

Many Australasian elapid species play the ecological role occupied by nonvenomous whipsnakes on other continents, presumably because frogs and lizards are abundant in numbers and in species, whereas small mammals are neither common nor seasonally predictable in the "land down under." Several species of Australasian elapids constrict as well as envenom their prey, behavior that might be especially useful in subduing large, slippery lizards (e.g., Eastern Brownsnakes [*Pseudonaja textilis*] sometimes eat Shingle-backed Skinks [*Trachydosaurus rugosus*]). Conversely, as is generally true with other elapids, among Australasian species there is a puzzling lack of predators on invertebrates (e.g., centipedes, earthworms) and fish.

Two species of taipans (*Oxyuranus*) resemble African mambas (*Dendroaspis*) in their large size and slender build, extremely toxic venom, and lifelong diet of rodents, as well as behaviorally, in biting and releasing prey rather than holding on after it is seized and in their willingness to bite an adversary. In some areas, these snakes have recently grown in number relative to other Australasian elapids as a result of human activities: Giant Toads (*Bufo marinus*) with toxic skin secretions, mistakenly introduced from the neotropics, have evidently killed many native elapids that eat ectotherms; at the same time, agricultural practices have led to increased rodent populations and thus abundant food for taipans.

Several genera of small burrowing Australian elapids, often resembling certain other Old World elapids (e.g., Asian coralsnakes [*Calliophis*] and New World coralsnakes), are noteworthy for their unusual diets and defensive displays. Narrow-banded Snakes (*Simoselaps fasciolatus*) are equipped with fairly typical elapid dentition and eat mainly skinks, whereas Northern Shovel-nosed Snakes (*S. roperi*) have stout, sawlike teeth on their pterygoid bones and specialize on a diet of reptile eggs. The Bandy-bandy (*Vermicella annulata*) eats only relatively huge blindsnakes (*Ramphotyphlops*); like that species, Northern Red-naped Snakes (*Furina ornatus*) and Curl Snakes (*Suta suta*) have unusual defensive repertoires. The meter-

Bandy-bandy (*Vermicella annulata*), Australia, in defensive display.

Curl Snake (*Suta suta*), Australia.

long Solomon Islands Coralsnake (*Loveridgelaps elapoides*) is banded black and yellow with a bright white head.

Three species of Australasian death adders (*Acanthophis*) are stocky, viper-shaped snakes that sometimes sidewind on sand or mud. Compared to most other elapids, they have long, curved fangs on relatively mobile maxillary bones. Death adders have exceptionally cryptic coloration, presumably associated with their ambush hunting style, and lure prey by wriggling

their contrastingly colored tails. They eat diurnal skinks, as well as birds and mammals. Like most vipers, death adders have a biennial reproductive cycle, evidently linked to similarly low rates of energy acquisition.

NEW WORLD VENOMOUS CORALSNAKES

Approximately 60 species of North, Central, and South American elapids are placed in 2 genera and collectively called New World coralsnakes. They range in length from <40 cm, in the Sonoran Coralsnake (*Micruroides euryxanthus*), to >1.5 m, in the Amazonian Coralsnake (*Micrurus spixii*) and Aquatic Coralsnake (*M. surinamensis*). They vary in shape from fairly stout Amazonian Coralsnakes and Broad-ringed Coralsnakes (*M. latifasciatus*) to attenuate Slender Coralsnakes (*M. filiformis*) and Andean Black-backed Coralsnakes (*M. narduccii*). New World elapids are generally similar to each other in overall appearance, with heads scarcely wider than their bodies, fairly short tails, small eyes, and shiny scales in fifteen rows. They might be closely related to certain morphologically and ecologically similar Old World groups (e.g., Asian coralsnakes [*Calliophis*]).

New World coralsnakes are best known for their stunningly beautiful colors. Bicolored species are usually ringed with black and either white, yellow, or some shade of red. Many-banded Coralsnakes (*Micrurus multifasciatus*) span those lighter hues within a population, although their head and tail bands are typically red. Some populations of Tuxtlan Coralsnakes (*M. limbatus*) have red and black rings, while others are bright orange with irregular black dorsal blotches. Most coralsnakes are tricolored, at least ventrally, with yellow-bordered black bands alternating with red bands; in some species the yellow bands are bordered by additional black bands. Depending on the number of black bands, those patterns are called monads

Sonoran Coralsnake (*Micruroides euryxanthus*), Arizona, envenoming a Plains Black-headed Snake (*Tantilla nigriceps*).

(e.g., Harlequin Coralsnakes [*M. fulvius*]), triads (e.g., Aquatic Coralsnakes [*M. surinamensis*]), and so forth. Dumeril's Coralsnake (*M. dumerili*) varies geographically, with some populations characterized by monads and others by triads. In some species (e.g., Oaxacan Coralsnakes [*M. ephippifer*], Carib Coralsnakes [*M. psyches*]), the red bands are partly or totally obscured by black dorsally, which makes the serpent appear bicolored. Andean Black-backed Coralsnakes (*M. narduccii*) are unicolored dorsally, with numerous oval red or yellow blotches on the belly.

New World elapids generally inhabit moist tropical locales, with six or more species found at some places in the Upper Amazon Basin. Sechura Desert Coralsnakes (*Micrurus tschudii*) and Sonoran Coralsnakes (*Micruroides euryxanthus*) live in arid regions, however, confining their surface activities to temporarily moist microhabitats. Most New World elapids are terrestrial, often diurnal predators on snakes and other elongate vertebrates; some species are nocturnal hunters, whereas others vary their daily activity cycles with seasonal changes in temperature. Aquatic Coralsnakes (*Micrurus surinamensis*) feed on eels and occasionally other fish and have very different venom from that of other elapids; Allen's Coralsnakes (*M. alleni*) are also found near water and sometimes eat eels. Hemprich's Coralsnake (*M. hemprichi*) is unique among serpents in eating onycophorans, strange little invertebrates that look like a cross between an earthworm and a caterpillar, and the Andean Black-backed Coralsnake (*M. narduccii*) feeds on the broken-off tails of slender, almost limbless lizards. The defensive behavior of most New World coralsnakes (see the Special Topic in this chapter) includes flattening and writhing their bodies, hiding their heads under a loop of the body, elevating their coiled tails, and biting repeatedly. Cloacal discharge is accompanied by hemipenial protrusion in Southern Coralsnakes (*M. frontalis*) and by sharp, popping sounds in Sonoran Coralsnakes.

In Costa Rica, the Neck-banded Snake
(*Scaphiodontophis annulatus;* right)
closely resembles the venomous
Allen's Coralsnake (*Micrurus alleni;*
below right).

In 1862, three years after Charles Darwin and Alfred Russel Wallace published their theory of evolution by natural selection, Henry W. Bates attributed the color patterns of certain palatable Amazonian butterflies (known as mimics) to the fact that predators mistook them for noxious species (the models) with similar color patterns. In 1879, Fritz Müller pointed out that sometimes two noxious species also resemble each other, acting as both mimics and models, such that fewer individuals of either one would have to encounter predators. During those same decades, others proposed that various supposedly harmless snakes were Batesian mimics of venomous species, and over the next century herpetologists hotly debated that possibility. Controversy usually centered on New World coralsnakes and various brightly colored colubrids, a resemblance emphasized in 1867 by Wallace (as young men, he and Bates had explored the Amazon Basin together).

Whether the color patterns of certain venomous snakes serve as warning signals to predators and whether other serpents mislead predators by mimicking the dangerous species are separate issues. Initially the evidence for both questions was circumstantial—the remarkable similarities among New World coralsnakes and seemingly harmless serpents with colorful

In Central America, the Black Halloween Snake (*Pliocercus euryzonus;* left) is strikingly similar to the Many-banded Coralsnake (*Micrurus multifasciatus;* below left).

banded patterns. Among the objections were that nocturnal coralsnakes could not be seen by predators with color vision; that because they were deadly, a predator could not learn the meaning of their bright markings; and that mimetic color patterns (e.g., of California Mountain Kingsnakes [*Lampropeltis zonata*]) also occurred where there were no venomous models. Plausible alternative explanations were that coralsnake patterns could be cryptic amid the mosaic of tropical forest colors and that red-and-yellow snakes might rebuff predators simply by virtue of their novelty. Frequencies of supposed models and mimics in large snake collections from Panama and Brazil were used as evidence for and against the mimicry hypothesis, assuming that models would have to far outnumber mimics if predators were to avoid the latter. Mildly venomous rear-fanged snakes (e.g., coralsnake mimics [*Erythrolamprus*]) were proposed as models, since predators could survive their bites and later avoid both deadly and harmless coralsnake mimics.

In fact, many New World coralsnakes and their mimics are diurnal. Dozens of nonvenomous, rear-fanged, and deadly snake species have brightly ringed color patterns, which implies a range of possible model-mimic abundance ratios—because mimics can be fairly com-

Mussurana (*Clelia clelia*), Costa Rica; adults are a uniform dark gray, but juveniles like this one vaguely resemble venomous coralsnakes.

mon if the punishment for making a mistake with a venomous model is severe. The presence of harmless, brightly marked species in areas with no venomous models is not evidence against mimicry, since coralsnake patterns might have arisen in a mimetic ancestor and been retained for other ecological roles. Studies with captive predators show that aversion could stem from social experiences or selection for avoidance of particular color patterns rather than from individual learning. Rubber models that looked and moved like coralsnakes repelled Coatis (*Nasua narica*) and Collared Peccaries (*Tayassu tajacu*), both group-living mammals. Young Great Kiskadees (*Pitangus sulphuratus*) and Turquoise-browed Motmots (*Eumomota superciliosa*) attacked both green-and-blue-ringed and red-and-yellow-striped wooden models, but those tropical birds

gave alarm cries and flew away when offered red-and-yellow-ringed sticks.

Although we expect convergent color patterns in snakes with similar natural histories, species with the same ecological roles typically do not live together. Usually, given a pair of serpents with strikingly similar color patterns, either the two are found in different places and are not closely related (e.g., Emerald Tree Boas [*Corallus caninus*] in South America and Green Tree Pythons [*Morelia viridis*] in New Guinea) or at least one of them is venomous, just as predicted by the mimicry hypothesis. Moreover, parallel geographic variation in color patterns of venomous and nonvenomous snakes strongly implies diverse Batesian and Müllerian mimicry associations. From northern Mexico to Honduras, Red Halloween Snakes (*Pliocercus elapoides*) look

astonishingly similar to local elapids. For example, in southern Veracruz those rear-fanged serpents are spotted, like Tuxtlan Coralsnakes (*Micrurus limbatus*); elsewhere in southeastern Mexico, they resemble two venomous models, combining the long red bands of Variable Coralsnakes (*M. diastema*) with the extra black bands of Elegant Coralsnakes (*M. elegans*). In Central America, closely related Black Halloween Snakes (*P. euryzonus*) mimic the similarly bicolored Many-banded Coralsnakes (*M. multifasciatus*). Nonvenomous Milksnakes (*Lampropeltis triangulum*), Neck-banded Snakes (*Scaphiodontophis annulatus*), and several other colubrids also impressively parallel color patterns among dangerous coralsnakes throughout much of the former species' ranges.

Given limitless time, money, and patience, we eventually might observe a predator encounter a coralsnake, avoid it upon recognition of a specific color pattern, and then later not attack a mimic that it actually could have eaten without penalty. An enterprising student in a tropical biology course did the next best thing: having laid out arrays of clay snakes in a Costa Rican rain forest, he confirmed that resemblance to venomous coralsnakes deters some predators on small serpents. Olive-colored models were most often disturbed by predators, whereas those ringed in red, yellow, and black (resembling the Central American Coralsnake [*Micrurus nigrocinctus*]) were handled rarely; red models with black-and-yellow collars (suggesting juvenile Mussuranas [*Clelia clelia*]) were manipulated an intermediate number of times. Marks left on the models matched the size and shape of motmot beaks, which proved that those birds were indeed relevant predators. Models laid in triplets on white paper incurred similar rates of disturbance, so differences in detectability of the color patterns cannot account for the results.

Decisive captive and field experiments, as well as concordant geographic variation among models and mimics, strongly support both parts of the coralsnake mimicry hypothesis. Color patterns of New World coralsnakes do act as warning signals, and other serpents survive predator encounters because they resemble the more dangerous species. Along with less well studied cases (e.g., African egg-eaters [*Dasypeltis*] that resemble deadly vipers), these findings lend credence to the idea that venoms have had a pervasive influence on the evolution of snakes and other creatures with whom they coexist.

Yellow-bellied Seasnake (*Pelamis platurus*), Costa Rica, sculling with its oarlike tail.

SEAKRAITS AND SEASNAKES

Willoughby Lowe described this remarkable experience in his 1932 book, *The Trail That Is Always New:*

> Leaving Colombo we departed for Penang, and the voyage from now on became more interesting . . . To starboard lay the beautiful green island of Sumatra and to port the Malay Peninsula. The water now became very calm and oily in appearance. After luncheon on 4th May, I came on deck and was talking to some passengers when, looking landward, I saw a long line running parallel with our course. It must have been four or five miles off. We smoked and chatted, had a siesta and went down to tea. On returning to the deck we still saw the curious line along which we had been steaming for four hours, but now it lay across our course . . . As we drew nearer we were amazed to find that it was composed of a solid mass of sea-snakes, twisted thickly together. They were orange-red and black, a very poisonous and rare variety [Stokes's Seasnake (*Astrotia stokesii*)] . . . Some were paler in colour and as thick as one's wrist, but the most conspicuous were as thick as a man's leg above the knee. Along this line there must have been millions; when I say millions I consider it no exaggeration, for the line was quite ten feet wide and we followed its course for some sixty miles . . . It certainly was a wonderful sight. As the ship cut the line in two, we still watched the extending file of foam and snakes until it was eventually lost to sight.

As young children aboard a troopship to the Philippine Islands, my brother and I marveled at sharks, flying fish, and whales. We imagined ourselves sailors for three glorious weeks but saw no sea serpents in those warm tropical waters. Later travels mostly kept me away from the Pacific, and—more to the point—although I'd read Lowe's account, oceans and seasnakes failed to capture my imagination. I've never learned to scuba dive, snorkeling was a choking mishap, and my second time at sea in a small boat ended when we crashed in the dark on a Galapagos reef. These prejudices vanished abruptly in 1991, when I held a live Yellow-bellied Seasnake (*Pelamis platurus*) at Alejandro Solórzano's serpentarium in Costa Rica. Surprisingly eel-like with oddly rough scales, it had a dull black back and brilliant saffron underside. The nostrils were dorsal, the better to breathe while floating, and I recalled that this species lives on "slicks"—lines of calm surface water, a few meters wide, that form where two currents meet. Unlike the eyes of some other aquatic creatures, those of the Yellow-bellied Seasnake were decidedly lateral and thus well-suited for aiming a sideways strike at nearby fish.

Enthralled with the idea of a planktonic reptile, I soon journeyed with Alejandro to a place famed for its seasnakes. Bahía de Culebra ("snake bay") is bounded on its upper, western corner by Punta Mala ("bad point") and opens to the south above Punta Gorda ("fat point"). Our hangout was a murky little fishing village, still off the main tourist route and not so much run-down as casual about appearances. Reminiscent of other maritime communities, rank with the sights and smells of the sea, Playa Hermosa ("beautiful beach") was languorous and ever-so-vaguely wild when we first visited. One night, intrigued by familiar sounds from behind a little beachside bar, I found dozens of treefrogs singing among floating trash and cow pies in a freshwater lagoon. Two partially inundated outhouses guarded the shore, doors creaking in the breeze, and as I savored the frog chorus something oddly unsettling darted by my feet. A red-and-yellow crab scuttled backward down its burrow, carrying in one claw the faded head of a small doll.

Each morning we walked along the shining beach, passing open truck beds already full of huge fish, and arranged for a local boatman to take us on a seasnake hunt. We cruised past rocky islets crowned by columnar cacti and patrolled the slicks; we scrutinized lines of debris and saw the little fish on which our quarry feeds. Failing to find any, I tried to form a mental picture of what Yellow-bellied Seasnakes would look like in the glassy green water and mused about their strange lifestyle. Without protection, a light-skinned, temperate-zone creature like myself wouldn't last long under that sun, and maybe seasnakes wouldn't either, without black exteriors and a midday retreat to cooler depths. Some desert lizards even have their body cavities lined with black pigment, and I wondered if the interiors of Yellow-bellied Seasnakes likewise are safeguarded from intense ultraviolet radiation. Later that year, another visit to Playa Hermosa was equally pleasant, thought-provoking, and unsuccessful.

The Fogdens and I returned to Bahía de Culebra in the summer of 1993, once again looking for seasnakes. A Seattle tourist showed us the beached one he'd bashed with a rock, fearing children would be bitten, and I photographed its discolored carcass, lest that be our only encounter. We searched the bay for two hours the first morning, in mildly choppy water that was at first olive-green and later almost black, as storm clouds gathered. The few slicks we crossed were short and lacked debris. Our boatman remarked that *culebras del mar* were much more common there in December, when the waters were colder, and I wondered if I'd ever find any seasnakes. Over dinner that night we laid plans for future work and played with a tame Crab-eating Raccoon (*Procyon cancrivorus*) that wandered among the restaurant tables.

Early the next morning our mood was optimistic, and Patricia suddenly hugged Michael as they waded to the boat. Blue sky and sun burst through clouds as we reached the mouth of the bay. We then worked our way north for a few kilometers and methodically checked several slicks. The ocean rolled gently, like an enormous inky water bed, and the three of us stood easily while scanning for snakes. During the first hour and a half, two sea turtles raised their heads off our bow, then dove. We chuckled as a shrimp trawler passed, its rigging sagging under scores of seabirds, all facing forward. With about forty minutes of boat time left,

I thought about borrowing someone's slides so I could at least daydream accurately about how seasnakes look in slicks.

As Punta Mala loomed ahead, signaling our impending return to shore, Michael said, "There's one!" Straining to glimpse a bright stripe in the water, I worried I'd miss it and never get another chance. We circled past the first Yellow-bellied Seasnake and took photographs, and soon I caught sight of another one. Their colorful undersides and black-spotted tails were obvious from several meters away, and within fifteen minutes we found six more. They ranged from 30-cm juveniles to adults almost 1 m long, all swimming separately and slowly on the same slick. The seasnakes seemed to sense the approaching boat, undulating faster or lifting their heads above the water as if to view us, and sometimes swimming backward. Eventually they all plunged, wriggling like brilliant yellow streamers, and disappeared from our view.

In 1519, six years after Vasco Núñez de Balboa first crossed Panama and gazed on the Pacific Ocean, a Spanish expedition sailed along the west coast of what is now Costa Rica. Reporting on the voyage, Gaspar de Espinosa accurately described "innumerable" black-and-yellow serpents—"the largest of them fatter than a big toe"—in waters he named Bahía de Culebra. Almost five centuries later, in that same bay, nothing I'd ever seen presaged the exhilarating sight of those creatures. Now literature about them has a fresh appeal and the western Pacific beckons, rich with dozens of species of deep-diving, bottom-crawling seasnakes. Perhaps someday I'll even learn to dive.

A few Australasian filesnakes (*Acrochordus*), homalopsine colubrids, and New World natricine colubrids have solved or sidestepped the difficulties associated with living in saltwater. None of those taxa has speciated widely in marine environments, however, and only a few turtles, crocodiles, and lizards live there. Conversely, as many as three independent invasions of the ocean have occurred within Elapidae, all in the Indo-Australian region, and one of those has been modestly successful in terms of the abundance of individuals, the richness of species, and overall biological diversity. As a result, of the approximately 75 species of turtles, crocodiles, lizards, and snakes that spend at least parts of their lives in the sea—less than 1.5 percent of all "reptiles"—more than half are cobra relatives.

Seakraits (*Laticauda*) and several genera of seasnakes are well known in regions where they dwell, for both their venomous bites and their roles in human customs and commerce. Tremendous food and leather industries involving these creatures are major sources of livelihood in the Philippine Islands and some other Asian countries, as well as potential threats to the long-term survival of marine snake populations. Seakraits and seasnakes are especially interesting to biologists because of their adaptations for marine life and their diverse roles in ocean ecosystems.

SYSTEMATICS AND DIVERSIFICATION OF MARINE ELAPIDS

The approximately 55 species of marine elapids are currently placed in 14 to 16 genera, all restricted to warm tropical and subtropical waters. There is no strong evidence either for monophyly of all these snakes or of any groups of marine elapid genera other than seakraits

(*Laticauda*) or for monophyletic groupings among the terrestrial elapids that are most closely related to seasnakes. Accordingly, Laticaudinae and Hydrophiinae (or their respective family names) are misleading as separate taxa; instead, I retain seakraits and seasnakes in the Elapidae, with cobras and their relatives, and discuss certain possibly monophyletic, informal subdivisions among marine elapids.

Sometimes called "amphibious seasnakes," the 6 species of seakraits (*Laticauda*) return to land—or at least the edges of it—to bask, mate, and lay eggs. Most seakraits are beautifully banded, cylindrical snakes whose only obvious, external indicator of marine life is a flattened tail. They reach a maximum length of about 2 m. Five species of *Laticauda* occur in coral reef habitats and commonly are found on exposed reefs, under nearby rocks and logs, and occasionally even in trees. One rather divergent seakrait is found only in a brackish lagoon in the Solomon Islands (see p. 242).

The remaining marine elapids, all viviparous and more specialized for ocean life than seakraits, fall into three groups. Three species in the "primitive" genera *Ephalophis, Hydrelaps,* and *Parahydrophis* are restricted to mangroves and mud flats around estuaries; they morphologically resemble terrestrial elapids (e.g., they have broad ventral scales) more than do other seasnakes. Another lineage, sometimes lumped with seakraits (*Laticauda*) and perhaps independently derived from terrestrial elapids, consists of two genera (*Aipysurus* and *Emydocephalus*) and about 9 species of relatively small (no more than 50 cm) "reef seasnakes." The remaining seasnakes, here called "hydrophiines," are more highly modified and include roughly 40 species in about 10 genera. Hydrophiine seasnakes range in total length from as little as 50 cm in several species to 2.75 m in the Yellow Seasnake (*Hydrophis spiralis*). The most massive among them, Stokes's Seasnake (*Astrotia stokesii*), has a midbody girth of up to

Yellow-bellied Seasnake (*Pelamis platurus*), Costa Rica; note the dorsally placed nostrils and lateral eyes.

26 cm at a length of 1.8 m. Beaked Seasnakes (*Enhydrina schistosa*) inhabit estuaries, various species of *Hydrophis* forage on muddy ocean bottoms, and other *Hydrophis* and some *Disteira* occasionally dive to more than 50 m. The Yellow-bellied Seasnake (*Pelamis platurus*) inhabits slicks, narrow strips of calm water where two ocean currents meet, and feeds on small fish that congregate at the accumulated debris in these drift lines.

MORPHOLOGICAL AND PHYSIOLOGICAL ADAPTATIONS

Snakes are better suited at the outset than most terrestrial vertebrates for invading marine environments. Lateral undulatory locomotion works well in water, and the low metabolic rates conferred by ectothermy, coupled with an elongate, serpentine lung, facilitate prolonged submergence. Powerful venoms, legacies of their terrestrial ancestors, are used by some seakraits (*Laticauda*) and seasnakes for feeding and defense. Beyond those attributes, marine elapids exhibit various unique specializations for life in the ocean. Body shape varies from cylindrical in seakraits, primitive seasnakes (e.g., *Hydrelaps darwiniensis*), and reef seasnakes (e.g., *Aipysurus*) to absurdly tapered, with slender foreparts and tiny heads, in some microcephalic *Hydrophis*. All marine elapids have flattened tails, and all seasnakes have at least moderately flattened bodies as well; in some hydrophiine seasnakes the body is markedly compressed laterally. Hydrophiine seasnakes have elongate neural spines on the tail vertebrae that support their caudal "oar," whereas in other marine elapids the tail fin is simply skin and connective tissue, without skeletal supports. Seasnakes have dorsally directed nostrils, whereas seakraits have lateral nostrils, like terrestrial serpents; the nostrils of all marine elapids are valvular. All these snakes have laterally or dorsally placed eyes with round pupils, and the

Olive Seasnake (*Aipysurus laevis*) has photoreceptors on its tail. An extension of tissue behind the rostral scale (in seakraits and reef seasnakes) or the rostral scale itself (in other seasnakes) fits into a notch at the front of the lower jaws, thus sealing the tongue's opening when the mouth is closed underwater.

All seakraits (*Laticauda*) and many seasnakes (e.g., most *Hydrophis*) have light and dark rings or bandlike blotches, although in several species dark pigment obscures the color patterns dorsally. Only the Yellow-bellied Seasnake (*Pelamis platurus*) is striped. The dorsal scales of marine elapids may be keeled or smooth; they are usually imbricate, rarely hexagonal and juxtaposed. Seakraits and primitive seasnakes (e.g., *Hydrelaps darwiniensis*) have enlarged ventral scales, as in terrestrial colubroids. The ventral scales of most other seasnakes are not well differentiated, scarcely if at all larger than the dorsal scales; Stokes's Seasnake (*Astrotia stokesii*) has a midventral pair of scales enlarged to form a longitudinal keel on the belly. Seakraits, primitive seasnakes, and reef seasnakes (e.g., *Aipysurus*) retain the one-to-one correspondence of vertebrae and ventral scales typical of most other snakes; in keeping with their independence from terrestrial locomotion, hydrophiines have especially variable ventral scale counts, which show no close concordance with numbers of vertebrae. Eyelash Seasnakes (*Acalyptophis peronii*) have spinose head scales, and male Turtle-headed Seasnakes (*Emydocephalus annulatus*) have a distinctive rostral spine.

Seawater is roughly 3.5 percent salt, a concentration about three times as high as in vertebrate body fluids; thus, water loss is a potentially serious problem for marine organisms. Marine elapids solve the problems of retaining freshwater, excluding saltwater, and excreting excess salt with skin modifications and a novel head gland. Their skin is more impermeable to salt than that of terrestrial snakes, yet it allows water to pass into their bodies much more readily than it is lost. Seakraits (*Laticauda*) and seasnakes have a salt gland beneath the tongue that concentrates sodium, which they then excrete when protruding their tongues.

Although some seakraits (*Laticauda*) and seasnakes might occasionally dive to more than 150 m, bottom-feeding species are typically found well within 100 m, and most species are active at depths of less than 30 m. Sleeping or resting snakes remain submerged longer than active ones. Submergence time is usually a half hour or less (e.g., in seakraits) but can extend to more than two hours in some hydrophiines. Marine elapids manage these feats with relatively minor anatomical and physiological adjustments compared to their terrestrial ancestors. The single lung is even longer than that of other snakes, and in addition the posterior, "saccular," portion is muscularized, so that stored air can be pumped forward into the vascular part for respiration. Seasnakes have much higher rates of respiration through their skin than seakraits and terrestrial snakes, although even hydrophiines must periodically surface to breathe. Seakraits exhibit both terrestrial respiration, characterized by frequent regular breaths, and aquatic respiration, or long periods of apnea followed by rapid multiple breaths. Seasnakes often perform only a single, audible breath before diving, and they quickly become distressed and unable to breathe on land.

Seakraits (*Laticauda*) and seasnakes shed their skins at two to six week intervals, more frequently than most terrestrial species. Those that regularly travel along the bottom of the ocean loosen their old skins by rubbing their heads against the substrate, as do terrestrial snakes. Floating on the surface, a Yellow-bellied Seasnake (*Pelamis platurus*) sheds its skin by forming complex body knots and crawling among its own coils, and as a result the cast-off skin is often knotted. These snakes also remove barnacles in a similar fashion.

The maxillary dentition of marine elapids varies from almost colubrid-like in the Yellow-bellied Seasnake (*Pelamis platurus*), with numerous teeth behind small anterior fangs, to cobra-like in several other hydrophiines, with a few small teeth following relatively large anterior fangs. Although yields are relatively low in quantity, the venoms of seakraits and seasnakes usually consist of highly concentrated neurotoxins. Some marine elapid venoms contain powerful myotoxins, substances that destroy muscle cells, but otherwise they include little in the way of tissue-destructive components.

FOOD AND FEEDING

Marine elapids eat mainly fish that are sedentary or dwell in holes and crevices, rather than prey that must be actively chased. Most seakraits (*Laticauda*) and some hydrophiine seasnakes usually feed on eels. Beaked Seasnakes (*Enhydrina schistosa*) eat primarily catfish, while Olive Seasnakes (*Aipysurus laevis*) are generalized predators on various fish, fish eggs, cuttlefish, prawns, and crabs. Prey can be relatively large in the case of eel-eaters and catfish specialists, whereas Yellow-bellied Seasnakes (*Pelamis platurus*) feed frequently on a variety of small fish.

Although details vary, most seakraits (*Laticauda*) and seasnakes are probably similar in foraging behavior. Chemoreception and sensitivity to nearby movements, rather than vision, seem to play major roles in locating prey. Seakraits swim slowly over a reef, investigating small holes for eels and partly or completely disappearing for brief periods, then emerge and continue to forage. Some reef seasnakes (e.g., *Aipysurus duboisii*) forage in water only 2–15 cm deep, so shallow that at times they crawl rather than swim. Another eel-eater, the Black-headed Seasnake (*Hydrophis melanocephalus*), locates burrowing prey by swimming slowly over the sand and entering each hole it encounters. In that species and other microcephalic hydrophiines, the small head and narrow neck penetrate crevices or burrow in sand while the larger body and tail rise vertically in the water. Most fish-eating seasnakes forage when their prey is inactive in crevices, either diurnally or nocturnally, and sometimes trap small fish in their lairs with a body coil before seizing them with the mouth. The Eyelash Seasnake (*Acalyptophis peronii*) eats mainly gobies, located by probing its head into their burrows, and sometimes ingests pairs of males and females in their joint retreat.

Only two species deviate from the approach-and-seize hunting behavior typical of most marine elapids. Beaked Seasnakes (*Enhydrina schistosa*) cruise near the ocean floor, striking laterally at catfish within a few centimeters or even when bumping into them. As a floating object, the Yellow-bellied Seasnake (*Pelamis platurus*) is attractive to small fish and seizes them with a sideways strike or by rapidly swimming backward.

Like many other elapids, seakraits (*Laticauda*) and most seasnakes rely on venom to subdue relatively large prey. They either hold the prey in their mouths until it ceases to struggle or release and then reseize it after it has been immobilized. Seakraits can swallow relatively large eels in as little as seven seconds. Beaked Seasnakes (*Enhydrina schistosa*) turn their heads backward while swimming, thereby using the flow of the water to help keep a catfish aligned in the mouth during swallowing. Most marine elapids ingest prey headfirst, but Black-headed Seasnakes (*Hydrophis melanocephalus*) and some other eel-eaters occasionally swallow their elongate prey tail-first. Seakraits bask on emergent reefs after feeding and sometimes, like terrestrial vipers, expose only the part of the body that contains food to sunlight. Yellow-bellied Seasnakes (*Pelamis platurus*) and perhaps other hydrophiines rest on the surface, per-

Turtle-headed Seasnake (*Emydocephalus annulatus*), Australia. This species eats only fish eggs and has largely lost its venom apparatus. (Photograph by Harold Cogger)

haps basking to enhance digestion after feeding, whereas Turtle-headed Seasnakes (*Emydocephalus annulatus*) and some other species coil in and under corals when not active.

The most dramatic modifications for feeding are found in the White-spotted Seasnake (*Aipysurus eydouxi*) and Turtle-headed Seasnake (*Emydocephalus annulatus*), specialists on the buried eggs of gobies, blennies, and other bottom-dwelling fish. These species possess a "geniomucosalis" muscle, found elsewhere only in blindsnakes and perhaps associated with suction feeding. *Aipysurus* and *Emydocephalus* have consolidated lip scales, which make the border of the mouth more rigid. The maxillary, palatine, and dentary teeth are greatly reduced or lacking in those seasnakes (as in African egg-eaters [*Dasypeltis*]), and their venoms are much less toxic than those of other marine elapids. Turtle-headed Seasnakes have pointed snouts (the basis for their common and generic names), accentuated in males and perhaps used for probing the substrate for goby nests (see also p. 241).

MORTALITY, DEFENSE, AND MUTUALISMS

With the possible exception of Yellow-bellied Seasnakes (*Pelamis platurus*), even dangerously venomous reptiles have enemies. White-breasted Sea-Eagles (*Haliaetus leucogaster*) commonly catch seakraits (*Laticauda*) and seasnakes on the surface, leaving the remains of their prey beneath their perches. Several species of sharks and other fish occasionally eat seasnakes, and Tiger Sharks (*Galeocerda cuvieri*) often do so. Like terrestrial species, marine elapids probably respond to threats primarily with locomotor avoidance, biting only when that fails. About 10 to 25 percent of seasnakes in waters off northern Australia have major scars, often to the tail, which suggests that they have successfully escaped some predator. Seakraits virtually never bite humans when handled; other species vary from usually inoffensive to those, such as the

Olive Seasnake (*Aipysurus laevis*),
Australia, with barnacles on its head.
(Photograph by Harold Cogger)

Beaked Seasnake (*Enhydrina schistosa*) and Stokes's Seasnake (*Astrotia stokesii*), that bite readily. Seakraits and perhaps other brightly banded species are avoided by some sharks, and other fish sometimes reject seasnakes they are offered as food. Certain seemingly defenseless eels are stunningly similar to seakraits in coloration, which implies that the snakes' coloration serves as a warning signal to predators and that the eels are mimics. Most seasnakes react to divers with curiosity, approaching from distances of up to 10 m; during the breeding season, however, some species rapidly and persistently pursue swimmers, biting repeatedly if touched.

The Yellow-bellied Seasnake (*Pelamis platurus*) is one of very few vertebrates with no regular enemies, even among predators that often eat other marine elapids. Seabirds occasionally seize and drop Yellow-bellied Seasnakes, and the latter's scars and missing tails—much less frequent than in other seasnakes—suggest that they sometimes also escape from biting predators. Their relative immunity from predation evidently results from a combination of factors. Yellow-bellied Seasnakes have small fangs and infrequently bite when handled, but their venom is exceptionally toxic. The bicolored black-and-yellow pattern, especially the strikingly spotted tail, is easily visible and functions as a continuous warning display. Pacific Ocean fish rejected Yellow-bellied Seasnakes as prey under experimental conditions, whereas naive Atlantic Ocean predators ate and then regurgitated them. Pacific fish favored other meat over the flesh of that species and found the skin even less acceptable, which suggests that postingestive chemical cues, in addition to visual signals, are involved. Naive egrets and herons treated drab-colored terrestrial snakes and eels as prey but fled from Yellow-bellied Seasnakes, so those avian predators too regard the latter as dangerous.

Seakraits (*Laticauda*) are commonly infested with chigger mites and ticks, as are some terrestrial snakes. Like many other strictly marine creatures, seasnakes are subject to fouling by algae, bryozoans, barnacles, and other organisms. Algae are most prevalent on reef sea-

Yellow-bellied Seasnakes (*Pelamis plat-urus*), Costa Rica, stranded on a beach; individuals like this one are doomed unless someone releases them in the ocean.

snakes and other species that forage in shallow waters, where light is available for photosynthesis. One species of barnacle is even completely restricted to seasnakes and causes its hosts little or no harm. Seasnakes attempt to rid themselves of fouling organisms by knotting behavior and frequent skin shedding, and perhaps by permitting fish to graze their algal infestations.

SOCIAL BEHAVIOR, REPRODUCTION, AND POPULATION BIOLOGY

Seakraits (*Laticauda*) mate on beaches and in caves, where they form large seasonal aggregations. Seasnakes probably all pair in the water, but little is known about their mating behavior. During courtship a male Turtle-headed Seasnake (*Emydocephalus annulatus*) swims slightly above and behind the female, rubbing her head and neck with his snout, his back pressed against hers, and attempts to push his tail under hers. Olive Seasnakes (*Aipysurus laevis*) lie on the ocean floor while copulating, whereas Yellow-bellied Seasnakes (*Pelamis platurus*) mate on the surface, the male wrapping coils of his posterior body around the female.

Gravid females of marine elapids are rare in collections, which suggests that they retreat to secluded places to lay eggs or give birth. Seakraits (*Laticauda*) oviposit communally in caves in exposed coral reefs, their eggs sometimes partially submerged in freshwater. Unlike marine turtles, crocodilians, and lizards, seasnakes are viviparous, giving birth to their young in the ocean. Neonate Yellow-bellied Seasnakes (*Pelamis platurus*) usually emerge tail-first, wrapping their posteriors around the female's body to pull themselves free, then remove remnants of the placenta by knotting behavior similar to that used in shedding. Beaked Seasnakes (*Enhydrina schistosa*) congregate in estuaries for birth, and the young remain there for several months before migrating to deeper waters.

The reproductive cycles of seakraits (*Laticauda*) and seasnakes vary among species at a single site and within species at different sites. Several seasnakes are highly seasonal, synchronized breeders, while some seakraits oviposit throughout the year. Individual female Olive Seasnakes (*Aipysurus laevis*) reproduce less frequently than every year. Seakraits and seasnakes tend to have smaller clutches and litters than do terrestrial elapids of comparable size, and their developing eggs or young are placed farther forward in the body cavity—both traits that minimize the impact of pregnancy on locomotion in water. At the extremes for marine elapids, the Stocky Seasnake (*Lapemis curtus*) and Viperine Seasnake (*Praescutata viperina*) produce three or four large young in a litter, whereas Beaked Seasnakes (*Enhydrina schistosa*) give birth to an average of eighteen relatively small offspring.

Relatively little is known about the population biology of seakraits (*Laticauda*) and seasnakes, beyond the extraordinary abundances of several species. Beaked Seasnakes (*Enhydrina schistosa*) suffer high juvenile mortality (80–90 percent), presumably from predators, but survivors mature rapidly and first reproduce at about two years of age. Olive Seasnakes (*Aipysurus laevis*) live at least ten years in nature. Singularly adapted for life in currents and unable to crawl on land, Yellow-bellied Seasnakes (*Pelamis platurus*) are helpless when current direction shifts; beach stranding is probably their greatest, perhaps only major, source of mortality.

Typical reptiles cannot survive long in saltwater, and hydrophiine seasnakes are incapable of locomotion and prolonged respiration on land. Nevertheless, as many as three lineages of cobra relatives have bridged these seemingly disparate habitats and successfully invaded the western Pacific Ocean. How might terrestrial attributes and ecologies have translated, over evolutionary time, into obligate marine creatures? What accounts for the proliferation of seasnakes in certain regions and their complete absence elsewhere?

Colonization of the sea through a series of intermediate lifestyles is a plausible scenario, illustrated by the pioneering habits of certain living snakes. Among North American natricine colubrids, some Common Gartersnakes (*Thamnophis sirtalis*) and Western Terrestrial Gartersnakes (*T. elegans*) catch fish in rocky intertidal pools, while Salt Marsh Watersnakes (*Nerodia clarkii*) routinely forage in brackish water, and coastal Common Watersnakes (*Liophis miliaris*) feed on saltwater fish in the Atlantic littoral of Brazil. Homalopsine colubrids and primitive seasnakes (e.g., *Hydrelaps darwiniensis*) retain several characteristics of terrestrial species (e.g., wide ventral scales) and hunt on estuarine mud flats; reef seasnakes (e.g., *Aipysurus*), more specialized for marine life, still forage in water so shallow they sometimes resort to terrestrial crawling locomotion.

Historical factors probably account for the adaptive radiation of marine elapids in Indo-Australian waters and almost total lack of these animals elsewhere. Seakraits (*Laticauda*), primitive seasnakes (e.g., *Hydrelaps darwiniensis*), reef seasnakes (e.g., *Aipysurus*), and hydrophiines co-occur only off northern Australia, which suggests that they originated in that region. The Straits of Malacca, inhabited by 27 species of seakraits and seasnakes, perhaps was also a staging ground for speciation in those groups. In fact, only the broad continental shelves of northern Australia and southeastern Asia offer great expanses of shallow, warm water with little seasonal variation in temperature—ideal conditions for ectotherms involved in risky evolutionary experiments. Moreover, the reefs, trenches, peninsulas, landbridges, and islands of Australasian seas provide a richly varied landscape for speciation by isolation and the diversification of ecological niches.

Marine elapids are occasionally found in rivers and lakes and twice have colonized freshwater to the point of speciation. Lake Taal in the Philippine Islands has an endemic seasnake (*Hydrophis semperi*), and two sympatric seakraits (*Laticauda*) inhabit a brackish lagoon on Rennell Island in the Solomons. The latter situation is particularly interesting as a microcosm for the evolution of specialized marine snakes. Like other amphibious members of its genus, the widely distributed Yellow-lipped Seakrait (*L. colubrina*) is brightly colored, regularly rests on land, eats eels, infrequently flicks its tongue, submerges for less than five minutes, and is oviparous. The Rennell Island Seakrait (*L. crockeri*) is closely related to the Blue-banded Seakrait (*L. laticaudata*) yet resembles hydrophiine seasnakes in several respects: it is dwarfed and has an obscured color pattern, is never seen on land, flicks its tongue constantly while foraging, feeds on gobies, submerges for six to eleven minutes, and reputedly gives birth to live young.

Having originated repeatedly as substrate-foraging predators in warm, western Pacific waters, marine elapids have been unable to disperse globally because of deep open seas and polar currents. Although a few individual seakraits (*Laticauda*) have reached the Pacific Coast of

Middle America, evidently only the Yellow-bellied Seasnake (*Pelamis platurus*) survives waif dispersal with sufficient frequency to colonize the eastern Pacific. The latter species occasionally washes up on the coast of Siberia and as far south as New Zealand and Cape Horn but can neither survive long nor reproduce in such cold areas. Its absence in tropical Atlantic waters implies that seasnakes had not reached the eastern Pacific by about four million years ago, when the Panamanian landbridge formed, separating those two oceans.

VIPERS, ADDERS, AND PITVIPERS

14

With their misty canyons, bright cascades, and opulent biological diversity, the uplands of Latin America, equatorial Africa, and southern Asia are especially poignant for naturalists: traveling among them, we seek answers to grand intellectual puzzles and wonder about far more personal issues; faced with the escalating destruction of tropical forests, we race to publicize irreplaceable treasures. Late one afternoon in 1982, four of us emerged on a ridge above Costa Rica's Rio Peije, skirting the river's dark gorge, and rejoined familiar outer worlds of wind and sky. As dusk fell and we set up camp, forest colors changed from green to rich purples and scarlet, then merged into obscurity. Nearby trees flattened as they dissolved into blackness, and the stars, so preternaturally sharp, evoked surprising memories of cooler climates. For two weeks our multinational team had backpacked up countless muddy slopes, documenting flora and fauna on one of the last unexplored altitudinal transects in Central America. Agriculture and logging threatened this provisional *zona protectora,* between La Selva Biological Station and a highland national park; our survey would bolster conservation efforts. Later that evening, staring wearily into a small cooking fire and remote from twentieth-century problems, I was spellbound by the translucent fangs of an imaginary, jade-colored serpent.

Like most other organisms, snakes increase sharply in number of species from polar regions to the equator. Although large and topographically complex, California harbors only 30 species of snakes and no more than 15 at one site. Where the Amazon Basin mingles with Andean foothills, 75 species inhabit a chunk of Peruvian rain forest no larger than an average city in the United States. Ecuador has three times as many species of vipers as all of western Europe, and the snakes of that small country are more diverse in size, shape, and lifestyle than their temperate counterparts. We know why there is such richness of tropical species, at least in broad outline: Much of the earth's land surface is near the equator, where superabundant water and sunlight favor plants and the animals that feed on them; more kinds of insects are eaten by more kinds of frogs and lizards, and they in turn are prey for more kinds of snakes, hawks that eat snakes, and so forth. Beyond such matters of area and sustenance, these regions have repeatedly been shattered by mountain uplifting, volcanic eruptions, and long-term climate cycles, which have promoted speciation and new sorts of organisms. On smaller scales of space and time, earthquakes and fallen trees create gaps in forests, thus renewing plant succession, scrambling the accompanying animal communities, and allowing more species to persist in one place.

Arboreal green vipers are exquisite reminders of orogenies, volcanism, and fluctuating sea levels that have shaped tropical biotas. The ancestor of Middle American palm pitvipers (*Bothriechis*) split from Mexican horned pitvipers (*Ophryacus*) when early Tertiary oceans overran the Isthmus of Tehuantepec. Subsequent uplifting, from the Chiapan highlands in southern Mexico to the Cordillera de Talamanca in Costa Rica, isolated lowland Eyelash Pitvipers (*B. schlegelii*) from the progenitor of today's montane species. A prehistoric seaway, the Nicaraguan Depression, then separated highland prototypes in northern and southern Middle America, and still later Eyelash Pitvipers dispersed into South America when receding waters exposed a Panamanian landbridge. Finally, earthly rumblings and Pleistocene climatic cycles disrupted cloud forest habitats, yielding six species at higher elevations. Some African bush vipers (*Atheris*) paralleled their New World analogs and speciated during fragmentation of the ancient Eastern Arc ranges, whereas many Asian tree pitvipers (*Trimeresurus*) differentiated when flooding of the Sunda Shelf left ancestors of modern organisms on Borneo, the Philippines, and other Pacific islands.

Turning from geologic events to the lives of individuals, we encounter intriguing ecological questions. Bright yellow Eyelash Pitvipers, called *oropeles* (gold skins), occur in litters with normal Eyelash Pitvipers and, unlike those leaf-colored snakes, are easily seen in tropical forests. Some African bush vipers and Philippine tree pitvipers (*Trimeresurus*) also are yellow or orange, as are the young of Emerald Tree Boas (*Corallus caninus*) and Green Tree Pythons (*Morelia viridis*). Such snakes would be eliminated by predators if their conspicuous colors were not advantageous, and our few encounters suggested that nectar-feeding birds might be attracted to the flowerlike *oropeles*: we saw one gold Eyelash Pitviper hunting on a flowering *Heliconia* plant, and the Fogdens photographed another lunging at a hummingbird. Exploring that idea, Wade Sherbrooke—an ingenious specialist in animal coloration who every spring dons lipstick and persuades hummingbirds to drink sugar water from his mouth—placed painted rubber snakes among flowering plants at La Selva. Unexpectedly, hummingbirds hovered attentively near the green models and ignored the yellow ones, as if the supposedly cryptic snakes were recognized as danger, whereas *oropeles* were treated as meaningless yellow objects. Sherbrooke's observations suggest that yellow snakes indeed might be mimicking flowers, but their ambush relies on stealth rather than attraction.

Bites from arboreal vipers are uncommon and usually not fatal, but the prospect of running into these creatures at face height is unnerving. A Ugandan friend, so at home in the mountains that he walked barefoot up trails that left me gasping, so sophisticated in natural history that he discerned gorilla travel from bent grass, reacted emphatically when I caught a Lowland Bush Viper (*Atheris squamiger*): "That snake jumps from the trees and you die instantly!" Costa Ricans refer to Eyelash Pitvipers with an aboriginal word for "a devil whose bite is fatal," also the title of a popular short story. Carlos Salazar Herrera's "La Bocaracá" is the simple, compelling tale of a tormented and nervous man. Returning from work in the fields, Jenaro Salas finds his wife unconscious, with small scratches on one hand. Nearby are Jenaro's screaming small son, with "eyes like two questions," and the smashed, writhing body of an Eyelash Pitviper. Just as the panic-stricken man runs from the house, Tana rouses and yells, "Nothing happened, nothing happened . . . " In fact, the little boy had appeared in the doorway, smiling and carrying the deadly serpent as if it were a beautiful flower. Singing and cajoling, concealing her terror from the child, the mother had transferred the snake to her own hands, killed it with a rock, and fainted. Riding off in a cloud of dust to seek help, Jenaro was the only real victim, burdened as always by his anxious personality.

Vipers perhaps arose on windswept Eurasian steppes more than thirty million years ago, but today they achieve extremes of species richness, lifestyle, and beauty in wet equatorial forests. Now human overpopulation and greed are devouring tropical habitats at astonishing rates, threatening the existence of countless plants and animals. Among arboreal vipers worldwide, only the common Eyelash Pitviper in Honduras is listed by the Convention on International Trade in Endangered Species; others, including the rare March's Palm Pitviper (*Bothriechis marchi*) in that country, receive no special protection. On the bright side, global leadership is changing, the biodiversity crisis receives ever more attention, and gaudy serpents capture the imagination of urbanites: a thriving Central American artists' collective is called Bocaracá, Eyelash Pitvipers are on national park posters, and the emerald head of a palm pitviper graced a recent cover of the Sierra Club's national magazine. The *zona protec-·tora* we explored in 1982 has been saved, thanks to international funding and Costa Rica's famous environmental resolve.

Today an expanded Braulio Carrillo National Park reaches from montane cloud forest to the lowlands at La Selva, showcasing some of the most magical vistas on earth. Among those luxuriant ridges and valleys live the Jaguars (*Panthera onca*), Red-eyed Treefrogs (*Agalychnis callidryas*), Bare-necked Umbrellabirds (*Cephalopteris glabricollis*), and other creatures usually sought by nature tourists. Three species of arboreal pitvipers also grace the park, challenging visitors to adopt the respectful caution of forest dwellers, to behave less like careless intruders. The odd nuance of a nearby mossy frond, under idle scrutiny during a lunch break, finally reveals itself as a catlike pupil surrounded by green coils. Among many bright flowers, one abruptly unfurls, suspended by its prehensile tail, and flashes an open-mouth defensive display. Alert to the presence of these small serpents, we may even appreciate their mysterious beauty and cherish such chance meetings.

Except for stiletto snakes (*Atractaspis*), all venomous serpents with highly movable, enlarged, anterior hollow fangs are in the Viperidae. Collectively called vipers, there are about 230 species in 28 genera of adders, Old World vipers, and pitvipers. Vipers have achieved modest morphological and behavioral diversity; arboreal lifestyles and other ecological themes recur within the group, but only Cottonmouths (*Agkistrodon piscivorus*) are semiaquatic, and only some night adders (*Causus*) have snouts modified for digging. Nonetheless, viperids are found on all major, nonpolar landmasses except Australia and occur farther north (the European Adder [*Vipera berus*], to 69°N) and farther south (the Patagonian Lancehead [*Bothrops ammodytoides*], to 47°S) than any other serpents. Vipers are among the largest vertebrate predators in some terrestrial ecosystems, and often they are remarkably common.

Rhombic Night Adder (*Causus rhombeatus*), southern Africa.

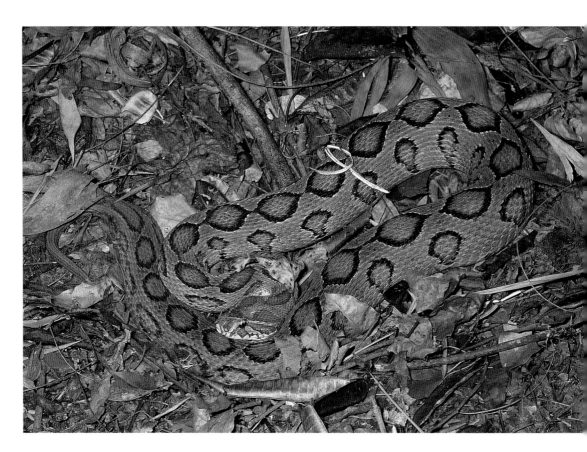

Russell's Viper (*Daboia russellii*), India.

Although these serpents are usually characterized as heavy-bodied and sedentary, most arboreal vipers are rather slender, and some terrestrial species travel several kilometers in the course of an active season. Most members of this group subdue, swallow, and rapidly digest exceptionally heavy and bulky items, even at relatively low temperatures. Vipers immobilize and process large prey by deeply injecting tissue-destructive venoms, and their phenomenal feats of ingestion are facilitated by unusually long quadrate bones and mandibles.

Relationships among viperid genera and higher taxa are not well resolved, but herpetologists typically focus on the four lineages discussed below. Night adders (*Causus*) are restricted to Africa, whereas geographic distributions of the other three taxa overlap in Eurasia and imply that the Viperidae first might have differentiated in the latter area. This chapter summarizes each of those lineages, then provides an overview of the natural history and diversification of vipers.

NIGHT ADDERS (CAUSUS)

Typically less than 80 cm long, 6 species of African night adders (*Causus*) have smooth scales, large head plates, and round pupils; they thus are rather different from other viperids in overall appearance. Night adders do, however, resemble North American hog-nosed snakes (*Heterodon*) and false pitvipers (*Xenodon*) in their blustery defensive behavior and specialized diets. Most night adders are spotted or blotched in shades of brown, but the Velvety Night Adder (*C. resimus*) is a lovely green snake. The snout is upturned in some species and is perhaps used for digging up their anuran prey. Striped Night Adders (*C. bilineatus*) eat highly aquatic clawed frogs (*Xenopus*), which suggests that this species might enter water more frequently than its relatives. Two species of *Causus* have long venom glands, extending onto the trunk and beneath the skin; night adder bites are painful but not life-threatening to humans.

OLD WORLD ADDERS AND VIPERS (VIPERINAE)

Sixty-six species in 10 genera of viperines are variously called adders or Old World vipers. About 20 species of small to medium-sized, terrestrial Eurasian species of *Vipera* share certain primitive colubroid features (e.g., large head scales in some taxa, as in many colubrids and elapids), superficially look like Eurasian pitvipers of the genus *Gloydius,* and among living forms might best represent the behavioral ecology of early vipers. They vary in length from less than 50 cm in the Meadow Viper (*V. ursinii*) to about 1 m in the Long-nosed Viper (*V. ammodytes*) and several other species. The color patterns of Eurasian Adders often include one or more wavy stripes on a lighter background, but sexual dichromatism and color pattern polymorphisms are common in these snakes. Probably owing to live-bearing habits and the ability to digest large items in cool climates, European Adders (*V. berus*) range as far north as the Arctic Circle. Several species of larger, heavy-bodied viperines in southwestern Europe, Asia, and northern Africa are placed in two other genera. Among them are the Levantine Viper (*Macrovipera lebetina*) and Russell's Viper (*Daboia russelli*), each reaching at least 1.8 m in some areas. Unlike smaller European viperines, these large species usually have small head scales and blotched color patterns.

Thirteen species in the exclusively African viperine genus *Bitis* are exceptionally stout, terrestrial creatures. These include the well-known Gaboon Adder (*B. gabonica*), Rhinoceros

Rhinoceros Adder (*Bitis nasicornis*),
equatorial Africa.

Horned Viper (*Cerastes cerastes*),
northern Africa. Like several other small
desert vipers, this species often uses
sidewinding locomotion.

Asian Saw-scaled Viper (*Echis carinatus*),
India, in defensive warning display.

Adder (*B. nasicornis*), and Puff Adder (*B. arietans*), surely the most heavy-bodied living snakes; wild-caught Gaboon Adders have exceeded 1.8 m in length and 10 kg in mass! Several smaller members of this genus are sidewinding adders of southern African deserts. Like the North American Sidewinder (*Crotalus cerastes*), some of these smaller *Bitis* also have "horns" —raised supraocular scales or spines, or erect fringes over the eyes. Among the smallest of all vipers, adult Peringuey's Adders (*B. peringueyi*) reach only 30 cm in length. This species is specially modified for desert life, with dorsally placed eyes and a black or white tail that facilitate caudal luring of lizards from beneath the sand.

Several additional genera of Afro-Asian viperines that inhabit arid regions and are typically less than 1 m long are perhaps not all closely related. Among them, Horned Vipers (*Cerastes cerastes*) resemble Sidewinders (*Crotalus cerastes*) in size, color pattern, and the presence of "horns" over their eyes; the smaller Avicenna Viper (*Cerastes vipera*) is more like Peringuey's Adder (*Bitis peringueyi*) in appearance. Found in southwest Asian deserts, Macmahon's Viper (*Eristicophis macmahoni*) has a scoop-shaped rostral scale and peculiar, whorl-like rows of dorsal scales. Saw-scaled vipers (*Echis*) are amazingly abundant in arid parts of Africa, the Middle East, western Asia, and India; the venom of some of these irascible serpents contains powerful anticoagulants, and despite their small size, perhaps no other snakes cause so many human fatalities. Members of these genera sidewind, but some also climb well and are occasionally found in shrubs.

Feathered Bush Viper (*Atheris hispidus*), equatorial Africa.

Fea's Viper (*Azemiops feae*), eastern Asia.

African arboreal vipers and their supposed relatives form a small but highly interesting group. Nine species of bush vipers (*Atheris*), usually brightly colored and less than 1 m in length, inhabit equatorial African forests. Eyelash Bush Vipers (*A. ceratophorus*) are stunningly similar to New World Eyelash Pitvipers (*Bothriechis schlegelii*) in their fringed head scales and polymorphic color patterns. Lowland Swamp Adders (*A. superciliaris*) and Hind's Adders (*A. hindii*) are terrestrial and have been assigned at times to *Vipera* or *Bitis*. The poorly known Uzungwe Viper (*Adenorhinos barbouri*) from Tanzania, closely related to arboreal species of *Atheris,* is unique among vipers in having a short, rounded head and perhaps eating soft-bodied invertebrates; like some blunt-headed colubrids that specialize on slugs and snails, this species has no mental groove.

FEA'S VIPER (AZEMIOPS FEAE)

Fea's Viper (*Azemiops feae*) is found only in mountainous regions of Burma, southern China, and Vietnam. Still known only from a few dozen specimens, the largest less than 1 m long, Fea's Viper was originally described in the nineteenth century as an elapid. Although it was once regarded as the most primitive living viper, recent biochemical studies suggest that *Azemiops feae* is closely related to pitvipers. In any case, broad purplish black bands and narrow orange rings, smooth scales, and the strangely rounded, flattened head of this snake create a decidedly un-viperlike appearance. Their few bites to humans have not had serious consequences.

PITVIPERS (CROTALINAE)

Pitvipers (Crotalinae) are distinguished by the paired infrared-sensing facial pits for which they are named, as well as by structural and behavioral attributes for confronting rather than fleeing from adversaries (see the Special Topic in this chapter).

With the exception of a few arboreal forms, most pitvipers have an emergent tail spine that produces a buzzing sound when vibrated against leaves, grass, and other flimsy substrates—and from which evolved the rattle in one group of crotalines. Pitvipers include 16 genera and 157 species, but relationships among major groups are poorly understood. Old World crotalines encompass several small lineages of relatively primitive terrestrial pitvipers, as well as the species-rich, ecologically diverse genus *Trimeresurus*. The larger adaptive radiation of New World pitvipers probably stems from one or more invasions from Asia, across a Bering landbridge.

Eurasian pitvipers include almost 50 species in 6 genera. Ten species of *Gloydius* include the well-known Japanese Mamushi (*G. blomhoffii*), Himalayan Pitviper (*G. himalayanus*), and other small denizens of temperate forests, grasslands, deserts, and montane rocky environments. By contrast, Malayan Pitvipers (*Calloselasma rhodostoma*), three species of Sri Lankan hump-nosed pitvipers (*Hypnale*), and Hundred-pace Pitvipers (*Deinagkistrodon acutus*) occur in tropical forests, and individuals of the last species reach more than 1.5 m in length. Thirty-one species of *Trimeresurus* are confined to Asia, India, and numerous islands in the western Pacific Ocean. Within this highly diverse assemblage, the Mountain Pitviper (*T. monticola*) and several other species are small and terrestrial, whereas Habus (*T. flavoviridis*) are nocturnal climbers, with large males reaching 2.2 m in length; there are several green

Right: Eyelash Pitviper (*Bothriechis schlegelii*), Costa Rica, with infrared pits clearly visible between the nostrils and eyes.

Opposite: Santa Catalina Island Rattlesnake (*Crotalus catalinensis*), Mexico, typically lacking a rattle.

Boa Constrictors (*Boa constrictor*) and some Old World vipers simply use thermally sensitive nerve endings in their facial scales to localize warm objects, whereas various other boas, pythons, and pitvipers concentrate those receptors in one or more well-defined depressions. Each of the paired receptor organs of a pitviper is a double chamber, divided by a membrane and positioned against a prominent cavity in the maxillary bones; perhaps the crotaline structure evolved by the fusion of two adjacent labial scale pits, like those seen along the lips of some living boas and pythons. The pit membrane is a double sheet of epidermis, separated by connective tissue, tiny blood vessels, and free nerve endings from the trigeminal nerve. A small pore in front of the eye connects the inner pit chamber with outer air, whereas the outer chamber forms the visible pit on the snake's face.

An anatomist guessed in 1869 that crotaline pits were sense or-

gans, but their remarkable thermoreceptive properties were not demonstrated until the 1930s. A blinded pitviper can perceive a mouse 10°C warmer than its surroundings from 70 cm away and guide the direction of its strike within about 5°. Crotaline pits have long served as model systems for studies of infrared receptor physiology, and we now know that the spatial arrangements of receptors in pits are recreated and integrated with visual information in the brain's tectum. Pits are probably infrared imaging devices rather than simply thermal receptors.

Although some snakes do use thermal cues in feeding, recent findings challenge the assumption that crotaline pits evolved solely for localizing prey. Pitvipers generally resemble most other viperids in diet and in predatory behavior, and Puff Adders (*Bitis arietans*) and Russell's Vipers (*Daboia russelli*) are as sensitive to and dependent on thermal cues as crotalines; together these

facts suggest that pitvipers achieve no feeding advantage with their namesake structures. Recall, however, that female pitvipers (unlike viperines) accompany their eggs or newborn young rather than abandon them, and that most crotalines confront adversaries with defensive sounds and other displays. Pits might provide nonvisual information about the size and shape of an adversary, perhaps crucial for a nocturnal pitviper faced with striking at, threatening, or fleeing from a looming animal. Pitviper color patterns are consistent with the idea that pits arose with confrontational parental and defensive behavior; whereas European Adders (*Vipera berus*) and some other viperines are striped, a pattern that generally correlates with locomotor escape tactics, pitvipers typically are crossbarred or blotched, patterns that facilitate camouflage and static defenses.

An explanation for crotaline pits based on their defensive role has in-

teresting implications. In terms of antipredator behavior, bush vipers (*Atheris*) gape with fangs fully extended, whereas Eyelash Pitvipers (*Bothriechis schlegelii*), Cottonmouths (*Agkistrodon piscivorus*), and other crotalines that gape as a threat gesture do so with fangs retracted. That difference is puzzling if pits function only in feeding, but it makes sense in light of an evaluate-and-engage defensive strategy: given that the sides of a viper's snout are folded up when rising maxillary bones erect the fangs (photo, p. 277), a pitviper that gaped like a bush viper would not be able to "see" with its pits. Of course pits might also help in striking prey and searching for warm spots to bask or hibernate, and they might play other roles as well. In any case, decades of experimental neurobiology must be reevaluated if these infrared receptors respond differently to large looming carnivores than to small perpendicularly moving rodents.

The rattle is used only for defense and arose in the context of an established antipredator role for noisy tail vibration. The popular scenario whereby the first rattlers warned away trampling herds of American Bison (*Bison bison*) is probably incorrect, since most species of rattlesnakes are not found in habitats with grazing herd animals, and those rattlers found in grasslands are not relatively primitive. Two facts, however, support an alternative theory, consistent with a widely held suspicion that some small montane species of the Mexican Plateau are remnants of early rattlesnake divergence. First, the rattle perhaps arose in areas lacking an appropriate substrate (e.g., dry grass or leaves) against which the primitive pitviper tail spine could vibrate to produce sound; among the habitats frequented by *Crotalus* and *Sistrurus*, rock slides best fit that description. Second, Rock Rattlesnakes (*C. lepidus*), Twin-spotted Rattlesnakes (*C. pricei*), and other species that

live in talus often sound off before they are discovered and after they disappear (this inspires one way of searching for them, by tapping rocks with a stick). Of course, rattling from hiding might be related to small size or some other factor, but it also might repel predators that probe into crevices. Coatis (*Nasua narica*), Ringtails (*Bassariscus astutus*), and Black Bears (*Ursus americanus*) are common in the rocky haunts of montane rattlesnakes, and all seek small, hidden vertebrates with their sensitive hands and snouts.

Evolutionary loss of the rattle has occurred twice on islands in Mexico's Sea of Cortés, in close relatives of the Red Diamond Rattlesnake (*Crotalus ruber*) of the adjacent Baja California Peninsula. In Santa Catalina Island Rattlesnakes (*C. catalinensis*), the lobes of the basal segments are so reduced that subsequently shed segments are not retained to form a functional rattle; the tail tip is less altered in San Lorenzo Island Rattlesnakes (*C. r. lorenzoensis*), but about half of the adults lack a rattle and others have only a few segments. The surprising loss of this complex adaptive structure might have resulted from random changes in genes controlling formation of the basal segment; reduced selection for defensive adaptations in predator-free environments; or—since rattles click against vegetation, even when the tail shakes slowly—selection for stealth in snakes that climb bushes, perhaps hunting for birds.

Temple Pitviper (*Tropidolaemus wagleri*),
southeastern Asia.

tree pitvipers among mainland and island members of the genus as well. The beautiful Temple Pitviper (*Tropidolaemus wagleri*), distinguished from other Asian pitvipers by keeled chin scales, also is arboreal.

Among New World pitvipers, those with large head scales include the Copperhead (*Agkistrodon contortrix*), still an abundant snake over much of the eastern United States, and the tropical Cantil (*A. bilineatus*). Speciation is evident among populations of the latter, exemplified by the distinctive Castellana (*A. b. howardgloydi*) and other isolated subspecies. Most *Agkistrodon* are mainly terrestrial and usually no more than 1 m long, but the semiaquatic Cottonmouth (*A. piscivorus*) reaches 1.88 m. New World pitvipers with variously fragmented head scales include more than 30 species of terrestrial lanceheads (*Bothrops*); that genus encompasses large snakes, well known in Latin America for their bite (e.g., Terciopelos

Castellana (*Agkistrodon bilineatus howardgloydi*), Costa Rica.

[*B. asper*], Jararacas [*B. jararaca*]), as well as shorter, stocky snakes (e.g., Patagonian Lance-heads [*B. ammodytoides*], Urutus [*B. alternatus*]). Among other neotropical viperids, 3 species of jumping pitvipers (*Atropoides*) are exceptionally heavy-bodied and approach the large African adders (*Bitis*) in relative girth. Other terrestrial taxa include thickset hog-nosed pitvipers (9 species of *Porthidium*), the more slender montane pitvipers (3 species of *Cerrophidion*), and the infamous Bushmaster (*Lachesis muta*). Palm pitvipers (7 species of *Bothriechis*) and forest pitvipers (7 species of *Bothriopsis*) are independently evolved arboreal groups of more slender serpents, and the former is closely related to the Mexican Horned Pitviper (*Ophryacus undulatus*).

The rattle is so similar among 30 species of *Crotalus* and 3 species of *Sistrurus* that it surely arose only once, in the common ancestor of those snakes. This structure evolved by changes in the shape of the tail spine that are characteristic of most other pitvipers, and in fact the tail tip of a Patagonian Lancehead (*Bothrops ammodytoides*) resembles rather closely that of a newborn Pigmy Rattlesnake (*S. miliarius*). Rattle segments are added at the base of the structure during skin shedding, with each segment hanging loosely on the lobes and grooves in the one that precedes it; end segments thus are the oldest and often break off with age. Highly

Side-striped Palm Pitviper (*Bothriechis lateralis*), Costa Rica.

Western Diamond-backed Rattlesnake (*Crotalus atrox*), Arizona.

McGregor's Pitviper (*Trimeresurus mcgregori*), Philippine Islands.

specialized tail-shaker muscles vibrate the interlocking rattle segments at about fifty-five cycles per second, producing a buzzing noise that varies in quality with the size of the snake. Generally stout-bodied, rattlers range from Pigmy Rattlesnakes and Twin-spotted Rattlesnakes (*C. pricei*) less than 60 cm long to Eastern Diamondbacks (*C. adamanteus*) and Western Diamondbacks (*C. atrox*) exceeding 2.3 m. Four or more species occur at some desert and rocky montane localities, but among these serpents only Timber Rattlesnakes (*C. horridus*) inhabit woodlands of the northeastern United States and only Neotropical Rattlesnakes (*C. durissus*) are found in the dry forests and savannas of Central and South America.

THE NATURAL HISTORY AND DIVERSIFICATION OF VIPERS

Compared to elapids, viperids have been conservative in structural and behavioral diversification. Although cosmopolitan as a group, most adders, Old World vipers, and pitvipers prefer open habitats or at least are most abundant where basking is easy. Many terrestrial vipers climb occasionally, and prehensile-tailed arboreal specialists have evolved independently on each continent where this group occurs. Unrelated sidewinders inhabit deserts of northern and southwestern Africa, the Middle East, and North America; in the dunes of

Sidewinder (*Crotalus cerastes*), Mexico.

southern Spain, Hog-nosed Adders (*Vipera latastii*) use that locomotor mode as well, although they occur elsewhere in non-sandy habitats. Probably all vipers rely on cryptic coloration and immobility to avoid discovery by predators, then react to direct threats with coiling and striking. Beyond those widespread traits, night adders (*Causus*) and African adders (*Bitis*) respond to adversaries with body inflation, loud abrupt hissing, and rapid, exaggerated strikes. Species of *Cerastes* and *Echis* rub their serrated, obliquely keeled lateral scales against each other to produce a rasping sound when disturbed; some bush vipers (*Atheris*) have similarly modified scales, which suggests that all those snakes might have had a common ancestor with that scalation and defensive behavior. Like pitvipers, Fea's Vipers (*Azemiops feae*) vibrate their tails and sometimes gape if threatened.

In contrast to other highly venomous snakes and regardless of habitat, most viperines and pitvipers have remarkably similar diets and consume several types of prey. Juvenile viperids typically eat large invertebrates (especially centipedes), frogs, and lizards, then switch as soon as size permits to feeding on mammals. Interspecific dietary differences reflect mainly local availability of particular mammals, especially mice, rats, and rabbits. Of course, larger snakes can swallow larger prey, and Puff Adders (*Bitis arietans*) sometimes eat dik-diks (*Madoqua*) and African Hedgehogs (*Atelerix frontalis*), while Gaboon Adders (*B. gabonica*) occasionally take Royal Antelope (*Neotragus pygmaeus*), Black-and-Rufous Elephant Shrews (*Rhynchodon petersi*), and even Brush-tailed Porcupines (*Atherurus africanus*)! At the other extreme of diet breadth, even adult Meadow Vipers (*Vipera ursinii*) feed mainly on arthropods; narrow diets are not confined to small vipers, as Bushmasters (*Lachesis muta*) eat only rodents throughout life. Viperines and pitvipers sometimes take exceptionally heavy prey relative to their own size—the largest item recorded for any snake was a whip-tailed lizard (*Cnemidophorus*) weighing 156 percent of the mass of the juvenile Common Lancehead (*Bothrops atrox*) that had eaten it. Night adders feed almost exclusively on frogs, mainly toads (*Bufo*), and the only recorded natural prey of Fea's Viper (*Azemiops feae*) was a Common Gray Shrew (*Crocidura attenuata*) that weighed as much as the juvenile snake that contained it.

Vipers hunt in ways that minimize the energetic costs of feeding and reduce danger from other predators, a feasible lifestyle because just a few large food items per year permit them to grow and reproduce. Night adders (*Causus*) actively search for toads (*Bufo*), but most other viperids hunt by strategic mobile ambush—using chemical traces of their prey to seek out good places for sitting-and-waiting. Hundred-pace Pitvipers (*Deinagkistrodon acutus*) sometimes hide in leaves under fruiting trees to catch birds, whereas Timber Rattlesnakes (*Crotalus horridus*) coil perpendicular to fallen logs or vertically against tree trunks, striking from concealment at passing deer mice (*Peromyscus*) and squirrels. Prairie Rattlesnakes (*C. v. viridis*) migrate several kilometers from winter dens, then remain for weeks in grassland patches with high concentrations of rodents. Vipers occasionally feed on nestling birds and mammals, presumably encountering these sedentary prey while searching for ambush sites; examples include Mexican Dusky Rattlesnakes (*C. triseriatus*) eating Volcano Rabbits (*Romerolagus diazi*), Prairie Rattlesnakes taking Mountain Cottontails (*Sylvilagus nuttalli*), and Twin-spotted Rattlesnakes (*C. pricei*) eating Yellow-eyed Juncos (*Junco phaeonotus*).

Visual and infrared cues guide a viper's strike, after which prey is usually released; arboreal species often hold on to prey, however, as do terrestrial vipers feeding on birds and lizards. As the bitten mammal bounds off, its dying heartbeats circulate the venom, beginning digestion of the prey even before it has been ingested. The viper relocates its victim by following the scent trail; it can chemically distinguish trails of envenomed mice from others, as well as determine the direction in which the prey has traveled. Although trailing prey might be dangerous, a viper's cryptic coloration and slow movements probably reduce the likelihood of detection by other predators.

The feeding biology of Cottonmouths (*Agkistrodon piscivorus*) is especially flexible. This often common species eats cicadas, fish (e.g., adult Yellow Bullheads [*Ictalurus natalis*]), frogs, salamanders (e.g., Lesser Sirens [*Siren intermedia*]), turtles, baby American Alligators (*Alligator mississippiensis*), birds, lizards, serpents (even an adult Timber Rattlesnake [*Crotalus horridus*]), and mammals (e.g., Eastern Moles [*Scalopus aquaticus*], Muskrats [*Ondatra zibethicus*], Cottontails [*Sylvilagus floridanus*])—but never toads (*Bufo*). One massive Cottonmouth contained four bird eggs, an immature Great Egret (*Casmerodius albus*), and an adult

Glossy Ibis (*Plegadis falcinellus*)! These semiaquatic pitvipers usually pull captured fish on land for swallowing and might even cooperate in herding stranded fish in small drying pools. Although other vipers occasionally eat carrion, Cottonmouths seem particularly attracted to long-dead prey; some individuals on small islands in the Gulf of Mexico even feed primarily by lying directly under the nests of waterbirds, snatching fallen chicks, as well as fish regurgitated by the parents.

Modes of reproduction among vipers are unusually variable taxonomically. Night adders (*Causus*) lay eggs, presumably retaining this mode from ancestral colubroids. Other oviparous viperids include the Levantine Viper (*Macrovipera lebetina*) and its close relatives, the Palestine Viper (*Vipera palestinae*), Malayan Pitviper (*Calloselasma rhodostoma*), Hundred-pace Pitviper (*Deinagkistrodon acutus*), Bushmaster (*Lachesis muta*), and perhaps the Colombian Lancehead (*Bothrops colombianus*), whereas large African adders (*Bitis*), bush vipers (*Atheris*) and their close relatives, most Eurasian viperines, and almost all pitvipers are viviparous.

Even close relatives differ in their modes of reproduction. The Horned Viper (*Cerastes cerastes*), Painted Saw-scaled Viper (*Echis coloratus*), East African Saw-scaled Viper (*E. pyramidum*), and Macmahon's Viper (*Eristicophis macmahoni*) lay eggs, but the Avicenna Viper (*Cerastes vipera*) and Asian Saw-scaled Viper (*Echis carinatus*) are live-bearing. Within *Trimeresurus*, Habus (*T. flavoviridis*) and at least ten other species lay eggs, whereas the Mangrove Pitviper (*T. purpureomaculatus*) and at least ten more species in that genus bear live young. Herpetologists usually assume that oviparity is primitive and that viviparity has thus arisen multiple times within Viperidae, but the repeated occurrence of egg-laying species deep within this radiation implies that oviparity might have re-evolved several times as well.

Viperines abandon their newly laid eggs or neonates immediately, as do most other snakes, but night adders (*Causus*) might remain with their eggs during incubation, and maternal care probably characterizes pitvipers as a group. Oviparous crotalines attend their incubating eggs, and some become more aggressive during that period. Mountain Pitvipers (*Trimeresurus monticola*) oviposit within a pile of bamboo, whereas Bushmasters (*Lachesis muta*) nest in large rodent burrows, and Malayan Pitvipers (*Calloselasma rhodostoma*) simply coil around their eggs in a depression on the tropical forest floor. Copperheads (*Agkistrodon contortrix*), Eastern Diamond-backed Rattlesnakes (*Crotalus adamanteus*), and at least a dozen other species of temperate pitvipers remain with the young for a week or so after birth, until they shed their natal skins and disperse. Since she is not free to encounter prey, a female pitviper risks declining health by protecting her litter during this vulnerable period, which suggests that this practice is not coincidental. Postnatal care in live-bearing pitvipers might be derived from egg-guarding, as still practiced by some crotalines; viviparous tropical pitvipers have not been found with their offspring, so perhaps maternal behavior has been lost in those groups.

Vipers exhibit a fascinating blend of differences from and similarities to other snakes in various aspects of behavioral ecology. Like pythons and boas, many vipers hunt by ambush, feed infrequently on large prey, and rely on camouflage and aggressive defense to avoid predation; some members of each group remain with their eggs or young for extended periods. Determining whether such similarities were retained by vipers from earlier macrostomatan ancestors or convergently acquired will require improved resolution of phylogenetic relationships among higher snake taxa.

Vipers contrast with most other colubroids, as well as boas and pythons, by possessing powerful, tissue-destructive venoms and occupying cold temperate environments. Unlike most other snakes, many vipers use multiple habitats during an annual cycle, often migrating over long, individually consistent routes year after year; among numerous examples, some European Adders (*Vipera berus*) crawl over snow to traditional mating sites in early spring, and neonate Prairie Rattlesnakes (*Crotalus v. viridis*) and Timber Rattlesnakes (*C. horridus*) use the chemical trails of adults to localize winter dens. Finally, some island viper populations reach unusually high densities, probably by virtue of low-energy lifestyles and reduced mortality. Shedao Island Pitvipers (*Gloydius shedaoensis*) off the coast of China, Golden Lanceheads (*Bothrops insularis*) off the coast of Brazil, and Cycladean Vipers (*Macrovipera schweitzeri*) in the Aegean Sea all feed primarily on migratory birds, ambushing them from low vegetation. Clearly, we have barely scratched the surface of physiological and behavioral complexity in these venomous snakes.

PART THREE: SYNTHESIS

In the end, we will conserve only what we love, we will love only what

we understand, we will understand only what we are taught.

Baba Dioum, Senegalese conservationist

Temple Pitviper (*Tropidolaemus wagleri*),
southeastern Asia.

EVOLUTION AND BIOGEOGRAPHY

Biogeographers decipher the vagaries of life in space and time, the history of plants, animals, and their surroundings. Broadly speaking, these relationships are fairly straightforward. Food, water, and other necessities limit individuals, whereas species distributions expand, divide, and contract with long-term geologic and climatic changes; inhospitable areas separate populations, and distinctive lineages later evolve from the remnants of once-continuous gene pools. The fine details of biogeography vary endlessly, by contrast, and are much harder to come by. Birds, large mammals, and other highly mobile creatures can cross hostile expanses, so their ranges often coincide with the limits of tropical biotas, savanna vegetation, and other continental patterns. Informed by science and inspired by obvious grandeur, we even symbolize such spacious landscapes with the likes of Scarlet Macaws (*Ara macao*) and African Lions (*Panthera leo*). Smaller, less vagile organisms often cannot disperse widely and thus better indicate the former extent of isolated habitats on local and regional scales. In this vein, thanks to the labors and romantic musings of field biologists, an attractive little rattlesnake exemplifies the history of disjunct ecosystems in the southwestern United States and northern Mexico.

The Sierra Madre Occidental runs 1,200 km along the west flank of Mexico's Central Plateau, with outlying fragments in southern Arizona and New Mexico. The range was formed by widespread volcanism in the early Cenozoic era, and subsequent faulting and uplifting created numerous ridges and valleys in the Sierra; its massive Pacific escarpment rises to 3,000 m, dissected by tremendous canyons known locally as *barrancas*. Tarahumaras, Yaquis, and other indigenous people have long occupied this rugged country, leaving such evocative place-names as Chuhuichupa ("valley of the mists") and Huachucas ("thunder mountains"). The Sierra has a distinctively tropical feel, even at its northern end, where desert and grasslands surround montane islands. In the Chiricahuas a pungent odor of Creosote Bush (*Larrea tridentata*), washed up from nearby lowlands by a late afternoon cloudburst, mingles with the raucous din of Thick-billed Parrots (*Rhynchopsitta pachyrhyncha*) settling in for the night. Four species of wildcats found on our Costa Rican study site also have prowled Arizona within recent centuries, and Jaguars (*Panthera onca*) still straggle across from Mexico. Iridescent green trogons and more than a dozen kinds of hummingbirds are there too, typical of wet forests to the south. These are ancient legacies in a region cloaked with arid-adapted tropical plant groups, the Madro-Tertiary Geoflora, that evolved long ago in marginally humid areas.

For at least twenty million years, this also has been the home of rattlesnakes, descendants of ancestral pitvipers that probably crossed the Bering landbridge from Asia. Famous among relict reptiles of the northern Sierra Madre, Ridge-nosed Rattlesnakes (*Crotalus willardi*) reach 65 cm in length and are named for a raised edge along the snout. Dorsal blotches are always present, but individuals may be reddish brown to gray and match local habitat colors. Four of the five well-differentiated geographic races have prominent facial stripes, whereas the New Mexico subspecies has only faint markings. Carl Kauffeld devoted a chapter to Ridge-nosed Rattlers in *Snakes and Snake Hunting* (a book I read over and over in high school), portraying the species as alpine, based on a pair he had found in the Huachucas. At the time of his 1941 trip, *C. willardi* was known elsewhere in the United States only from the Santa Rita Mountains, and Kauffeld's colorful prose launched throngs of collectors in search of these supposedly rare creatures. No wonder, then, the excitement when I finally saw one in the field: a barely glimpsed movement in leaf litter, beside a fallen oak limb; momentary disbelief at the lacquerlike head markings, sure signs of my good fortune; and then, "Hey, a *willardi*! I found a *willardi*!!"

Earlier, more humble ramblings about Ridge-nosed Rattlers came from a man whose abundant joy in life encompassed carpentry, music, photography, pistols, friends, and all of nature. Howard K. Gloyd's many publications on pitvipers are enduring tributes to scholarship and successful fieldwork, but several visits to the Huachucas in the 1930s seemed, in his words, "under the auspices of an evil jinn":

I have spent more time and lost more pounds in pursuit of this herpetological will-o'-the-wisp than in hunting any other single species. Why this particular jinx should attach itself to me I can not reason out; but I am not a philosopher.

On the third day . . . [a companion] returned from a hike down the trail and, after reporting his success with spider collecting, nonchalantly produced a tightly tied cloth bag and inquired: "What will you give me for a *willardi?*"

"I am in the market for a *willardi,*" I answered dryly. Fearing a trap, I refused to warm up prematurely. He had me dangling, and knew it.

"Do they come in browns?" he asked, prolonging the suspense. I could stand it no longer. I snatched the bag, opened it, and looked down upon a small coil of velvety brown, a turned-up snout, a delicate "white enamel" line diagonally across a scaly cheek.

"Yes, *willardi* comes in browns and this is it!" I exclaimed. "*Where* did you find it?"

"*Right in the trail,* near the Hamburg Mine," was the answer! Delighted as I was to have the snake—truly a thing of beauty if you appreciate ophidian pulchritude—I could not help muttering: "There ain't no justice."

Another unsung player in the Ridge-nosed Rattlesnake saga is ornithologist Joe T. Marshall. Marshall had studied with Alden H. Miller, Joseph R. Grinnell's successor as founding director of Berkeley's Museum of Vertebrate Zoology. Miller espoused Grinnell's pioneering research theme of variation within and among populations along environmental gradients, and that perspective inspired Marshall's dissertation on Song Sparrows (*Melospiza melodia*) in 1948. After finishing graduate school, he investigated bird distributions and habitats at the northern end of the Sierra Madre Occidental; a renowned walker, Marshall mapped plant assemblages from coniferous forests at high elevations down through pine-oak and evergreen oak woodlands, thence into the grasslands and desert scrub that separated adjacent mountain ranges. An unusually broad biologist as well (his scientific papers extend to gibbon calls and Asian rat systematics!), while surveying plants and birds Marshall collected snakes too; he found Arizona Ridge-nosed Rattlers (*Crotalus w. willardi*) at new sites in the Santa Ritas and Huachucas, and first recorded them for Cerro Azul and the Sierra de los Ajos in the neighboring Mexican state of Sonora.

Marshall's role unfolded as I savored his 1957 monograph, *Birds of Pine-oak Woodland in Southern Arizona and Adjacent Mexico,* seeking to understand Southwestern biogeography and imagining how wonderful backwoods travel must have been then. About the same time I ran across mention in Laurence M. Klauber's massive tome on rattlesnakes of a letter from Marshall, about a *Crotalus willardi* in northern Mexico eating a Wilson's Warbler (*Wilsonia pusilla*). Wondering if that snake actually might have been a New Mexico Ridge-nosed Rattlesnake, first reported in 1961 as an isolated population in the Animas Mountains, I called Marshall at the Smithsonian Institution. Within a minute or two he'd checked field notes from forty years earlier and confirmed my hunch, that he had collected the first specimen in the Sierra San Luis of what would much later be named by others *C. w. obscurus* and placed on the Endangered Species List. He then added matter-of-factly that he'd eaten one of the little rattlers—and, in answer to my obvious next question—"Because I was hungry!" Later I discovered that the San Luis snake with a bird in its stomach (but not the one he ate) is reported near the end of Marshall's monograph, in the Wilson's Warbler account, although overlooked by subsequent herpetologists who copied his maps.

A lot has been learned in the decades since Gloyd and Marshall explored the Sierra Madre Occidental. *Crotalus willardi* occurs more widely in oak woodlands than previously supposed, yet during the Pleistocene it probably survived dry interglacial climates only at higher, cooler elevations. Several teams have embarked on long-term field research, so in spite of restricted distribution and supposed rarity, these lovely brown serpents may soon be relatively well known. All this new information and appreciation still build on earlier studies by people like Joe Marshall, who grabbed a few supplies and headed into the mountains, confident he could drink from the creeks and eat a snake or two if food ran low.

E volution, the central organizing principle of biology, encompasses two irrefutable facts. First, organisms have histories, reflecting their immediate descent from parents and thus their derivation from ancestral generations over longer periods of time. Second, living things depend on and are affected by their surroundings; to persist, reproduce, and thus have histories, they must survive in particular environments. Evolution involves several processes, evident at different levels of analysis. At the most basic level, new life forms are initiated by genetic

Short-nosed Vinesnake (*Oxybelis brevirostris*), Costa Rica.

mutations and by their effects on developmental mechanisms that make whole, functioning organisms. Shifts in the proportions of alternative genetic combinations, and thus ultimately in the kinds of organisms that make up a population, occur as a result of natural selection, linkage to other genes that are themselves subject to selection, and chance (genetic drift). On grander scales of time and space, continents move, climates change, predators invade previously simpler faunas, and so forth; those "macro" phenomena promote immigration, extinction, and speciation, thereby leading to regional and global variation in biodiversity. These last topics blend contemporary ecology with paleontology, collectively the focus of biogeography.

Evolutionary biology involves comparisons among extinct and living organisms, to infer changes in their attributes through time; research on embryonic tissue interactions, on shifts in developmental timing, and on other mechanisms that produce new kinds of organisms; and studies of differential reproduction and mortality among individuals in populations, the process of natural selection. Snakes are prominent in modern studies of evolutionary and biogeographic mechanisms, at levels of analysis ranging from genes to individuals and landscapes, and some broad patterns in serpent diversification are emerging. This chapter summarizes the origin and radiation of snakes, some ecological correlates of those historical patterns, and mechanisms of evolutionary change.

MESOZOIC LIZARDS AND THE EARLY EVOLUTION OF SNAKES

Tropical climates and tremendous evolutionary activity characterized the Mesozoic (250–65 million years ago [mya]), during which flowering plants, mammals, and dinosaurs originated. Those organisms played no known role in early serpent evolution, but several key aspects of snake biology ultimately reflect the prior divergence of two major squamate lineages in the Jurassic period of that era (>150 mya; Fig. 3 illustrates relationships among those and other relevant groups of lizards and snakes). Iguanians (iguanas, chameleons, and their relatives) seize and taste food with a fleshy, bloblike tongue; in contrast, scleroglossans (autarchoglossans and geckos) grasp food with their jaws and transport odor molecules to their vomeronasal organs with the tongue. Iguanians emphasize visually guided behavior and have never evolved body elongation and limb loss, whereas scleroglossans have repeatedly achieved specialized chemosensory abilities, elongation, and limb reduction. Within Autarchoglossa, snakes are closely related to Anguimorpha, which includes knob-scaled lizards (Xenosauridae), alligator lizards (Anguidae), beaded lizards (Helodermatidae), and monitors (Varanidae); varanids resemble serpents in several respects, especially in having long forked tongues. Serpents also share some features with Scincomorpha, and within that other major clade of autarchoglossans, giant neotropical tegu lizards (*Tupinambis*) independently evolved a snakelike tongue. The oldest fossil snake is from the Cretaceous period, about 95 mya; however, because other scleroglossans diverged in the Jurassic, serpents might well have originated then too, more than 140 mya. Ancient fossil snakes, most blindsnakes, and most basal alethinophidians occur on southern continents, which implies that serpents perhaps first radiated on Gondwanaland, at a time when other autarchoglossans were diversifying on Laurasia.

Characteristics of the first snakes—of the most recent common ancestor of blindsnakes and alethinophidians—included an elongate body, greatly reduced limbs and eyes, chemosensory hunting, and use of cloacal scent glands for defense against predators. A traditional view holds that the first snakes were structurally and ecologically like Asian pipesnakes (*Cylindrophis*), feeding infrequently on heavy, slender vertebrates. This hypothesis was inspired by *Dinilysia patagonica,* a well-preserved, 2-m-long Cretaceous snake or snakelike squamate whose skull resembles that of pipesnakes in some respects. A more likely alternative is that the earliest snakes were small creatures and, like limbless anguimorphs and blindsnakes, primarily ate tiny arthropods. Anguimorphs, the closest relatives of snakes, have ranged in length from <25 cm in some alligator lizards (*Elgaria*) to 6 m in an extinct Australian monitor lizard (*Megalania prisca*) and even larger in prehistoric aquatic mosasaurs. Like blindsnakes and most living basal alethinophidians, however, limbless anguimorphs are invariably less than

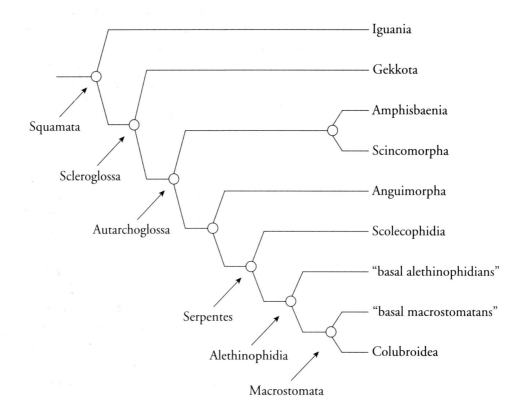

Fig. 3

Phylogenetic relationships among major
groups of lizards and snakes.

1 m long and usually much shorter (e.g., California Legless Lizard [*Anniella pulchra*]). Australian flap-footed lizards (Pygopodidae) evolved a limbless lifestyle from gecko ancestors, and their history independently parallels the second view of snake origins. Most pygopodids resemble limbless anguimorphs and blindsnakes in behavioral ecology, and among flap-footed lizards, only species of *Lialis* ingest relatively large prey; as is the case for snakes, that specialization arises well within a limbless clade, rather than at its base.

Instead of presenting a single key innovation, snakes initially combined limbless locomotion with highly specialized, tongue-mediated searching behavior. Other groups with only slender bodies or only long forked tongues have not matched in species richness even the smaller of two basal branches in serpent evolution; among extant squamates, there are 300 species of blindsnakes, versus about 40 species of monitors and 5 of tegus (*Tupinambis*) with snakelike tongues. Blindsnakes, however, despite considerable anatomical variation, have not differentiated much in behavior and ecology. On the other basal branch of serpent evolution, innovations in the skull of alethinophidians enhanced flexibility and gape, while constricting behavior permitted some early snakes to immobilize massive prey and feed less frequently— a dramatic shift in foraging strategy. One blindsnake (*Acutotyphlops subocularis*) and all shield-tailed snakes eat earthworms, a plausible transitional diet between the tiny invertebrates that most scolecophidians eat and the heavy, elongate vertebrates eaten by other basal alethino-phidians. And, judging from the behavior of their living relatives, mimicry of noxious, elon-gate invertebrates might have facilitated the evolution of more vulnerable, surface-dwelling lifestyles in early alethinophidians. Dwarf pipesnakes (*Anomochilus*), Asian pipesnakes (*Cylin-drophis*), Red Pipesnakes (*Anilius scytale*), and shield-tailed snakes expose alternating dark and light ventral blotches during their writhing defensive displays, and those rapidly moving, repetitious markings might evoke the appearance and behavior of centipedes and millipedes.

Several lineages of basal macrostomatans had appeared by the end of the Mesozoic and beginning of the Cenozoic eras (65 mya); these were boa-like snakes with cranial modifica-tions that permitted them to swallow substantially more diverse prey than are taken by Asian pipesnakes (*Cylindrophis*) and other basal alethinophidians. Long quadrate bones and lower jaws, enhanced movement between the tips of the mandible, and loosened connections among bones of the skull enabled these serpents to ingest bulky prey; they thus could add relatively large crocodilians, birds, and mammals to their diets (see the Special Topic in this chapter).

Like pythons and other surviving basal macrostomatans, those snakes achieved consider-ably more variation in size and habits than have basal alethinophidians. At least one Mesozoic boa-like snake (*Madtsoia bai*) reached 10 m, rivaling the largest living serpents in length. By the Eocene epoch (50–35 mya), a radiation of primarily fossorial erycine boas was widespread in the Northern Hemisphere; the habits of living boas and pythons suggest that by the early Cenozoic their basal macrostomatan ancestors occupied arboreal and aquatic niches as well. Most early snake lineages died out by the Oligocene epoch (35–25 mya) or, like the Neotropi-cal Sunbeam Snake (*Loxocemus bicolor*) and Asian pipesnakes, are represented today by only one or a few species. Among ancient groups surviving into the Pleistocene epoch (<2 mya), the most surprising is represented by *Wonambi naracoortensis*. This 5-m-long, heavy-bodied Australian serpent belonged to the Madtsoidae, a group of boa-like creatures otherwise last found in the early Eocene (50 mya).

CENOZOIC INNOVATIONS AND THE COLUBROID EXPLOSION

Whereas tropical biotas and drifting continents characterized the Mesozoic, large-scale changes in habitats and climate marked the end of that era and the first half of the Cenozoic (beginning 65 mya). Huge seaways obliterated parts of some continents and divided others; landbridges formed between continents, disappeared under water for millions of years, and then reappeared. Climates gradually became more seasonal as immense tracts of grassland and other open habitats blanketed much of the Northern Hemisphere; large dinosaurs went ex-tinct, and new groups of vertebrates appeared. Rodents originated in Asia near the beginning of the Cenozoic, and modern families of those mammals and of birds arose in the Eocene (55–35 mya) and Oligocene (35–25 mya) epochs. Small and medium-sized carnivores diver-sified during that period, as did modern groups of eagles, falcons, and hawks. Lemur- and tarsier-like primates arose in the Eocene, and apelike species appeared by the early Miocene epoch (25 mya). Along with the rodents, primates, carnivores, birds, and other groups that now are often their most important ecological associates, colubroids radiated so explosively in the Miocene that the Cenozoic has been called "the age of snakes." Today colubroids make up fully 80 percent of the world's snake fauna.

Brown Sand Boa (*Eryx johnii*), India, displaying its heavily scarred tail tip. Sand boas and their relatives are found today in Eurasia, Africa, and North America; they also have a rich fossil record.

Fossilized remains of plants and animals show that organisms with particular traits indeed lived at certain times in the past, and they sometimes also yield tantalizing clues to ancient lifestyles. Most fossils probably belonged to the many populations of extinct organisms that connected ancestors and their descendants through geologic time, rather than to actual ancestral populations; nonetheless, fossils provide valuable evidence of ancient relatives and transitional characteristics, as well as chronologies for evolutionary scenarios. Snake fossils, although relatively uncommon and usually incomplete, offer many important insights about the evolution of these animals.

The fossil record of serpents suffers from several unfortunate biases. Many modern snakes are terrestrial or arboreal, yet waterborne sediments are among the best environments for preservation and fossilization of carcasses. Snake fossils, especially geologically older ones, often come from deposits that once were ocean, lake, and stream bottoms; they thus might falsely imply that most truly ancient serpents were aquatic. Moreover, fossils are

not common in tropical regions, and so far they mainly afford glimpses of extinct taxa in Europe and North America. Perhaps most important, snake skeletons—unlike the skeletons of turtles and some other vertebrates—are especially fragile and easily destroyed. Predation is a common cause of serpent mortality, and their enemies often crush snake skulls (following the beeping transmitter of a telemetered Costa Rican Terciopelo [*Bothrops asper*], we found a fossil-in-the-making—her skeleton lay scattered in leaf litter, the foreparts represented by only a few vertebrae and skull fragments). Finally, a rotten snake floats with its head hanging down, so the skull bones gradually fall out as connective tissues disintegrate; the skin acts as a sleeve for the body, however, and might retain the vertebral column and ribs in place until they are covered in sediments. Predation and postmortem processes leave us mainly with fossil snake vertebrae, usually disarticulated or in short chains of a few bones. Identification

of individual vertebrae is often difficult, and those of living blindsnakes are so similar to each other that we cannot assign fossils to families, let alone genera and species.

There are wonderful exceptions to the poor fossil record of snakes, illuminating major evolutionary patterns as well as behavior within populations. For decades, *Dinilysia patagonica* from the Cretaceous period of Argentina has inspired and challenged efforts to understand snake origins. The single well-preserved skull exhibits such a blend of snakelike, lizardlike, and unique characteristics that herpetologists still argue about whether this creature was indeed a snake. An extensive Old and New World record of fossil erycines shows that modern Rubber Boas (*Charina bottae*) and sand boas (*Eryx*) retain an ancient antipredator specialization, their reinforced, clublike tail skeletons, perhaps associated with the origin of that widespread and diverse clade. Other fossils document the last few million years of serpent history, in-

A fossil aggregation of two genera of Oligocene sand boas (an *Ogmophis* on the left, two *Calamagras* on the right), Wyoming.

cluding a rich record of speciation and extinction among Eurasian cobras (*Naja*) and vipers. New fossil serpents are discovered ever more frequently, and with luck and careful study they will play special roles in our further understanding of snake evolution.

There are several truly spectacular snake fossils, most of them from a former municipal garbage dump near Frankfurt am Main. The Messel oil shale deposits preserve an incomparable array of middle Eocene plants and animals, including leaves and berries, a frog with eggs in its abdomen, a horse with seeds in its gut, and perfectly intact bat skeletons. Among the first snakes discovered there was a beautifully preserved, boa-like serpent with a small crocodilian in its stomach, testimony to headfirst consumption of a rela-

tively large item roughly 50 mya. Other impressive serpents among the Messel fossils include a meterlong erycine with a perfect skull and fine details of a rudimentary pelvic girdle and femurs. Part of the windpipe is visible in a smaller specimen, perhaps related to dwarf boas, and a basal alethinophidian (superficially similar to Asian pipesnakes [*Cylindrophis*]) contains three or four young in its abdomen.

My favorite fossil snakes are preserved together, a stunning aggregation of two genera of common early erycines from Oligocene deposits in Wyoming. Approximately 32 million years old, these sand boa relatives were in a rich sedimentary fossil assemblage with an amphisbaenian, marsupials, camels, and various other vertebrates. An *Ogmophis* about 50 cm long lies beside

two entwined *Calamagras,* each of the latter about half the size of the larger snake; both genera are represented by extremely well preserved skulls. We should be cautious about interpreting fossilized behavior, because animals usually do not die while acting normally. These Wyoming snakes, however, are in horizontal postures that look like those of living serpents—not twisted in death throes or jumbled in a heap. Rubber Boas sometimes occupy colonial winter shelters and various colubrids and vipers often den together, so perhaps the Wyoming fossil erycines died during a similar period of seasonal inactivity. That chunk of Oligocene mudstone and its fossil skeletons imply that a social interaction typical of diverse living serpents extended well back into the early stages of snake evolution.

Western Brownsnake (*Pseudonaja nuchalis*), Australia.

Although serpents perhaps originated on the ancient southern continent of Gondwana-land, major events in colubroid evolution began on the old northern landmass, Laurasia. Colubrids and perhaps colubroids arose in Asia, the former group in the Oligocene (>30 mya) or earlier and the latter perhaps as long ago as the late Cretaceous (>65 mya). Some modern colubrid, elapid, and viperid genera appeared by the early Miocene (25 mya), and many fossil snakes from the late Pliocene epoch (3–2 mya) are indistinguishable from their modern counterparts.

Colubroids are marked by three syndromes, all maybe interrelated and associated with an early radiation in open habitats. First, these snakes generally move faster, have more stamina, and exhibit more diverse locomotor repertoires than boas and other basal macrostomatans; this breadth is correlated with more diverse hunting and defensive tactics in colubroids as well as with changes in their musculoskeletal systems. Second, further loosening of the jaw apparatus in ancestral colubroids allowed the inner rows of teeth (those on the palatoptery-goid bar) to play an ever-greater role in prey ingestion, which in turn liberated the maxillary bones for new feeding and defensive roles. Consequently, and in contrast to boas and other more basal snakes, colubroids have specialized maxillary dentitions for coping with a wide array of prey. Only Colubroidea includes species with toothless mouths for eating bird eggs, needlelike teeth for extracting snails from their shells, and other such modifications. Lastly, the origin of venom injection in colubroids profoundly affected the subsequent history of snakes. Venoms not only facilitate otherwise improbable feeding biologies, but through mimetic protection they permit some nonvenomous species to pursue weaponless, highly vulnerable lifestyles (e.g., eating snails or bird eggs) that are not even exhibited by venomous species.

The Mottled Slug-eater (*Sibon longi-frenis*; above), Costa Rica, perhaps mimics the Eyelash Pitviper (*Bothriechis schlegelii*; left), Costa Rica, here shown in open-mouth threat display.

Boas, pythons, and other basal macrostomatans diminished in diversity during the Oligocene and early Miocene, their decline broadly contemporaneous with the rise of colubroids. Whether that pattern reflects some sort of competitive superiority in advanced snakes would be difficult to prove; increasingly seasonal climates might also have played a role in the demise of more basal serpents. In any case, at least some modern members among the latter snakes thrive best in the absence of colubroids. Today diverse assemblages of boas and dwarf boas occur only on Caribbean islands (totaling 25 species) and of pythons only in Australia (totaling 19 species), both regions that lack their behavioral and ecological counterparts, the vipers. The only species-rich group of basal macrostomatans in temperate regions is the fossorial Erycinae (14 species), and those snakes occupy subsurface habitats rarely used by viperids.

Modern snake faunas were largely in place by the beginning of the Pliocene (5 mya), and two subsequent phenomena are noteworthy. First, the Pleistocene epoch (10,000 to 2 million years ago) was distinguished by periodic changes in global climate and the extent of polar ice caps. Ranges of snakes variously shrunk, fragmented, expanded, and coalesced as ice sheets spread and contracted over large portions of continents. Examples of Pleistocene distributional shifts in North America include the movement of Worm Snakes (*Carphophis amoenus*) into Florida (where they are no longer found) during glacial maxima and the expansion of Western Hog-nosed Snakes (*Heterodon nasicus*), Bull Snakes (*Pituophis catenifer sayi*), and other prairie species into the north-central United States since the last glacial retreat (see p. 288). In Europe, the Aesculapian Snake (*Elaphe longissima*) today is found only as far north as central Germany, but within the past seven thousand years its range extended to Denmark.

Second, humans entered the New World from Asia about twelve thousand years ago, and within a few hundred years several dozen species of large mammals and birds went extinct. Although our role in destroying the Pleistocene megafauna was surely substantial, until the twentieth century no New World snake species had disappeared as a result of human activities. By contrast, people migrated to Australia tens of thousands of years ago, and it is possible that a large, slow-moving snake (*Wonambi naracoortensis*) might have been hunted to extinction.

MECHANISMS OF SNAKE EVOLUTION

Changes in developmental timing have played obvious roles in snake evolution. Retention of embryonic or juvenile characteristics, for example, can yield novel behavior and anatomy for individuals of adult size and reproductive status. A brightly colored tail tip enhances caudal luring by Copperheads (*Agkistrodon contortrix*) and many other species, but usually both the color and the practice of luring are lost as soon as young snakes reach a size at which they can switch from frogs and lizards to rodents as prey. Two-striped Forest Pitvipers (*Bothriopsis bilineata*) reach adult lengths of 60–90 cm, but unlike other comparably large arboreal vipers they retain the distinctively marked tail, luring behavior, and diet of juveniles. The divided maxillary bones of Round Island boas, once regarded as a major anatomical novelty, also might result from shifts in the timing of existing developmental mechanisms. Each snake maxilla begins as two centers of bone formation, so the unusual structure in adult bolyeriids probably results when those embryonic segments fail to fuse.

Natural selection, proposed jointly by Charles Darwin and Alfred Wallace, shapes the attributes of populations, because variation among individuals affects their survival and repro-

Two-striped Forest Pitviper (*Bothriopsis bilineata*), Peru; note the distinctive tail color, retained for caudal luring in adults of this species.

duction in particular environments. Natural selection has played a major role in diversification among species and higher taxa, although it acts primarily on individuals within populations. For example, different color morphs of European Adders (*Vipera berus*) incur advantages and disadvantages whose importance varies among habitats. Black Adders are more easily seen by predators than are striped individuals, but the former thermoregulate more effectively and thus grow faster; as a result, black Adders are more numerous in populations for which high body temperature and rapid growth are at a premium. Common Gartersnakes (*Thamnophis sirtalis*) vary within and among populations in their tolerance of the neurotoxic skin secretions of Rough-skinned Newts (*Taricha granulosa*), but individuals that routinely feed on those salamanders have heritable increased resistance to their chemical defense. Patternless Lake Erie Watersnakes (*Nerodia sipedon insularum*) are cryptic on island beaches, whereas cross-banded Northern Watersnakes (*N. s. sipedon*) are difficult to see among rocks and vegetation on the adjacent mainland. Some gene flow occurs as watersnakes migrate between those populations, but birds more frequently eat the conspicuously colored immigrants; each morph is thereby favored in its respective habitat, so that snakes are usually banded on the mainland and unicolored on the islands.

New species are a product of evolution, and the time required for diverging lineages to attain independence undoubtedly varies greatly (see the Appendix for further comments on speciation). Differences in external appearance and lifestyle that are typical of closely related

Neotropical Rattlesnake (*Crotalus durissus*), Costa Rica. Rattlesnakes arose in North America, and only this species is widespread in South America.

species in sympatry can accumulate fairly rapidly, as shown by some snakes in recently glaciated regions and on islands. Massasaugas (*Sistrurus catenatus;* see p. 288) must have colonized the northeastern United States and southern Canada after glaciers receded at the end of the Pleistocene, within roughly the last ten thousand years; Eastern Massasaugas (*S. c. catenatus*) in that area now are larger, have more dorsal blotches and ventral scales, and feed more on frogs and rodents than do Western (*S. c. tergeminus*) and Desert Massasaugas (*S. c. edwardsi*). Likewise, the distinctive Golden Lancehead (*Bothrops insularis*) of Queimada Grande diverged from mainland Jararacas (*B. jararaca*) only about 11,000 years ago, when that island separated from the nearby Brazilian coast. Cycladean Vipers (*Macrovipera schweitzeri*) and Levantine Vipers (*M. lebetina*) are much older, having diverged at least 5 mya when rising Aegean waters isolated the Cyclades Archipelago from mainland Eurasia.

Regardless of their ages, distinctive lineages may later overlap without interbreeding, proof that speciation is complete. Western Diamond-backed Rattlesnakes (*Crotalus atrox*) and Red Diamond Rattlesnakes (*C. ruber*) originated several million years ago, when a northward extension of the Sea of Cortés divided their common ancestral population in what is now southern California. Today those two species are in narrow contact in that area but retain their distinctive appearances and temperaments; Western Diamondbacks usually are quick to rattle, whereas Red Diamond Rattlesnakes generally are mild mannered.

A few major themes pervasively influence snake evolution. Body elongation, limblessness, and chemically mediated behavior characterized the origin of serpents, and those attributes profoundly affect many other aspects of their biology. Within Alethinophidia, a few successive modifications for feeding and locomotion underlie much of the diversification of that large group, with further, often parallel, specializations accompanying radiation of particular lineages. The repeated origin of similar characteristics in independent lineages might reflect similar selective pressures, as well as constraints peculiar to those organisms—after all, the ancestor of Asian flyingsnakes (*Chrysopelea*) had no forelimbs to modify into wings!

Sometimes such convergent similarity involves a particular biological role, such as the snouts used for digging by patch-nosed snakes (*Salvadora*) and Shield-nosed Cobras (*Aspidelaps scutatus*), the salt-excretion glands of Australasian filesnakes (*Acrochordus*) and marine elapids, and the defensive spitting adaptations of the Rinkhals (*Hemachatus haemachatus*) and some other cobras (*Naja*). Snakes in which more diverse series of features have evolved in parallel, reflecting more complex ecological demands, include Emerald Tree Boas (*Corallus caninus*) and Green Tree Pythons (*Morelia viridis*), each with similar overall shape, age-related shifts in coloration, and perching behavior in rain forest canopies; Black Mambas (*Dendroaspis polylepis*) and Taipans (*Oxyuranus scutellatus*), both unusually large, brown, fast-moving rodent-eaters with extremely toxic venoms; and vipers and death adders (*Acanthophis*), stout-bodied, venomous ambush predators with low-energy lifestyles. Convergence in color pattern and antipredator response is especially widespread in serpents, probably because the simplified body plan of these limbless animals affords relatively few options for defense and because mimicry of venomous species is common.

Geographically distinct snake assemblages often resemble each other because of convergence among distantly related serpents, whereas regional and continental differences underscore other, equally interesting, patterns. For example, tropical faunas are predictably similar and contrast with those from elsewhere; they consistently harbor far more total species as well as more big-headed climbers, cylindrical burrowers, and other kinds of serpents that are rare or lacking in temperate regions. Variation among those tropical faunas, however, reflects their divergent evolutionary histories and unique environmental characteristics. Colubrinae, Dipsadinae, and Xenodontinae are primarily North, Central, and South American radiations, respectively, and each of those groups emphasizes different lifestyles. From Mexico to Argentina, ecological communities vary according to the contribution each of those colubrid lineages makes to local snake faunas; South American spiders and centipedes enjoy a paucity of arthropod-eating colubrines (the Sonorini) on their continent, Central American streams and ponds lack the xenodontine watersnakes (*Helicops*) so ubiquitous in the Amazon Basin, and most New World rain forests would be rather different at night without the dipsadine radiation of snail-eaters (*Dipsas*), cat-eyed snakes (*Leptodeira*), and their relatives.

Habitats influence evolutionary vagaries as well. For instance, although neotropical parrotsnakes (*Leptophis*) and African bushsnakes (*Philothamnus*) may launch themselves from tree limbs, only flyingsnakes (*Chrysopelea*) specialize in gliding and parachuting. Asia also harbors the only specialized "flying" frogs and lizards, as well as more species of gliding mammals than elsewhere, a difference correlated with rain forest structure: Asian canopies reach 60 m rather than 30–40 m in height, as in Africa and the neotropics, and far fewer lianas connect trees at lower levels of Asian forests than elsewhere.

Right: Emerald Tree Boa (*Corallus caninus*), South America. Opposite: Green Tree Python (*Morelia viridis*), New Guinea.

The elongate shape common to all snakes might underlie two striking global patterns among squamate reptiles, namely, that there are no large temperate lizards and no large viparous lizards. Temperate lizards usually weigh <50 g and are <40 cm in length (e.g., Long-nosed Leopard Lizards [*Gambelia wislizenii*]), whereas temperate snakes commonly approximate 1 m in length and 500 g in mass (e.g., Timber Rattlesnakes [*Crotalus horridus*]; Black Ratsnakes [*Elaphe o. obsoleta*] in the northeastern United States even exceed 2 m). The largest viviparous lizard (the Solomon Islands Skink [*Corucia zebrata*]) is <1 m long, and most are far smaller, whereas live-bearing Green Anacondas (*Eunectes murinus*) reach at least 6 m, and even temperate North American watersnakes (*Nerodia*) exceed 1 m and are viviparous.

Recall now that live-bearing squamates enhance the developmental rate of their embryos by behavioral thermoregulation and that snakes can do this while otherwise concealed in a

crevice; they simply extend a loop of their body—the part that contains the developing embryos (or a full stomach, when digesting)—into the sun. Perhaps the squat form of most large lizards precludes avoiding predators while thermoregulating, whereas even fairly large snakes hasten the birth of their young or rapidly digest large meals without exposing most of themselves to enemies and high temperatures. Exceptions might prove the rule in this case, because the largest temperate lizard (the European Glass Lizard [*Ophisaurus apodus*]) and the largest temperate viviparous lizard (the Slowworm [*Anguis fragilis*]), both Eurasian anguimorphs, are in fact elongate and limbless.

SNAKES AND OTHERS: PAST, PRESENT, AND FUTURE

Rattlesnakes always evoke strong reactions, ranging from disgust and fear to awe and even admiration. Historically, they were the *cascabeles* (jingle bells) of seventeenth-century Spanish explorers' logs, the sacred messengers of Hopi ceremonies, and the "horned rattlers" of cowboy tales. "Don't Tread on Me" and the Timber Rattlesnake (*Crotalus horridus*) on an American Revolutionary War flag symbolized independence and defense of one's home against a threat. Today these enigmatic, sometimes beautiful creatures are widely regarded as villains and persecuted in ways long since outlawed for those more "lovable." No other North American vertebrates inspire fear so out of proportion to their real threat to human welfare or are so underappreciated relative to their merits. At the same time, more than half of all pitviper species and about a third of all rattlesnake species are currently vulnerable to extinction. The precarious status of these animals is due primarily to habitat destruction and persecution by humans; their salvation ultimately depends on changes in our attitudes toward snakes, and that is a matter of education.

Students easily appreciate vultures and badgers once they learn to value venomous snakes. For this reason I use a Western Rattlesnake (*Crotalus viridis*) with an unusual background as the "flagship" in my natural history courses. Seventeen years ago, I found the rattler basking on the rock wall of a Berkeley day care center, surrounded by curious children. As I gently lifted the velvety brown reptile to safety, its only responses were a sweeping arc of the glistening black tongue and a shifting of coils to maintain balance on my snakehook. Every spring since then, with the local rattler resting or crawling slowly nearby, I lecture on the biological, cultural, and aesthetic aspects of our interactions with these remarkable animals.

Prejudices evaporate as my students come to see rattlesnakes as an accumulation of special adaptations to their environment. Like all snakes, rattlers flow over the ground without limbs; as vipers, rattlesnakes are uniquely able to subdue, swallow, and digest up to a third or more of their annual energy needs in a single prodigious meal. Like all snakes, they rely heavily on chemical signals in hunting for prey and locating mates; as pitvipers, rattlesnakes see the world in a unique mixture of infrared images and more typical vertebrate vision. Finally, blessed with their rattle and a special tail-shaker muscle, these creatures possess one

16

of the most dramatic warning devices in all of nature. Thus introduced to the special qualities of my reptilian companion, even students who have been apprehensive about harmless salamanders clamor for a closer look. With a little luck we find rattlers on class field trips and ponder again the roles of snakes in our lives.

I remind my students that rattlesnakes thrive amid some of the world's most spectacular scenery, and that without these animals our North American landscape would be less vibrant, more tame. If we have a rational perspective on the role of rattlers in natural ecosystems and their threat to humans, finding one can be as precious as seeing a wildcat or raptor. Never really extravagant, these intricately specialized predators are beautiful in subtle, meticulous ways that emerge with lingering contemplation. Idly puzzling over the adaptive significance of jet-black tails, we also discover that the unusual color pattern of Black-tailed Rattlesnakes (*Crotalus molossus*), often likened to beadwork, stems from the fact that each body scale has only a single hue. After meeting Massasaugas (*Sistrurus catenatus*) here and there, I noticed a resemblance to local soil color superimposed on the universal blotched markings. These inconspicuous little rattlesnakes have a distinctly reddish tinge in Arizona, whereas those on the loamy plains of north-central Texas are grayish brown.

Soon after the spring semester ends in Berkeley, exhilarated and exhausted by five months of teaching vertebrate natural history and herpetology, I resume field research. Back in southern Arizona, there is no distinction between "work" and recharging mental batteries; with the welcome respite from phone calls and committee meetings, memos and deadlines, I unwind. Squatting down in desert grassland on a hot June morning, sifting parched volcanic soil through my fingers, I try to imagine the rattlesnakes waiting for rain. Two months from now, violent thunderstorms will transform the desert, and its newly lush surface will swarm with a mind-boggling array of insects, toads, and other creatures. Massasaugas will emerge too, living reminders of the changing fortunes of prairie ecosystems over the past ten thousand years. On a distant horizon, jagged peaks of the Chiricahua Mountains rise abruptly from the surrounding landscape. Up there on some lichen-covered ledge, among the trogons and hummingbirds that traditionally attract nature lovers, a fragment of life's mystery will glint off the yellow backs of a pair of mating Blacktails. Surely all of this is worth saving.

The outlook for pitvipers in the twenty-first century is mixed. Timber Rattlesnakes face extinction over significant parts of their range, and some Oklahomans and Texans make a gruesome, public spectacle of slaughtering Western Diamondbacks (*Crotalus atrox*). If the more dire predictions about global warming prove true, isolated patches of prairie will dry up like water drops on a hot griddle, and Massasaugas, dependent on seasonally moist grasslands, will disappear from much of the southwestern portion of their range. Conversely, some populations of Arizona and New Mexico Ridge-nosed Rattlesnakes (*C. willardi*) might survive climate change in higher, forested parts of their distribution and later reinvade lower elevations. Like Western Rattlesnakes in Berkeley and Jararacas (*Bothrops jararaca*) in São Paulo, a few species of large venomous snakes may even thrive in urban areas.

Earlier inhabitants of the New World were well aware of rattlesnakes, so foreboding and yet somehow attractive. Pre-Colombian peoples generally tolerated rattlers, although attitudes varied from aversion to reverence, and aboriginal lore encompassed accurate natural history as well as mythological exploits. Often rattlesnakes were regarded as agents of justice, avenging the deaths of their own relatives and punishing religious infractions by humans.

Connections between rattlers and weather also prevailed in Native American cosmology, perhaps because a serpent's strike resembles a lightning bolt. With increased understanding, we too might appreciate venomous snakes as living symbols of wilderness; perhaps we can even afford them a place in modern landscapes. I look forward to a day when park rangers will proudly guide visitors to a Timber Rattlesnake den, when nature tourists will travel to Arizona or Costa Rica with hopes of seeing several species of pitvipers on one trip.

Snakes play diverse roles in many terrestrial, aquatic, and marine ecosystems, as well as in human culture. All serpents are directly involved with at least a few other animals as predators and prey, and most snakes probably interact with numerous other organisms as competitors and mutualists. Human relationships with serpents undoubtedly have preceded written history by several million years, extending culturally and perhaps genetically back through our common ancestors with the African apes and beyond. Now, in the late twentieth century, people variously treat snakes as vermin; commodities in the food, leather, and pet trades; components of religious ceremonies and wilderness appreciation; and subjects of scientific research. Our growing population and increasing degradation of the environment threaten the persistence of many forms of life, and some attributes of snakes make them especially vulnerable to extinction. In this chapter I discuss relationships between snakes and other organisms, including humans, and then explore prospects for our joint future.

SNAKES AS ECOLOGICAL ASSOCIATES

Snakes live alongside many other organisms, with which their interactions range from life-and-death to insignificant. Massasaugas (*Sistrurus catenatus;* see the Special Topic in this chapter), for example, take refuge in burrows with centipedes, lizards, and small mice—all part of their diet—as well as with other invertebrates, snakes, and rodents that are neither predators nor prey. Many snakes capitalize on environmental modifications by other organisms; they lay eggs in termite mounds and abandoned rodent burrows, they crawl in tunnels formed by rotting tree roots or constructed by other animals. Beyond such obvious and straightforward associations, however, serpents also influence the structure, function, and evolution of ecological communities in numerous less direct ways.

Some snakes potentially compete with many raptors and mammalian carnivores for food, and because serpents can eat relatively huge prey, their ecological analogues among other predators often are surprisingly large. Brown Vinesnakes (*Oxybelis aeneus*), Boa Constrictors (*Boa constrictor*), Groove-billed Anis (*Crotaphagus sulcirostris*), Double-toothed Kites (*Harpagus bidentatus*), and Jaguars (*Panthera onca*) all sometimes eat Green Iguanas (*Iguana iguana*); although only the big cats and boas take the largest lizards, all these predators feed on the same populations. Likewise, by preying on ungulates, African Rock Pythons (*Python sebae*) and Asian Rock Pythons (*P. molurus*) overlap the diets of African Lions (*Panthera leo*) and Tigers (*P. tigris*), respectively; most vipers take the same rodents eaten by sympatric hawks, owls, foxes, weasels, and other predators.

Snakes have affected as well as been influenced by some other creatures through predator-prey races—long-term processes in which a prey population evolves novel protection, certain

"PICTURES OF THE SUN": MASSASAUGA RATTLESNAKES (*SISTRURUS CATENATUS*) AND CONSERVATION

A few miles south of the Chiricahua Mountains, sprawling grasslands and small, seasonally wet depressions of Arizona's San Bernardino Valley form an isolated outpost for the Massasauga (*Sistrurus catenatus*), a rattlesnake I first studied as a teenager in Kansas and Texas. Less than a meter long and so cryptic that an adult can be invisible from a few centimeters away, the Massasauga was known by some Native Americans as "pictures of the sun" for the large, silver-edged spots that decorate its back. The range of this enigmatic little snake stretches from Arizona to New York, an area so large it only can be visualized with a map or from many kilometers above the earth. Nevertheless, the Massasauga is a candidate for the United States Endangered Species List and in real need of protection; its fragile status, unlike that of some other venomous snakes, results from overall environmental degradation rather than direct persecution.

Transcontinental snake species often occupy several habitats, but Massasaugas everywhere inhabit seasonally moist grasslands and forest edges. This unusual distribution also characterizes the Western Hognosed Snake (*Heterodon nasicus*) as well as several other animals and plants, inspiring discovery by ecologists in the 1930s of a "postglacial peninsula" of prairie floras and faunas that extended into northeastern North America as Pleistocene ice sheets retreated. Sadly, because grasslands have usually been converted to agriculture and underemphasized by conservationists in the United States, most Massasauga populations are now small and isolated. Massasaugas in the eastern parts of the species range migrate seasonally between wet bogs and dry forest edges, and effective conservation and management must encompass both habitats.

Until recently, almost nothing has been known on which to base the conservation of desert-grassland Massasaugas in the southwestern United States and northern Mexico. At night these rattlesnakes cross highways in Arizona and New Mexico a few times a year, and several have been spotted by day at the mouths of rodent burrows. By examining museum specimens (mostly road kills), I learned that southwestern Massasaugas eat centipedes, small lizards, and mice. They produce smaller litters and smaller young than the larger Massasaugas of the northeastern United States and southern Canada. Judging from documented records, Massasaugas have a much more limited distribution in the Southwest than typically portrayed.

Northeastern Massasaugas are associated with wet bottomlands and hibernate in water-filled crayfish burrows, evidently sitting for weeks with their heads extended vertically above the water level. Desert Massasaugas live in areas where only a few thousand years ago large playas, or shallow lakes, were prevalent. My captives spent hibernation in temperature-controlled chambers and, unlike other Arizona rattlesnakes in the same room, sometimes rested for days in their water bowls with foreparts extended vertically. Their behavior is reminiscent of hibernating northeastern Massasaugas, and I suspect that those snakes actually inhabit the ancestral niche, which they had reoccupied after the glaciers retreated a few thousand years ago.

Now Massasaugas barely hang on in the arid Southwest, with most of their former range converted from grassland to agriculture or invaded by mesquite thickets as a result of other environmental insults. Perhaps the little rattlers persist during dry weather by taking refuge in the burrows of Banner-tailed Kangaroo Rats (*Dipodomys spectabilis*), a microhabitat with little year-round variation in temperature and humidity. Those rodents are far too large for desert-grassland Massasaugas to eat, but their tunnels also

A Massasauga Rattlesnake (*Sistrurus catenatus*) beside the mound of a Banner-tailed Kangaroo Rat (*Dipodomys spectabilis*), Arizona.

harbor pocket mice (*Perognathus*), Lesser Earless Lizards (*Holbrookia maculata*), and giant centipedes (*Scolopendra*)—all known prey of the small rattlesnakes.

As I finished this book, a doctoral student initiated detailed research on Arizona Massasaugas. Aided by radiotelemetry, Andrew Holycross

may determine if fragile, moist bottomland areas are crucial to their survival and help develop methods for assessing the presence or absence of this species in areas it has historically occupied. If the presence of these snakes in arid regions is tied to humid rodent burrows, understanding that relationship will

be crucial to effective conservation and management. Information on the Massasauga and other species in ephemeral habitats is especially urgent if current predictions of global warming prove accurate.

Dumeril's Boa (*Boa dumerili*), Madagascar. Like many snakes, this species serves as predator and prey of other local vertebrates.

Right: Orange-naped Snake (*Furina ornata*), Australia, during defensive threat display. Opposite: Common Flap-footed Lizard (*Pygopus lepidopodus*), Australia, a nonvenomous species whose defensive display resembles that of the Orange-naped Snake and certain other dangerous elapids.

predators then specialize in circumventing that defense, and so forth. Among many likely examples, Asian sunbeam snakes (*Xenopeltis*), mock vipers (*Psammodynastes*), and several other groups of serpents have independently evolved modified teeth that overcome the slippery scales of skinks. North American ratsnakes (*Elaphe*) are specialized for climbing and thereby capture baby birds, but Red Cockaded Woodpeckers (*Picoides borealis*) respond by pecking holes that ring their nest trees with serpent-proof resin; the devastation of Guam's previously enemy-free avifauna by introduced Brown Treesnakes (*Boiga irregularis;* see p. 179) indirectly implies that elsewhere, in the intruder's natural range, birds have means of avoiding nest predators. Among snakes, venoms probably often differ in ways that facilitate particular dietary themes, and their components are countered by biochemical responses in some potential predators and prey. Common Kingsnakes (*Lampropeltis getula*) and Mussuranas (*Clelia clelia*) are immune to the venoms of pitvipers, on which they sometimes feed, whereas Neotropical Opossums (*Didelphis marsupialis*) are occasionally eaten by Terciopelos (*Bothrops asper*) but are surprisingly resistant to their bites.

Venomous snakes serve as models for visual, chemical, and auditory mimicry, not only by other serpents but by a wide assortment of invertebrates and other vertebrates as well—a phenomenon that perhaps reflects the ease with which snakes can be crudely imitated by simple changes in other organisms. The larva of a Central American sphingid moth (*Hemeroplanes triptolemus*) looks rather similar to the head of a palm pitviper (*Bothriechis*), caterpillars of the Malayan Hawk Moth (*Panacra mydon*) bear a striking resemblance to some tree pitvipers (*Trimeresurus*), and the pupa of a Costa Rican nymphalid butterfly (*Dynaster darius*) looks amazingly like a small Terciopelo (*Bothrops asper*). In each of those insects the similarity to a viper extends to fake eyes, a "postocular" dark stripe, and crudely accurate head scalation.

Snake stones, India, used for fertility offerings.

Siberian Chipmunks (*Eutamias sibiricus*) anoint themselves with the skin secretions of rat-snakes (*Elaphe*), thereby using foreign odors to repel snakes, shrews, and mammalian carnivores that prey on small rodents. Burrowing Owls (*Athene cunicularia*) hiss loudly when disturbed, sounding remarkably like the rattlesnakes with which they sometimes share an underground refuge.

SNAKES AND PRIMATES

Humans have always regarded snakes with a mixture of inquisitiveness and fear, of awe and revulsion, but whether these conflicting tendencies reflect genetically based predispositions or cultural traditions has long been controversial. In laboratory experiments and field observations, primate responses to snakes range from instant terror through curiosity and mobbing to immediate consumption; that some naive nonhuman primates react negatively to snakes suggests an inborn response with obvious survival value. Human reactions to serpents are influenced by early experience and the behavior of adults; nonetheless, people in Western societies acquire ophidiophobia (from the Greek word *ophidion,* "a snake") much more readily than fear of guns or other modern objects more dangerously prevalent than snakes—a tendency that likewise suggests a retained, adaptive response to elongate creatures.

Like the bird-watchers described in the introductory essay to Chapter 10, even some nature lovers have second thoughts when it comes to serpents. Albert Schweitzer, renowned for his reverence for "all" life, kept a rifle to shoot snakes. Visitors surveyed at a national park in the eastern United States overwhelmingly supported the protection of "all wildlife" but were much less enthusiastic about a later question as to whether snakes in the park should be killed. On the other hand, some individuals and even entire cultures (e.g., inhabitants of New

Dreamtime painting by the contemporary Australian aborigine Brogas Tjapanardi. (Fogden collection)

Guinea rain forests) treat snakes with diffidence or even respect. Whatever the underlying mechanisms, attitudes toward snakes are steeped in an old and complex history that predates our origin as humans.

The wide range of responses by primates is understandable, given that snakes can represent extreme danger, valuable food, or uncertainty. Today serpents serve as predators and prey for creatures as different as tarsiers, lemurs, bush babies, monkeys, and humans. Prehistoric kitchen middens in Florida contained the remains of Eastern Diamond-backed Rattlesnakes (*Crotalus adamanteus*); modern subsistence hunters regularly eat African Rock Pythons (*Python sebae*); and probably some people have always eaten serpents, even venomous species that otherwise posed a deadly threat. But beyond these associations with fear and hunger, limbless reptiles are linked with emotional reflection and symbolism. Serpents appear in some of the earliest known art, at least thirty thousand years old, and over the intervening millennia their elongate form has been carved on bone implements, chipped and painted on the walls of rock shelters, sculpted on marble porticoes, and deified in countless legends and ceremonies. Serpents represented peace and healing to the early Greeks and Romans, and a pair of entwined snakes still symbolizes the medical profession. Brazilian money accurately depicts a Mussurana (*Clelia clelia*) eating a Jararaca (*Bothrops jararaca*), and other snakes are emblematized in countless equally mundane ways.

Snakes are prominent in the cosmology of many paleolithic, aboriginal, and modern cultures, although the nature of human spiritual relationships with snakes varies tremendously. In addition to the familiar Judaeo-Christian tale of Eve's temptation by a serpent in the Garden of Eden, snakes commonly appear in the belief systems of Eastern religions; they dominate the drug-induced visions of some South American Indians and play a major role in the creation legends of Australian aborigines. Even today, several hundred people in the

Indian Cobra (*Naja naja*) with a snake charmer. The snake rises and sways to accompany the man's movements and is not responding to the music.

southeastern United States handle Copperheads (*Agkistrodon contortrix*) and Timber Rattle-snakes (*Crotalus horridus*) as part of fundamentalist Christian services, and entire tribal groups in Asia work as snake charmers and snake catchers.

Serpents are sometimes treated positively in contemporary human cultures. Madagascan Ground Boas (*Boa madagascariensis*), Indigo Snakes (*Drymarchon corais*), and some other harmless species are tolerated by rural people because they eat rodents. In more urban settings, reptile houses are among the most popular exhibits in zoos, and visitors routinely crowd around the snake cages; ice hockey, baseball, and football teams are named after vipers, as is an upscale nightclub in Los Angeles. However, in Western societies serpents are largely used for economic benefit, in ways that bode poorly for their wild populations. Leather goods from snake skins sell for hundreds of millions of dollars annually on a global basis, and even the pet trade in serpents has reached multimillion dollar levels in Europe and the United States; in one recent year almost 200 million dollars in snake products and live snakes were imported into this country. Despite claims of promoting public safety and venom research, rattlesnake roundups in the southern United States are fueled mainly by financial considerations, bringing hundreds of thousands of dollars into each of several small towns in a single weekend.

Members of the Irula tribe in India support themselves by collecting snakes for venom research and by catching rats that plague farmers.

SNAKES AND SCIENCE

Systematics defines the fundamental units and relationships among living things, and natural history chronicles the lifestyles of organisms in relation to their environments. These twin pillars of classical biology—enhanced by radiotelemetry, DNA sequencing, and other sophisticated techniques—lay the foundation for modern ecological and evolutionary studies, as well as for nature appreciation and conservation. Without studies of distribution and variation, we would be unable to distinguish among kinds of organisms, and we would have no field guides; without detailed natural history, we could not marvel at spectacular adaptations to particular lifestyles, and wildlife management would run the risk of grave errors. Studies of snake biology thus play important roles in understanding those animals and in addressing interesting questions with broad implications for science and society.

Comments on the form and habits of snakes appeared early in recorded history, but serious studies of anatomy and natural history began only about two hundred years ago. Many of the first scientific reference works were based on one or a few specimens from widely scattered localities, and what little was mentioned about habits often was vague or based on unreliable hearsay—provoking Spencer Fullerton Baird of the Smithsonian Institution to complain in 1853 that "workers all carefully avoid the subject of Ophidians, each waiting for the others to

Bushmaster (*Lachesis muta*), Costa Rica, being captured by the author for studies of behavior and ecology.

make the first step." Extensive studies of variation within and among populations started in the first decade of this century, and long-term field research on marked individuals began shortly thereafter. Serpents attracted increasing attention over the next fifty years, although until recently many scientists still regarded them as uninteresting; even the first large symposium about these animals, in 1975, had a distinctly pioneering flavor. Now hundreds of lay naturalists and professionals study everything from natural selection and chemosensory behavior to the molecular systematics of higher taxa. As examples of the range of backgrounds and accomplishments encompassed by snake biologists, I highlight five individuals whose significant contributions perhaps have not been as widely recognized as they deserve to be.

Frank Wall (1868–1950), a British medical officer, was born in Sri Lanka, schooled in England, and then stationed in India and nearby countries. Colonel Wall combined a zeal for collecting with meticulous natural history observations, and he once named a snake after himself (*Bungarus walli,* now a synonym of the Indian Krait [*B. caeruleus*])—something no other herpetologist has ever done! Wall's most important contribution was "A Popular Treatise on the Common Indian Snakes," published between 1905 and 1919, in twenty-nine parts in *Journal of the Bombay Natural History Society*. These accounts, accompanied by fine color plates, include descriptions of scalation and color pattern as well as extensive firsthand notes on food, reproduction, and defensive behavior. Colonel Wall's more than two hundred publications, including several books, remain essential references on the biology of Asian snakes.

Henry S. Fitch (born 1909) received his Ph.D. with Joseph R. Grinnell at Berkeley in the late 1930s, and in 1995 he presented a conference paper on forty-seven years of snake population data from the University of Kansas Natural History Reservation. Fitch's dissertation was the first detailed study of species relationships among western North American gartersnakes (*Thamnophis*), a truly formidable task, and much later he undertook revisionary work on another difficult group, the Mexican *Anolis* lizards; above all else, however, Fitch exemplifies outstanding field natural history. He has authored nearly two hundred publications on organisms as diverse as spiders, snails, hawks, and rodents, seeking to understand their home ranges, food habits, and geographic patterns in life history. Fitch's central focus long has been snakes, and among his many monographic treatments of individual species, a 1960 opus on the Copperhead (*Agkistrodon contortrix*) is without peer in the literature on snake ecology.

Roger Conant (born 1909) spent his professional career in zoos, first in Toledo and later in Philadelphia; eventually serving as director of the latter institution, he has had a remarkable influence on North American herpetology. Conant's field guide to the amphibians and reptiles of the eastern United States, first published in 1958, provoked an explosion of interest in the topic and brought him into contact with countless young people and amateurs. Relentlessly encouraging, he answered thousands of letters, patiently explaining careers in herpetology and ferreting out details of his readers' new discoveries for subsequent editions of the field guide. Along the way, Conant conducted many scholarly studies of snake systematics, culminating in a monograph on Mexican watersnakes (*Nerodia*) and earning him a doctorate in 1971. After "retiring," he returned to collaboration with his longtime friend Howard K. Gloyd, and the two completed their massive book on the pitviper genus *Agkistrodon* and its taxonomic allies.

With no formal graduate training, Hymen Marx (born 1925) started his career as an assistant to Karl P. Schmidt at Chicago's Field Museum of Natural History. After several shorter papers, he published with fellow curator Robert F. Inger a landmark systematic study of reedsnakes (*Calamaria*). In the 1960s a live Macmahon's Viper (*Eristicophis macmahoni*) at nearby Brookfield Zoo provoked Marx's interest in viper relationships, an encounter that led over the next twenty-five years to a series of highly original, influential papers on snake evolution. Along with Brookfield's director, George B. Rabb, he first recognized the unusual attributes of a poorly known Tanzanian bush viper and established the genus *Adenorhinos* for what had earlier been described as *Atheris barbouri*. Others with special expertise enlisted in the wide-ranging snake phylogenetics project from time to time, with Marx providing the unifying perspective. Fea's Viper (*Azemiops feae*) held a special intrigue for Marx, and well into the viper research, he delighted in finding the eleventh known specimen of that extremely rare snake—within an old jar of unsorted Chinese frogs, in his own museum!

By now the reader might guess that only men study snakes, a reasonably accurate generalization until recently. Joan B. Procter (1897–1931) was an important exception and probably would have emerged as a major figure in herpetology had her life not been cut short by illness. She was an assistant at the British Museum (Natural History) and later Curator of Reptiles at the Zoological Society of London; her scientific papers include a detailed study of color pattern variation in the Common Lancehead (*Bothrops atrox*). Procter corresponded with a young Karl P. Schmidt soon after he arrived in Chicago, confiding that she had left the British Museum because conditions were unfavorable for a woman. Her later publications demonstrate a rich appreciation for reptiles as living creatures, and I imagine she would have been especially pleased that today field biologists use color patterns to identify free-living, individual Asian Rock Pythons (*Python molurus*), Ridge-nosed Rattlesnakes (*Crotalus willardi*), and other serpents. She surely would have enjoyed the emerging picture of behavioral complexity in snakes, and the fact that even venomous species are no longer studied only by men. Today a bust of Joan Procter graces the London Zoo's Reptile House.

CONSERVATION STATUS AND MANAGEMENT

Snakes have been killed by humans and other primates for millions of years, but with few exceptions, we have not caused the extinction of entire species of serpents until the past couple of centuries. About 200 species of snakes are currently listed as threatened or endangered or are otherwise afforded special protective status, but that number vastly underestimates the proportion of the world's snake fauna vulnerable to extinction in the near future. Beyond the fact that dozens of species remain undiscovered or known from only a single specimen (e.g., 7 species in Sri Lanka alone), the conservation status of most snakes has not been critically examined. In Europe, where the snake fauna is small and well studied, 19 of 27 species need active management; by contrast, the conservation status of almost all of South America's hundreds of species of snakes is unknown. Worldwide, only about 12 species of pitvipers are formally singled out for protection, but sustained persecution, habitat destruction, and intrinsically small distributions (e.g., the Aruba Island Rattlesnake [*Crotalus unicolor*] and other island populations) imply that perhaps 60 percent or more of the 157 species in that group are now threatened or endangered. In fact, many snakes might be especially vulnerable to extinction because of their slow growth rates, small clutch and litter sizes, and ordinarily high adult survivorship.

For most serpents currently listed as threatened or endangered, habitat destruction and fragmentation are undoubtedly the most severe and frequent culprits. Rattlesnake roundups and other forms of wanton killing, the leather trade, and the pet trade result in the deaths of millions of snakes annually, but the impact of those activities is generally focused on a few widespread species. Snakes that inhabit the North American Great Plains and other grasslands have sustained massive habitat loss during the past century, with some species already restricted to small remnant populations. Tropical forests everywhere are being destroyed at accelerating rates, and those extraordinarily rich habitats—especially in Madagascar, southeastern Asia (including the nearby archipelagoes), and South America—include more severely endangered snake species than anywhere else. Some temperate forests also are at risk, their destruction causing the precarious status of Louisiana Pinesnakes (*Pituophis ruthveni*) and several species of Eurasian vipers (*Vipera*), among others.

Often snake habitats are destroyed on a large scale—essentially cleared of all native plants and animals for agriculture, water impoundments, and other human activities—but sometimes the damage is more localized. Removal of sandstones for decorative gardens threatens the Broad-headed Snake (*Hoplocephalus bungaroides*), an Australian elapid restricted to rocky outcrops; in North America, unscrupulous collectors destroy many granite flakes and cap

Black-speckled Palm Pitviper (*Bothriechis nigroviridis*), Costa Rica. These snakes, and many other tropical highland species with small geographic distributions, are threatened with extinction as their habitat is destroyed.

rocks in their search for California Mountain Kingsnakes (*Lampropeltis zonata*), Rosy Boas (*Charina trivirgata*), and other commercially valuable species. Some paved highways pose particularly severe threats to snake populations, by interrupting travel routes and causing heavy mortality. On just one 44-km stretch of roadway in southern Arizona, traffic kills between 500 and 1,000 snakes each year, which amounts to the annual removal of more than a square kilometer of snake populations.

Invading species and other environmental modifications pose extremely serious threats to snake populations throughout the world. Lowered water tables, invading shrubby vegetation, and the resulting demise of native grasslands in southern Arizona have changed the distributions of some snakes and severely diminished the ranges of others. In the Indian Ocean, Round Island Ground Boas (*Casarea dussumieri*) formerly inhabited nearby Mauritius as well, but they disappeared from that larger island after rats arrived on shipping; likewise, deliberately introduced Mongoose (*Herpestes auropunctatus*) probably destroyed several West Indian snake populations. Fire Ants (*Solenopsis invicta*) and Nine-banded Armadillos (*Dasypus novemcinctus*), recent invaders of the southeastern United States, have annihilated many invertebrates and small reptiles, including snakes.

Circumstantial evidence illustrates how toxic substances are among the most insidious threats to snakes, and perhaps to all life on earth. Of two adjacent river valleys in southern Texas, one has never been subjected to extensive agricultural pesticides, whereas the other has

Gray-banded Kingsnake (*Lampropeltis alterna*), Texas. Like some other harmless species with beautiful colors, these snakes are popular pets.

been repeatedly treated with various highly toxic compounds. The first valley harbors a typical snake fauna, while the second lacks ratsnakes (*Elaphe*) and other egg-laying species; given their proven effects on reproduction in other vertebrates, a reasonable guess is that pesticides eliminated those Texas snakes.

An environmental catastrophe of global proportions is fast upon us, not just looming on some distant horizon. More than six billion people will burden the earth by the year 2000, and the irrevocable loss of many plant and animal species is under way now; new species of vertebrates are literally going extinct before their formal descriptions have been published. Simply stated, unlimited reproduction, technology, and greed have caused this biodiversity crisis, although a more detailed assessment would include complex social, political, and economic forces; only urgent, difficult measures will bring this tragic process under control. We must rapidly increase our understanding of nature and preserve chunks of landscape with as much as possible of what remains. We must drastically reduce pollution and other environmentally unsound by-products of technology, for which those of us in the large, excessively developed and consumptive countries are most responsible. We must eliminate overexploitation and special persecution of wild populations, increasingly serious problems for serpents; we need to better understand the ecology of highly altered environments, as these increasingly will provide most of the available wildlife habitat (someone should study urban rattlesnakes). All possible compassion for people cannot erase one final, overarching reality—that ever more of us are depleting increasingly scarce resources and further reducing the quality of life; we must control human population growth.

Our environmental crisis will not be solved quickly, and not everyone will like snakes; I prefer optimism to despair, however, and there are grounds for hope. In the 1930s, Boy Scouts

Round Island Ground Boa (*Casarea dussumieri*). Found on only one small island in the Indian Ocean, this species is threatened by habitat changes and introduced animals.

received bronze "Junior Conservationist" medals for killing Northern Watersnakes (*Nerodia sipedon*), believing they were thereby protecting trout; now there is a fine children's book about those much-maligned serpents. Today Arizona staunchly protects all of its small rattlesnake species, and the plight of the Timber Rattlesnake (*Crotalus horridus*) in eastern North America is gaining widespread recognition. As the millennium approaches, Uganda's Bwindi-Impenetrable Forest is a national park, and Brazilian wilderness reserves for indigenous peoples might protect a large portion of Amazonian snake diversity. The Nature Conservancy recently purchased the Gray Ranch, including a mountain range inhabited by the endangered New Mexico Ridge-nosed Rattlesnake (*C. willardi obscurus*), then sold it to a cowboy poet. Drum Hadley and his Animas Foundation are combining careful ranching with conservation on that spectacular property, hectare for hectare the biologically richest piece of land in the United States.

As so eloquently stated by the Senegalese conservationist Baba Dioum, we must learn about, understand, and appreciate nature in order to save her. If you agree with me that the world is richer for its serpents and other unpopular creatures, then use your money, your time, and your votes to encourage biodiversity research, habitat conservation, and population control. We are all teachers in one sense or another, whether in classrooms or over the backyard fence, so take what you know and introduce others to the marvelous world of snakes.

Rock Rattlesnake (*Crotalus lepidus*),
New Mexico.

There was no climate or soil . . . equal to that of
Arizona. . . . It is my land, my home, my father's
land. . . . I want to spend my last days there, and to
be buried among those mountains.

Geronimo

EPILOGUE: WHY SNAKES?

Natural history offers panoramas of space and time, an enlightening perspective on our own
lives. With this inspiration, field biology blurs the distinction between science and art, between what we know and how we feel. I am crouched at the front of a cave in the Chiricahua Mountains, sheltered from an afternoon cloudburst and wondering what Geronimo
thought about rattlesnakes. A small, dark pictograph on the back wall depicts a mounted
crown dancer or mountain spirit, an Apache symbol of weather and power; thunder shakes
the earth in long, low rolls, as if this fractured continental spine were rumbling in its sleep.
Looking up the forested, boulder-strewn slopes of Cave Creek Canyon, I remember that the
Sierra Madre Occidental lies across the nearby border with Mexico, and that over the centuries indigenous peoples have coexisted here with conquistadors and priests, drug dealers and
ecotourists. We have profoundly altered this region, all of us, as aptly conveyed by book
titles like *Cycles of Conquest* and *The Changing Mile*. Nonetheless, eight species of pitvipers
still live within 50 km of where I sit; usually unseen by humans, they are resting under li-
chenous granite slabs and rattling from talus slides, lying in ambush by streamside trails and
patrolling sandy desert flats. Whether or not Geronimo thought much about rattlesnakes, I
certainly have.

Despite thousands of years of human impact, these Sierra Madrean uplands and surrounding desert mix tropical exuberance with the low-key struggles of open country life—
and such distinctive attributes surely influence our individual worldviews. Deprived of lush
equatorial spectacles of life and death, would Gabriel García Márquez and other Latin
American authors have gained their distinctive style of magical realism? Without his up-
bringing in arid Algeria, would Albert Camus have found happiness in stark, perhaps ultimately unanswerable questions? Southern Arizona and New Mexico combine those usually
disparate landscapes, and I would rather confront life's riddles here than anywhere else,
studying nature with people I admire and care about. Wondering what it's like to be a serpent takes me to the edge of the explainable, and working with rattlesnakes, I often have
pondered the question from Norman Maclean that began this book. Why am I preoccupied
with snakes, especially venomous species? And why should others care about them?

At the time of our dinner, I uneasily brushed off Maclean's query with something about addressing important scientific problems by studying those creatures. More than a decade later, I realize that a more honest and complete answer would have to include parents who bought me books, took me to countless zoos and museums, and otherwise encouraged my interests. Surely the threads of my fascination with reptiles reach back into childhood, then fade to the Pliocene and beyond—our ancestors dodged pythons and large vipers on African savannas, and millions of years earlier we were bug-eyed prosimians, chomping off the heads of ancestral Asian coralsnakes. There are some specifics too, which were less obviously pertinent in 1983 than they are today. At that time I had hoped to write about working as an army medic among the guys who had made it back from Vietnam, chance assignments having spared me, and about how their shattered limbs and faces still sadden me. I wanted to articulate vivid memories of automobile accidents and fatal house fires; of shooting and stabbing victims in dark, smoky bars; and especially of the faintly sour milk smell of an unconscious child, of gently puffing into her mouth and pushing on her little chest, and finally collapsing in tears against a tile wall when the emergency room physician said she was gone. I imagined that my prose might honor a first lover and an influential teacher who each had been murdered, and that I would summon hope and gratitude for the squalling, healthy babies I had delivered. How obvious now that venomous serpents have been personal icons of danger, of life and death—as if in that crystalline moment when the fangs pierce another creature, I might finally understand my own fears and losses.

An answer to Maclean's question must also encompass the sheer joys of friendship and shared discovery, with which I am blessed many times over. David Chiszar, David Duvall, and I are lurching down a narrow ranch road in Wyoming's Haystack Mountains, intent on a distant rise. The two Daves are old buddies, well known for their clever experimental studies of reptile behavior, and now, full of pancakes and coffee, we banter easily about academic politics, the merits of field versus lab research, and pitviper biology. On this sunny morning after the first cold snap of September, several bloated, lightning-scorched cows remind us that the climate at 42° N and 2,100 m is hardly benign. Soon we are walking through rolling grassland and sagebrush, where Duvall and his graduate students are studying the behavioral ecology of Prairie Rattlesnakes (*Crotalus v. viridis*). Each snake we encounter is crawling toward a hibernaculum in an east-facing canyon, and eventually their collective routes lead us to other rattlers clustered around a small cleft in the rocks. In May these same snakes had fanned out over the surrounding plains, up to several kilometers from the den, and then spent the next four months hunting deer mice (*Peromyscus*) and seeking mates. The pregnant females bred last summer, and this year they retreated instead to a nearby side canyon where rocks for basking and shelter are plentiful. Today there are so many adults with newborn young that we walk gingerly to avoid stepping on little rattlers. My lifelong interest in snakes notwithstanding, the visual parallel between this birthing rookery and colonially nesting birds amazes me. We are wind-weary but still excited on the drive back to Laramie, and I will forever associate the two Daves with memories of those rattlesnakes and their litters.

Of course, art is in the eye of the beholder, so why should others appreciate serpents, especially venomous species? Snakes are natural puzzles, suggestive of things that haunt and inspire us; they have laconic expressions, subtle at best, with lidless eyes once dubbed "peepholes into hell." Snakes flow unhindered, as if free of friction or other restraint and thus beyond our control. We often regard our own natural functions with ignorance and shame, but snakes are overt tubes and thus expose the profound role of peristalsis in everyday life; they overpower and swallow other animals intact, sometimes passing prey into the stomach as visibly struggling lumps. Snakes are sinuous, supple eroticism exaggerated by paired, intricately ornamented sex organs. Their social behavior relies largely on tactile and chemical signals; it's as if I were blindfolded in an Andean marketplace yet able to distinguish dozens of friends amidst overripe fruit, roasting corn, urine in the gutters, and a seething mass of damp wool ponchos.

Even venomous snakes embody scaly symmetry and harmonious colors; many of these animals are profoundly cryptic and stingy with movement, so watching them teaches patience. Venomous serpents inspire outlandish legends and rational fears—after all, bites from Terciopelos (*Bothrops asper*) cause horrific symptoms, and a bush viper (*Atheris*) might well kill my Ugandan friend. Walking in their habitats fosters a special kind of respect for nature, an urgency that both humbles and enlivens us. Perhaps most important, with venomous snakes we contemplate violence and mortality without implications of real evil, devoid of anthropocentric traps laid by fur, feathers, and facial expressions; with pitvipers and their unusual infrared vision, we arrive at the very cusp of mystery, the illusive but tantalizing limits on empathy. And when we know snakes as living creatures in nature, venomous or otherwise, they can take on grander meaning. Sidewinders (*Crotalus cerastes*) and Peringuey's Adders (*Bitis peringueyi*) epitomize the demands of life on shifting desert sands, and flying-snakes (*Chrysopelea*) reflect the unique botanical characteristics of Asian rain forests. Each of the more than 2,700 species of snakes embodies special relationships with its environment, and the earth would be poorer without them.

Appreciating nature comes easily for field biologists; involved to the hilt, we rarely distinguish between work and play or debate the merits of science versus art. Four decades after Joe T. Marshall's work in the same region, Kelly and I are camped on a woodland flank of the Huachuca Mountains. Our study canyon is dotted with three species of oaks (*Quercus*), Mexican Pinyon (*Pinus cembroides*), Alligator Juniper (*Juniperus deppeana*), and manzanita (*Arctostaphylos*). These last plants are among my favorites, effervescent shrubs with countless small, lime-colored leaves and smooth, carmine bark. Working here over the last two summers, Fred Wilson, Dave Hardy, and I learned that pregnant Ridge-nosed Rattlesnakes (*Crotalus willardi*) favored particular rock shelters. Other telemetered individuals waited for the rains within bunchgrass clumps, and now, surprised by that finding, we explore the role of microclimate in their ecology. With a special probe I measure humidity at about 450 surface sites and Ridgenose-sized crevices, discovering that grass clumps are among the dampest places in the habitat of these small snakes. Intent on such matters, we pass mornings and late afternoons gathering data, breaking for lunch and a siesta in the midday heat. Evenings here are wild and magical, all the more so for things we have seen; the remains of a Mule Deer (*Odocoileus hemionus*) killed by a Mountain Lion (*Felis concolor*) lie a few dozen meters from our tent, and last night a herd of Collared Peccaries (*Tayassu tajacu*) snuffled so close we could see their piglike silhouettes in the moonlight. I fall asleep early, idly pondering whether predators ever catch rattlesnakes in rock piles.

Writing this book taught me a lot. Some unexpected patterns seem to be taking shape, such as the numerous similarities among pythons, boas, and vipers, and the pervasive role that mimicry might have played in snake evolution. The complex ways that serpentine muscles underlie locomotion now seem more within my grasp, as do venom biochemistry and pharmacology. Plenty of intellectual challenges remain too, from the role of moist habitats in the lives of individual snakes to relationships among climate, diet, reproduction, and social systems in these animals as a group. And in the course of writing about what I knew, by following Norman Maclean's advice, I learned about other things that I had not understood; lately my focus on venomous snakes has even weakened somewhat, loosened its grip, and my interests have broadened. During our 1993 trip to Brazil, I was more elated by Amazonian Watersnakes (*Helicops angulatus*) and their resemblance to unrelated North America counterparts (*Nerodia*) than by the Common Lanceheads (*Bothrops atrox*) we were studying. A few months later, back in Arizona, Kelly and I walked among towering red canyon walls to visit the cliff ruins called Keet Seel. Along the way we remembered that these same Kayenta sandstones yielded the oldest caecilian fossil, and upon entering the ancient Anasazi settlement we noticed a faded rock painting, unmistakably that of a serpent. Instead of much-sought-after Hopi Rattlesnakes (*Crotalus viridis nuntius*), however, the only limbless vertebrate we encountered in two days of hiking was a very agitated but harmless Gopher Snake (*Pituophis catenifer*).

Life is beautiful, at times almost peaceful, and I count myself extraordinarily lucky to be a teacher and field biologist. Just last week I became instant friends with a four-year-old while showing her a sleek, inoffensive Ball Python (*Python regius*) in my lab. Later, as we walked hand-in-hand beside a creek on the Berkeley campus and talked about why she liked snakes, Rachel beamed with happy confidence. As for the serpents themselves, we still can't say what it's like to actually *be* a Black-tailed Rattlesnake, much less a little Ridgenose. I must go farther and closer.

APPENDIX: SYSTEMATICS AND EVOLUTIONARY INFERENCE

PHYLOGENETIC CLASSIFICATION AND TAXONOMY

Several terms are central to phylogenetic systematics. A monophyletic group (species, genus, family, etc.), also called a "clade" or a "lineage," is one whose members have a common ancestor, which is contained, with all its descendants, in the group. Members of a paraphyletic taxon likewise have a common ancestor, but without all of its descendants being contained in the group; a paraphyletic taxon is a clade with one or more of its constituent taxa removed. A widely known example of paraphyly is the class Reptilia as traditionally defined, because it excludes birds. Members of a polyphyletic group have no common ancestor unique to them; also called "grades," such taxa reflect independently evolved similarity in one or more traits (e.g., anterior folding fangs in stiletto snakes [*Atractaspis*] and true vipers).

Historical relationships among populations, species, and higher taxa are inferred by comparisons of their characteristics with those of more distantly related groups. Higher taxa often have been arbitrary, based on vague traditions about "significant" similarities and differences (e.g., whether the anal scale is divided or single, as in North American ratsnakes [*Elaphe*] and kingsnakes [*Lampropeltis*], respectively) rather than evolution. Recovering phylogenetic patterns depends instead on overall differences in evolutionary directionality (or polarity) of their character states. A character is an attribute that is compared among organisms, and a state is the particular expression of that character; thus "rostral scale normal" and "rostral scale upturned" are alternative states of a character (snout shape) that describe variation among snakes. Primitive character states, such as rostral scale normal, are present in some members of the ingroup under study (e.g., snakes), as well as in closely related organisms, or outgroups (e.g., other squamates). Derived character states, such as the upturned snout of North American hog-nosed snakes (*Heterodon*), typify an ingroup and are lacking or at least uncommon in the outgroups (in this case, other snakes). Only derived character states are useful for recognizing monophyletic groups and deducing phylogenetic relationships. For example, the presence of four limbs characterizes a much larger, more inclusive group than squamates (namely, the tetrapods) and thus does not define squamates as a monophyletic taxon; the presence of four limbs is a derived state for tetrapods (their outgroups, lobe-finned fish and lungfish, lack true limbs) but a primitive state for those squamates that still possess limbs.

Phylogenetic systematists prefer monophyletic taxa, because such groups emphasize rather than obscure relationships among organisms. For example, the traditional definition of Reptilia (with birds removed) ignores common ancestry and thereby keeps us from rightly emphasizing the numerous similarities between birds and crocodilians. Phylogenetic classifications often will be dichotomous, reflecting evolutionary bifurcations; however, ambiguous inferences can preclude that goal, and perhaps sometimes more than two lineages arise simultaneously. In any case, it is neither practical nor necessary to name every branching point, or node, on a cladogram. We formally recognize only monophyletic groups that require discussion in terms of biogeography, adaptive radiation, and so forth. Some phylogenetic systematists believe that each taxon should pertain to larger taxa, but that these need not be ranked or redundant. Anomochilidae as a family for dwarf pipesnakes, for example, provides no additional information about relationships, since there are no fossil or living taxa included in it other than *Anomochilus*. By contrast, Acrochordidae encompasses both the living genus *Acrochordus* and other fossil taxa.

SPECIES AND SUBSPECIES

Biological species are reproductively continuous sets of populations that are isolated from other such populations. For example, external characteristics and biochemical genetic profiles suggest that Common Kingsnakes (*Lampropeltis getula*) across much of North America exchange genes among themselves (or at least have done so recently), yet these snakes do not interbreed with Mole Kingsnakes (*L. calligaster*) and other species with whose ranges they overlap. Two problems have been recognized by proponents of the biological species concept. Populations whose distributions overlap, such as Common and Mole Kingsnakes, are called "sympatric," whereas those not in geographical contact are known as "allopatric." By definition, then, the latter are not capable of interbreeding, regardless of their genetic relatedness (such is the case of Ridge-nosed Rattlesnakes [*Crotalus willardi*], now isolated from each other as a result of habitat fragmentation). Whether to refer to allopatric populations as a single species or several ones has usually hinged on whether the similarities and differences resemble the variation among populations of related organisms that do interbreed. Ring species are those in which some geographically contiguous populations interbreed, while others overlap with no evidence of gene exchange. The Milksnake (*Lampropeltis triangulum* [see p. 198]) is among the few postulated examples of a ring species in terrestrial vertebrates.

A more serious problem is that the biological species concept does not necessarily reflect the history of distinctive, evolving units in nature. Some phylogenetic systematists regard reproductive continuity as a retained primitive condition and propose instead that species are independent evolutionary lineages, sets of populations diagnosed by shared derived characteristics. For example, Ridge-nosed Rattlesnakes have been assigned to five allopatric subspecies (e.g., the Chihuahuan Ridge-nosed Rattlesnake, *Crotalus willardi silus*), but one or more of them might encompass a diagnosable set of populations and thus be a phylogenetic species. This approach emphasizes independently evolving lineages as the fundamental units of biodiversity. Conversely, there seems little point in recognizing subspecies that represent only vaguely overlapping geographic variation in one or a few characters (as in the case of some Milksnakes).

Chihuahuan Ridge-nosed Rattlesnake (*Crotalus willardi silus*), Mexico.

HOMOLOGY, HOMOPLASY, AND ADAPTATION

Several terms are especially important in discussing the evolution of similarities and differences among organisms and the relationships of those historical patterns to behavior and ecology. Homologous features are attributes, such as the erectile fangs of vipers, whose similarities stem from continuous presence during the shared evolutionary history of different organisms. Homoplasious character states are independently evolved similarities, resulting from either reversal to more primitive states (e.g., loss of the rattle in Santa Catalina Island Rattlesnakes [*Crotalus catalinensis*] and San Lorenzo Island Rattlesnakes [*C. ruber lorenzoensis*]) or convergent attainment of derived states (e.g., bright green colors in Emerald Tree Boas [*Corallus caninus*] and Green Tree Pythons [*Morelia viridis*]).

Adaptation, one of the most confusing and contentious terms in evolutionary biology, can denote either a process in populations or a character state that relates to environmental variables. In the first sense, individuals with particular characteristics cope better with environmental demands and are favored by natural selection; as a result, they increase their genetic contribution to future generations. In the second sense, the term connotes any feature of an organism that promotes survival and reproduction within a population, regardless of its evolutionary origin. For example, an elongate body might facilitate an arboreal lifestyle and somehow enhance reproductive success in species that climb, but snakes (and thus their primitively attenuate form) did not evolve in trees.

I prefer a stricter, phylogenetic concept that incorporates historical origins of attributes: Adaptations are features whose original presence conferred particular performance advantages; they thus can characterize individuals, species, and higher taxa. An exaptation, conversely, imparts a performance advantage not associated with the origin of that feature. For example, the infrared-imaging pits of crotaline snakes originally might have been an adaptation for assessing predators and only later served as an exaptation for feeding; the historical origins of performance advantages are inferred by comparing taxa that exhibit a particular trait with close relatives that lack it (see p. 254). These distinctions are significant for this book, because I do not emphasize adaptation in terms of natural selection and genetic processes within populations but focus instead on adaptations of species and higher taxa, important ingredients in any broad consideration of snake evolution.

INTRODUCTION

Linné is quoted in Goin et al. 1978, 11. Ditmars (1932) relates his encounter with a Bushmaster (*Lachesis muta*), and Campbell and Lamar (1989) summarize the literature on that species.

1. CLASSIFICATION AND GENERAL BIOLOGY

Aristotle's comment on snake tongues is quoted in Burghardt 1970, 245. For the history of herpetology, here and elsewhere in the book, I rely heavily on Adler 1989. For a single overview of reptile biology there is still none better than Bellairs 1969, and I often use his accounts as a start on particular topics; shorter but more up-to-date summaries are in Cogger and Zweifel 1992 and Zug 1993. Other references for this chapter are Andreadis and Burghardt 1993; Branch 1986; Burghardt 1978, 1985, 1988, 1991, 1996; Burghardt and Greene 1988; Chiszar et al. 1992; Collins 1990; Conant and Collins 1991; Cundall 1987; Cundall and Rossman 1993; Cundall et al. 1993; de Queiroz and Gauthier 1994; Dmi'el et al. 1990; Dodd 1988; Dowling and Savage 1960; Edgren 1953; Ford and Burghardt 1993; Forstner et al. 1995; Fox and Dessauer 1962; Gans 1974; Greene 1986, 1988a; Guyer and Donnelly 1990; Hedges et al. 1989; Inger and Marx 1962; M. K. Jackson and Reno 1975; Kluge 1991, 1993a, 1993b; Lillywhite 1987; Loveridge 1936; Lyman-Henley and Burghardt 1994; Macdonald 1974; Maddison and Maddison 1992; Marx and Rabb 1972; Mason 1992; McDowell 1979, 1987; J. B. Murphy and Barker 1980; Myers 1984; Pope 1941; Pough 1980, 1983; Rieppel 1988, 1991; Robinson and Hughes 1978; Sazima and Di-Bernardo 1991; Schaafsma 1980; Schwenk 1993, 1994; Shine 1991a; Stebbins 1985; Stuebing 1994; Tokarz and Slowinski 1990; Underwood and Kochva 1993; Weldon 1990; Young 1993; Zimmerman and Heatwole 1990; and Zweifel 1980.

2. LOCOMOTION AND HABITATS

Sir Richard Owen is quoted in Bellairs 1969, 108. My comments about "trophyism," in the introductory essay, should not be misconstrued as criticism of collecting specimens for museums (see Greene 1994a). The photograph I mention of a Puff Adder (*Bitis arietans*) striking is in Loveridge 1936. Nussbaum 1992 and Gans 1992 provide overviews of caecilians and amphisbaenians, respectively. For detailed discussions of limbless locomotion, see Cundall 1987; Edwards 1985; and Gans 1974. Other references for this chapter are Branch 1988; Bustard 1969; Conant and Collins 1991; Dial et al. 1987; Gans 1975, 1986; Greene and Roberts, MS; Greer 1983; Guyer and Donnelly 1990; Heyer and Pongsapipatana 1970; Jaensch 1988; Jayne 1982, 1986, 1988; Jayne et al. 1988; Jenkins and Walsh 1993; Lillywhite and Henderson 1993; Marx and Rabb 1972; Norris and Kavanau 1965; Pough 1969, 1980, 1983; Reinert 1993; Riley et al. 1986; Robinson and Hughes 1978; Shine 1986a; Thireau 1967; Wake 1993; Walton et al. 1990; and Wu et al. 1993.

3. DIET AND FEEDING

For reviews of the mechanics of feeding in snakes, see Gans's classic analysis (1961) and Cundall 1987, 1995. Other references for this chapter are Ananjeva and Orlov 1982; S. J. Arnold 1977, 1983, 1992, 1993; Beavers 1976; Broadley 1963; D. G. Brown 1990; Burghardt 1970, 1978; Camilleri and Shine 1990; Campbell and Lamar 1989; Cervantes-Reza 1981; Chiszar et al. 1990, 1992; Coleman et al. 1993; Cundall and Irish 1989; Drummond 1979, 1983, 1985; Duarte and Laporta-Ferreira 1993; Duvall et al. 1980; Eichholz and Koenig 1992; Fitch 1960, 1963a; Ford and Burghardt 1993; Forsman and Lindell 1993; Gans 1975; Gans and Oshima 1952; Gillingham and Rush 1974; Gillingham et al. 1990; Godley 1980; Greene 1976, 1983a, 1983b, 1984, 1986, 1988b, 1989a, 1992, 1994a, 1994b; Greene and Burghardt 1978; Greene and Santana 1983; Gregory et al. 1980, 1987; Gumbart and Sullivan 1990; Hailey and Davies 1986a; Halloy and Burghardt 1990; Hasegawa and Moriguchi 1989; Heatwole et al. 1978; Henderson et al. 1987; Hisaw and Gloyd 1926; K. B. Jones 1990; K. B. Jones and Whitford 1989; Kardong 1977, 1980, 1982, 1986; A.

Knight and Mindell 1994; J. L. Knight and Loraine 1986; Lind and Welsh 1994; Macdonald 1974; Marques and Puorto 1994; Mora 1991; Mori 1989, 1995; Mori et al. 1992a; Mushinsky 1987; Nalleau and Bonnet 1995; Pleguezuelos et al. 1994; Plummer 1977; Punzo 1974; Rathbun 1979; Reinert 1993; Reinert et al. 1984; Richmond 1944; Rodriguez and Reagan 1984; Rossman and Myer 1990; Saint Girons 1972; A. H. Savitzky 1978, 1980, 1981, 1983; Sazima and Puorto 1993; Schwaner 1989; Secor and Nagy 1994; Secor et al. 1994; Seib 1985; Shine 1977, 1991a, 1991b, 1994; Slip and Shine 1988a, 1988c; Takahashi 1981; Voris and Voris 1983; Webb and Shine 1993a; and Wendelken 1978.

4. VENOMOUS SNAKES AND SNAKEBITE

The quote about Terciopelo (*Bothrops asper*) bites is from Picado T. 1931, 14 (translation mine); the quote about Karl P. Schmidt's death is from Pope 1958, 281. Other references for this chapter are Assakura et al. 1992; Barr et al. 1988; Bellairs 1969; Bertelsen and Nielsen 1987; Bolaños 1984; Cadle 1982; Campbell and Lamar 1989; Chiszar et al. 1992; Daltry et al. 1996; Duvall et al. 1980; Elliott 1978; Emerson et al. 1994; Ferguson et al. 1984; Fowler and Salamão 1994; Furtado et al. 1991; Gans 1978; Gillissen et al. 1994; Glenn et al. 1983; Gómez and Savage 1983; Gopalakrishnakone 1986; Gopalakrishnakone and Chou 1990; Greene 1986, 1988b, 1992; Greene and Roberts, MS; Groves 1973; Gutiérrez et al. 1980, 1990, 1992; Hardy 1987, 1992, 1994a; Hayes 1995; Hayes and Galuska 1984; Hayes et al. 1992a, 1993; Hisaw and Gloyd 1926; K. Jackson and Fritts 1995; Jansen 1983; Jansen and Foehring 1983; Johnson 1988; Kardong 1979, 1980, 1982, 1986; Kardong and Lavín-Murcio 1993; Kardong et al. 1986; Klauber 1972; A. Knight and Mindell 1994; Kochva 1987; Kuch and Jesberger 1993; Mackessy 1988; Mackessy and Tu 1993; McDowell 1968, 1969, 1986, 1987; Mebs 1978; Minton 1974; Monge-Nájera 1994; Myers 1974; Pe et al. 1991; Pough and Groves 1983; Pough et al. 1978; Rodríguez-Robles 1994; Rodríguez-Robles and Thomas 1992; H. I. Rosenberg et al. 1985, 1986; Russell 1980; Salomão and Laporta-Ferreira 1994; A. H. Savitzky 1978, 1980, 1992; J. A. Seigel and Adamson 1983; Shine 1991a; Spawls and Branch 1995; Sutcliffe et al. 1994; Tan et al. 1993; Taub

1967; Thomas and Pough 1979; Underwood and Kochva 1993; Warrell 1991; Weinstein and Kardong 1994; and Wellman 1963.

5. PREDATORS AND DEFENSE

The introductory essay quotes from Wallace 1895, 71. The main text is based primarily on Greene 1988a and Pough 1988a; my ideas about widespread mimicry of venomous snakes have been especially influenced by Pough 1988a, 1988b; Savage and Slowinski 1992; Vitt 1992; and Y. L. Werner 1983. Other references for this chapter are Bierregaard 1984; Boinski 1988; Branch 1991a; Broadley 1974, 1987; Brodie 1989, 1993a; W. S. Brown 1993; Campbell and Lamar 1989; Carr 1940; Cooper and Greenberg 1992; Cox 1991; DeSilva 1990; Drummond and Wolfe 1981; Fitch 1960; Galat-Luong 1991; Gehlbach 1972; Gerhardt et al. 1993; Glaw and Vences 1994; Greene 1988b, 1989b; Greene and Hardy 1989; Haagner 1991; Hailey and Davies 1986b; Herzog and Burghardt 1988; J. F. Jackson 1979; J. F. Jackson et al. 1976; Lambiris 1967; Lorenz 1971; Macdonald 1974; W. F. Martin and Huey 1971; McDowell 1979; Mead and Van Devender 1981; Meinzer 1993; Mendelson 1992; Mori et al. 1992b; Myers 1986; Niemitz 1973; S. Rasmussen et al. 1995; Rubinoff and Kropach 1970; Sanchez-Herrera et al. 1981; Sazima 1992; Slowinski and Savage 1995; Weldon 1990; Wüster and Cox 1992; Wüster and Thorpe 1992; and Young 1991.

6. BEHAVIOR, REPRODUCTION, AND POPULATION BIOLOGY

The *Pennsylvania Journal* quote is from Medden 1929, 84–85. See W. S. Brown 1993 for a review of the biology and conservation status of Timber Rattlesnakes (*Crotalus horridus*). For reviews of social behavior, reproduction, and population biology of snakes, see Blackburn 1985; Carpenter 1977, 1984; Cooper and Greenberg 1992; Duvall et al. 1993; Fitch 1970; Ford and Burghardt 1993; Gillingham 1987; Graves and Duvall 1995; Halpern 1992; Mason 1992; Moore and Lindzey 1992; W. S. Parker and Brown 1987; R. A. Seigel and Ford 1987; Shine 1985, 1988a, 1988b, 1991a, 1992, 1993; Stewart 1992; and Whittier and Tokarz 1992. Reinert 1992 provides an excellent overview of radiotelemetry of snakes.

Other references for this chapter are Aldridge 1992; Aldridge and Semlitsch 1992; Bartecki and Heymann 1987; Bennion and Parker 1976; Blanchard 1930; Bothner 1974; Branch 1978; Breithaupt and Duvall 1986, 1990; Brodie et al. 1969; Burger and Zappalorti 1986; Burghardt 1988; Cadle and Chuna M. 1995; Charland 1995; Chiszar et al. 1991, 1993; Cobb 1990; Diller and Johnson 1982; Drummond 1989; Duarte and Laporta-Ferreira 1993; Duellman 1961, 1978; Duvall et al. 1985, 1993; Fitch 1960; Ford and Seigel 1994; Gibbons and Semlitsch 1987; Gloyd and Conant 1991; Goode and Schuett 1994; Graves et al. 1986; Greene 1970, 1990, 1994b; Gregory 1975; Gregory and Stewart 1975; Gregory et al. 1987; Guillette 1993; Hall and Meier 1993; Harlow and Shine 1992; Hayes et al. 1992b; Heatwole et al. 1978; Henderson et al. 1978; Hersek et al. 1992; Holmstrom 1981; Lipske 1979; Lyman-Henley and Burghardt 1994; Macartney et al. 1988; Madsen 1984; Madsen and Shine 1992; W. H. Martin 1992; Martins 1993; Minton 1966; J. B. Murphy and Barker 1980; Mushinsky 1979; Myers 1984; Nalleau and Bonnet 1995; Neitman 1992; Oliver 1956; Ota et al. 1991; Palmer 1992, 1993; W. S. Parker and Brown 1972; Platt 1969; Plummer 1990; Reichling 1990, 1995; Reinert and Zappalorti 1988a; Riley et al. 1985; Ripa 1994; Rodda 1992; Sazima 1992; Schuett 1992; Schwaner 1989; R. A. Seigel and Ford 1992; Shine 1986b; Solórzano 1989; Solórzano and Cerdas 1989; Stuebing 1988; Thorbjarnarson 1995; Treadwell 1962; Viitanen 1967; Y. L. Werner 1970; and Yaron 1985.

7. BLINDSNAKES

Quotations are from Bauer 1988, 45; Dixon and Kofron 1983, 263; Rose 1962, 316; and Wall 1918, 377. My comments on *Xenotyphlops grandidieri* are based on Wallach and Ineich 1996. No general overview of the biology of blindsnakes exists, but this chapter summarizes information in Bauer 1988; Branch 1988; Broadley and Watson 1976; Cagle 1946; Cox 1991; Cunha and Nascimento 1978; Dunn and Tihen 1944; Dyer 1990; Fox and Dessauer 1962; Gaulke 1995; Gehlbach and Baldridge 1987; Gehlbach et al. 1971; Greene 1988a; Groombridge 1979; Hibbard 1964; List 1966; McCoy 1960; McDowell 1974, 1987; Minton 1966; Nussbaum 1980a; Pienaar et al. 1983; Pitman 1974; Punzo 1974; Reid and Lott

1963; Richter 1955; Robb and Smith 1966; Schwartz and Henderson 1991; Shine 1985; Shine and Webb 1990; H. M. Smith 1957; Taylor 1922; Thomas 1965; Wallach 1993, 1994, 1995; Warner 1945; Webb and Shine 1992, 1993a, 1993b; and Wynn et al. 1987.

8. PIPESNAKES, BOAS, AND OTHER TRANSITIONAL GROUPS

Here and elsewhere I have relied primarily on Behrensmeyer et al. 1992 for information on geology and paleoecology. Other references for this chapter include Alvarez del Toro 1982; Barker and Barker 1994; Bhupathy and Vijayan 1989; Bloodworth 1989; Bogert 1969; Branch 1986, 1988; Branch and Hacke [sic] 1980; Campbell and Camarillo 1992; Chapman 1986; Cheney and Seyfarth 1990; Corn 1974; Cox 1991; Cundall and Irish 1986, 1989; Cundall and Rossman 1993; Cundall et al. 1993; Cunha and Nascimento 1978; Daniel 1983; Dixon and Soini 1977; Gans 1973, 1976; Gans et al. 1978; Gartlan and Struhsaker 1971; Glaw and Vences 1994; Greene 1980, 1983a, 1983b, 1994a; Greene and Burghardt 1978; Haacke 1982; Hardy 1994b; Harlow and Shine 1992; Haverschmidt 1970; Hay and Martin 1966; Hedges et al. 1992; Henderson 1994; Henderson et al. 1987, 1995; Heymann 1987; Irish and Alberch 1989; Iverson 1986; F. W. King 1962; Kluge 1989, 1991, 1993a, 1993b; Lax 1989; Lazell 1964; Lehmann 1970, 1971; Lillywhite 1991; McDiarmid et al. 1996; McDowell 1975, 1979, 1987; Merker and Merker 1995; Mora 1991; J. B. Murphy et al. 1978a, 1978b, 1981; Myres and Eells 1968; F. Parker 1982; Pope 1961; Preston-Mafham 1991; Rodriguez and Reagan 1984; Ross and Marzec 1990; Schwartz and Henderson 1991; Shine 1986b, 1986c, 1991a; Shine and Slip 1990; Slip and Shine 1988a, 1988b; Spawls 1991; Stanford 1994; Starin and Burghardt 1992; Strüssman and Sazima 1991; Stuebing 1994; Stuebing and Goh 1993; Sunquist 1982; Szyndlar and Schleich 1994; Thorbjarnarson 1995; Tolson and Henderson 1993; Trail 1987; Vinegar 1973; Vinson 1975; Vitt and Vangilder 1983; Wolf and Werner 1994; Zaher 1994; and Zhao and Adler 1993.

9. OLD WORLD COLUBRIDS

Kingdon 1989 provides a general summary of African landscapes and biotas,

and Matthiessen 1991 summarizes the environmental crisis in equatorial Africa. J. B. Rassmussen (1993) classifies the green Bwindi snake we found as *Dipsadoboa unicolor*. The Special Topic for this chapter quotes Marshall 1985: 260. Especially important recent references to African snakes are Branch 1988, 1991a; Broadley 1983; Cadle 1994; and Pitman 1974. Glaw and Vences 1994 provide an introduction to Malagasy snakes. Other references for this chapter are E. N. Arnold and Burton 1978; Böhme 1993; Burdick 1994; Butynski and Kalina 1994; Cadle 1987; Coleman et al. 1993; Cox 1991; Dunson et al. 1978; Fritts and McCoid 1991; Greene 1989a, 1989b; Gyi 1970; Hall et al. 1994; Hasegawa and Moriguchi 1989; Henderson and Binder 1980; Inger and Marx 1965; Jayne et al. 1988; Leviton et al. 1992; Lim and Lee 1989; Malnate 1953, 1960; Marx and Rabb 1972; Mori et al. 1992a, 1992b; Nussbaum 1980b; Ota et al. 1991; Pleguezuelos et al. 1994; Pope 1958; Preston-Mafham 1991; Raxworthy and Nussbaum 1994; Rodda and Fritts 1992; Saint Girons 1972; Savidge 1987; A. H. Savitzky 1981, 1983; Shine 1991a; M. A. Smith 1921, 1938, 1943; Steehouder 1984; Steward 1971; Sweeney 1971; Szyndlar 1991a, 1991b; Toriba 1987; Tweedie 1983; Wüster and Cox 1992; and Zhao and Adler 1993.

10. NEW WORLD COLUBRIDS

The quotes in the introductory essay are from Skutch 1980, 256–57, on which my treatment of his views is also based; I also quote Bonta 1991, 96, and allude to Fitch 1965. For discussions of and references to the complexity of colubrid relationships, see Cadle 1987, 1994; and McDowell 1987. Other references for this chapter include Alvarez del Toro 1982; Beebe 1946; Blair 1960; R. E. Bullock 1981; Cadle and Greene 1993; Campbell and Lamar 1989; Clark 1970; Cunha and Nascimento 1978; Dixon and Soini 1977; Duellman 1963, 1978; Eichholz and Koenig 1992; Ernst and Barbour 1989; Ferrarezzi 1993; Fitch 1949, 1963a, 1963b, 1965; Freiberg 1982; Gertler and Morales 1980; Greene 1979; Greene and Roberts, MS; Gregory et al. 1980; Henderson and Binder 1980; J. A. Jackson 1974; Laporta-Ferreira et al. 1986; W. F. Martin and Huey 1971; Martins and Oliveira 1993; Marx and Rabb 1972; Mount 1975; Myers 1974, 1982, 1986; Myers and Cadle 1994; Nickerson and Heringhi 1966; H.

Rosenberg 1955; Ruthven 1912; Salomão and Laporta-Ferreira 1994; Savage and Scott 1987; A. H. Savitzky 1978; Sazima 1989; Sazima and Abe 1991; Sazima and Puorto 1993; Scott et al. 1989; Seib 1985; Strüssmann and Sazima 1990; Underwood 1967; Vanzolini et al. 1980; Vitt 1980, 1983; Vitt and Vangilder 1983; Wellman 1963; Williams 1988; A. H. Wright and Wright 1957; and Zweifel 1954.

11. STILETTO SNAKES AND OTHER AFRICAN ENIGMAS

The introductory essay ends with a quote from Broadley 1978, 31. Information in this chapter comes from Bourgeois 1965; Branch 1988, 1991a, 1991b; Broadley 1978, 1994; Cadle 1982; Corkhill and Kirk 1954; Golani and Kochva 1988; Jennings 1994; Kloog et al. 1988; Kochva and Golani 1993; McDowell 1968, 1986, 1987; Shine 1985; Spawls and Branch 1995; Underwood and Kochva 1993; Wall 1906; Wilson 1976; and from my unpublished observations on captive and preserved stiletto snakes (*Atractaspis*).

12. COBRAS, CORALSNAKES, AND THEIR RELATIVES

The introductory essay quotes Broadley 1968, 432, and Carr 1994, 180. For an excellent introduction to the biology of Australian elapids, see Shine 1991a. This chapter is also based on Alvarez del Toro 1982; Angilletta 1995; Branch 1988; Broadley 1993b; Brodie 1993b; Burton 1950; Cadle 1987; Campbell and Lamar 1989; Chiszar et al. 1990; Cogger 1992; Cox 1991; Cunha and Nascimento 1978; Dixon and Soini 1977; Ernst 1993; Evans 1906; Greene 1984; Greene and McDiarmid 1981; Haagner 1991; Ionides and Pitman 1965; Kardong 1982; Kauffeld 1953; Kuch and Schneyer 1991; Leakey 1969; Leviton et al. 1992; Lloyd 1985; Loveridge 1928; Marx and Rabb 1972; McCarthy 1985; McDowell 1968, 1969, 1986, 1987; Oliver 1956; F. Parker 1982; Pough 1988a, 1988b; Roze 1989; Savage and Slowinski 1992; A. H. Savitzky 1978, 1985; Sazima and Abe 1991; Scanlon and Shine 1988; Shine 1985; Shine and Covacevich 1983; Slowinski 1994a, 1994b, 1995; M. A. Smith 1943; Soderberg 1973; Spawls and Branch 1995; Storr et al. 1986; Szyndlar and Rage 1990; Tweedie 1983; Wallace 1895, 70–73; Wüster and

Thorpe 1992; Young 1991; and Zinner 1971.

13. SEAKRAITS AND SEASNAKES

Lowe is quoted in Dunson 1975, 14; Espinosa, in Taylor 1953. Key references on marine elapids are the books by Dunson (1975) and Heatwole (1987) and the article by Voris and Voris (1983). Other references for this chapter are Caldwell and Rubinoff 1983; Cogger 1992; Cogger et al. 1987; Cox 1991; Dunson and Ehlert 1971; Dunson and Minton 1978; Gopalakrishnakone 1994; Gorman et al. 1981; Gregory 1978; Gritis and Voris 1990; Heatwole et al. 1978; Hecht et al. 1974; Kropach 1975; Lemen and Voris 1981; Mackessy and Tu 1993; Mao and Chen 1980; Marques and Souza 1993; McCarthy 1986, 1987a, 1987b; McDowell 1972, 1987; Pickwell 1971; Rubinoff and Kropach 1970; Shine 1988a, 1991a; Stuebing 1988; Takahashi 1981; Tu et al. 1990; Vallarino and Weldon forthcoming; Voris 1966, 1975, 1977, 1983; Weldon 1988; Young 1987; and Zimmerman and Heatwole 1990.

14. VIPERS, ADDERS, AND PITVIPERS

The systematics and ecology of Middle American palm pitvipers (*Bothriechis*) are discussed by Crother et al. (1992), and Rosenzweig (1995) discusses the causes of species diversity gradients. The short story "La Bocaracá" is in Salazar Herrera 1989. Major syntheses of pitviper biology include Campbell and Brodie 1992; Campbell and Lamar 1989, 1992; Gloyd and Conant 1991; and Klauber 1972. Other references for this chapter include Alvarez del Toro 1982; Andrén and Nilson 1981; Armstrong and Murphy 1979; Arnold and Burton 1978; Ashe and Marx 1988; Barr et al. 1988; Bothner 1974; Branch 1978, 1988; Brattstrom 1964; Breidenbach 1990; Broadley 1983; Brodmann 1987; D. G. Brown 1990; W. S. Brown 1993; T. H. Bullock and Cowles 1952; Butler et al. 1995; Cadle 1992; Cervantes-Reza 1981; Chiszar et al. 1992; Cock Buning 1993; Cox 1991; Cunha and Nascimento 1978; Duarte et al. 1995; Duvall et al. 1980, 1985, 1990, 1993; Ernst 1993; Fitch 1960; Gertler and Morales 1980; Graves and Duvall 1995; Graves et al. 1995; Greene 1992, 1994a, 1994b; Grismer et al. 1994; Groombridge 1986; Gumbart and Sullivan 1990; Hartline et al. 1978; Hayes and Galuska 1984;

Herrmann et al. 1992a, 1992b; Hoffmann 1988; Huang 1989; Kardong 1977, 1982, 1986, 1992; Kardong and Mackessy 1991; Kardong et al. 1986; Klusmeyer and Fausten 1994; A. Knight and Mindell 1993, 1994; A. Knight et al. 1992; Koba 1938, 1971; Liem et al. 1971; Lim and Lee 1989; Lowe et al. 1986; Madsen and Shine 1992; W. H. Martin 1992; Marx and Olechowski 1970; Marx and Rabb 1965, 1972; Mebs et al. 1994; Minton 1966; Molenaar 1992; R. W. Murphy et al. 1995; Nilson 1994; Nilson and Andrén 1986; Ottley and Murphy 1983; Pitman 1974; Reinert and Zappalorti 1988a, 1988b; Reinert et al. 1984; Ripa 1994; Robinson and Hughes 1978; A. H. Savitzky 1992; B. C. Savitzky 1992; Sazima 1992; Schwenk and Greene 1995; Sherbrooke 1996; Shine 1985, 1988b, 1994; Solórzano 1994; Spawls and Branch 1995; Tweedie 1983; Viitanen 1967; Wharton 1969; Young and Brown 1995; and Zhao and Adler 1993.

15. EVOLUTION AND BIOGEOGRAPHY

The long quote is from Gloyd 1940, 122. My favorite works on the United States–Mexico boundary region are Bowden 1991; Hastings and Turner 1965; Marshall 1957; Spicer 1962; and Weisman and Dusard 1986. Summaries of the fossil record of snakes are found in Cadle 1987; Rage 1987; and Szyndlar 1991a, 1991b. In addition to others cited earlier for individual taxonomic groups, references for this chapter are Albino 1993; Andrén and Nilson 1981; Armstrong and Murphy 1979; Barker 1992; Behrensmeyer et al. 1992; Bogert and Degenhardt 1961; Breithaupt and Duvall 1986, 1990; Brodie 1989; Brodie and Brodie 1990; Cadle and Greene 1993; Cooper et al. 1994; Cundall 1995; Cundall and Rossman 1993; Cundall et al. 1993; Duarte et al. 1995; Dudley and DeVries 1990; Estes 1983; Estes and Báez 1985; Estes et al. 1970; Forsman 1995; Forstner et al. 1995; Frazzetta 1970; Gehlbach 1981; Gilmore 1938; Greene 1983a; Grismer 1994; Herrmann et al. 1992b; Irish and Alberch 1989; Kauffeld 1957; R. B. King 1992; Klauber 1972; Kluge 1991; A. Knight and Mindell 1994; Kraus and Brown forthcoming; Ljungar 1995; Lowe et al. 1986; Madsen and Stille 1988; McDowell 1987; Mead and Bell 1994; R. W. Murphy and Crabtree 1985; R. W. Murphy et al. 1995; Novacek 1992; Parmley and Holman 1995; Patchell and Shine

1986; Rage 1987; Rage and Prasad 1992; Rage and Richter 1994; Rich et al. 1990; Rieppel 1991; Rodríguez-Robles and Greene 1996; A. H. Savitzky 1980; Schaal and Ziegler 1992; R. A. Seigel 1986; Shine 1986a; Shine and Covacevich 1983; Simpson 1933; Sweeney 1971; Szyndlar 1994; Szyndlar and Rage 1990; Szyndlar and Schleich 1994; Tanner 1985; Vanzolini and Heyer 1985; Vasile 1990; Vitt 1987, 1992; Webb and Shine 1993a; and C. Werner and Rage 1994.

16. SNAKES AND OTHERS: PAST, PRESENT, AND FUTURE

The quote under "Snakes and Science" is from Baird and Girard 1853, v. Important discussions of snake conservation are found in Dodd 1987, 1993; and Scott and Seigel 1992. Other references for this chapter include Adler 1989, 1994; Anonymous 1938, 1995; Arena et al. 1995; E. N. Arnold 1980; Ashe and Marx 1988; Bartecki and Heymann 1987; Bernardino and Dalrymple 1992; Bhupathy 1990; Boinski 1988; Bowers and Burghardt 1991; Burghardt et al. 1972; Bushey 1985; Butler and Shitu 1985; Butynski and Kalina 1994; Carr 1982; Case et al. 1992; Duarte et al. 1995; Duellman 1984; Felger and Moser 1985; Fitch 1987; Fleet et al. 1972; Francini et al. 1990; Gans 1995; Glaw and Vences 1994; Gloyd and Conant 1991; Gove 1991; Graham 1991; Greene 1988b, 1990, 1994a; Greene and Campbell 1992; Inger and Marx 1965; Inglish 1991; J. A. Jackson 1974; C. Jones 1969; Kellert and Wilson 1993; Kinkenborg 1995; Klemens and Thorbjarnarson 1995; Kobayashi and Watanabe 1986; Lanoie and Branch 1991; Lillywhite 1991; MacBride 1932; P. S. Martin and Klein 1984; Marx and Rabb 1965; Masataka 1993; McDowell 1987; Mellink 1995; Mendelson and Jennings 1992; Morris and Morris 1965; Nabhan 1995; Neill et al. 1956; North et al. 1994; Pough 1988a; Procter 1918; Quammen 1994; Rosen and Lowe 1994; Rowe et al. 1986; Schaafsma 1980; Scott et al. 1989; R. A. Seigel 1986; Shine 1986c; Shine and Fitzgerald 1989; Silva and Sites 1995; Skaroff 1971; Van Devender et al. 1994; Warwick 1990; Weir 1992; Weldon 1990; and P. C. Wright and Martin 1995.

EPILOGUE

Hastings and Turner 1965 and Spicer 1962 are the two titles I allude to. The

Prairie Rattlesnake (*Crotalus v. viridis*) site is described in Duvall et al. 1985.

APPENDIX

For discussions of phylogenetic systematics and its implications for modern biology, see Brooks and McLennan 1991; de Queiroz and Gauthier 1994; and Maddison and Maddison 1992. For alternative views on the concepts of species and subspecies in herpetology, see Frost et al. 1992 and Van Devender et al. 1992. E. N. Arnold 1994 and Losos and Miles 1994 provide excellent discussions of historical approaches to the study of adaptations.

Adler, K., ed. 1989. *Contributions to the history of herpetology.* Oxford, Ohio: Soc. Study Amph. Rept.

———. 1994. The remarkable career of Roger Conant. In *Captive management and conservation of amphibians and reptiles,* Contr. Herp. 11, ed. J. B. Murphy, K. Adler, and J. T. Collins, 17–23. Ithaca, N.Y.: Soc. Study Amph. Rept.

Albino, A. M. 1993. Snakes from the Paleocene and Eocene of Patagonia (Argentina): Paleoecology and coevolution with mammals. *Hist. Biol.* 7:51–69.

Aldridge, R. D. 1992. Oviductal anatomy and seasonal sperm storage in the southeastern crowned snake (*Tantilla coronata*). *Copeia* 1992:1103–6.

Aldridge, R. D., and R. D. Semlitsch. 1992. Male reproductive biology of the southeastern crowned snake (*Tantilla coronata*). *Amphibia-Reptilia* 13:219–25.

Alvarez del Toro, M. 1982. *Los reptiles de Chiapas.* Tuxtla Gutiérrez, Mexico: Inst. Hist. Nat. Est. Dept. Zool.

Ananjeva, N. B., and N. L. Orlov. 1982. Feeding behaviour of snakes. *Vert. Hungarica* 21:25–31.

Andreadis, P. T., and G. M. Burghardt. 1993. Feeding behavior and an oropharyngeal component of satiety in a two-headed snake. *Physiol. Behav.* 54:649–58.

Andrén, C., and G. Nilson. 1981. Reproductive success and risk of predation in normal and melanistic morphs of the adder, *Vipera berus. Biol. J. Linn. Soc.* 15:235–46.

Angilletta, M. J., Jr. 1995. Sedentary behavior by green mambas, *Dendroaspis angusticeps* (Elapidae). *Herp. Nat. Hist.* 2:105–11.

[Anonymous]. 1938. For better fishing—kill the watersnake. *Pennsylvania Board of Fish Comm. Combined Bienn. Repts.,* 71–75.

[Anonymous]. 1995. Freer Texas [*sic*] holds million dollar hunt. *N.C.S. [National Crotalus Society] News* 3 (2): 1, 8.

Arena, P. C., C. Warwick, and D. Duvall. 1995. Rattlesnake round-ups. In *Wildlife and recreationists: Coexistence through management and research,* ed. R. L. Knight and K. J. Gutzwiller, 313–24. Washington, D.C.: Island Press.

Armstrong, B., and J. B. Murphy. 1979. The natural history of Mexican rattlesnakes. *Spec. Publ. Univ. Kansas. Mus. Nat. Hist.* 5:1–88.

Arnold, E. N. 1980. Recently extinct reptile populations from Mauritius and Réunion, Indian Ocean. *J. Zool.* 191:33–47.

———. 1994. Investigating the origins of performance advantage: Adaptation, exaptation, and lineage effects. In *Phylogenetics and ecology,* ed. P. Eggleton and R. Vane-Wright, 123–68. San Diego: Academic Press.

Arnold, E. N., and J. A. Burton. 1978. *A field guide to the reptiles and amphibians of Britain and Europe.* London: Collins.

Arnold, S. J. 1977. Polymorphism and geographic variation in the feeding behavior of the garter snake *Thamnophis elegans. Science* 197:676–78.

———. 1983. Morphology, performance and fitness. *Amer. Zool.* 23:347–61.

———. 1992. Behavioural variation in natural populations. VI. Prey responses by two species of garter snakes in three regions of sympatry. *Anim. Behav.* 44:705–19.

———. 1993. Foraging theory and prey-size–predator-size relations in snakes. In *Snakes: Ecology and behavior,* ed. R. A. Seigel and J. T. Collins, 87–115. New York: McGraw-Hill.

Ashe, J. S., and H. Marx. 1988. Phylogeny of the viperine snakes (Viperinae): Part II. Caldistic analysis and major lineages. *Fieldiana (Zool.),* n.s., 52:1–23.

Assakura, M. T., M. G. Salomão, G. Puorto, and F. R. Mandelbaum. 1992. Hemorrhagic, fibrinogenolytic and edema-forming activities of the venom of the colubrid snake *Philodryas olfersii* (green snake). *Toxicon* 30:427–38.

Baird, S. F., and C. Girard. 1853. *Catalogue of North American reptiles in the museum of the Smithsonian Institution. Part I: Serpents.* Washington, D.C.: Smithsonian Inst.

Barker, D. G. 1992. Variation, infraspecific relationships, and biogeography of the ridgenosed rattlesnake, *Crotalus willardi.* In *Biology of the pitvipers,* ed. J. A. Campbell and E. D. Brodie Jr., 89–106. Tyler, Tex.: Selva.

Barker, D. G., and T. M. Barker. 1994. *Pythons of the world.* Vol. 1, *Australia.* Lakeside, Calif.: Advanced Vivarium Systems.

Barr, A. D., S. A. Wieburg, and K. V. Kardong. 1988. The predatory strike behavior of the mamushi, *Agkistrodon blomhomffii blomhoffii,* and the Malay pit viper, *Calloselasma rhodostoma. Jap. J. Herp.* 12:135–38.

Bartecki, U., and E. W. Heymann. 1987. Field observations of snake-mobbing in a group of saddle-back tamarins, *Saguinus fuscicollis nigrifrons. Folia Primatol.* 48:199–202.

Bauer, A. M. 1988. *Leptotyphlops occidentalis,* western worm snake: Size and distribution. *J. Herp. Assoc. Africa* 34:45.

Beavers, R. A. 1976. Food habits of the western diamondback rattlesnake, *Crotalus atrox,* in Texas (Viperidae). *Southw. Nat.* 20:503–15.

Beebe, W. 1946. Field notes on the snakes of Kartabo, British Guiana, and Caripito, Venezuela. *Zoologica* 31:11–52.

Behrensmeyer, A. K., J. D. Damuth, W. A. DiMichele, R. Potts, H.-D. Sues, and S. L. Wing. 1992. *Terrestrial ecosystems through time: Evolutionary paleoecology of terrestrial plants and animals.* Chicago: Univ. Chicago Press.

Bellairs, A. d'A. 1969. *The life of reptiles.* London: Weidenfeld and Nicolson.

Bennion, R. S., and W. S. Parker. 1976. Field observations on courtship and aggressive behavior in desert striped whipsnakes, *Masticophis t. taeniatus. Herpetologica* 32:30–35.

Bernardino, F. S., Jr., and G. H. Dalrymple. 1992. Seasonal activity and road mortality of the snakes of the Pa-hay-okee wetlands of the Everglades National Park, USA. *Biol. Conserv.* 62:71–75.

Bertelsen, E., and J. G. Nielsen. 1987. The deep sea eel family Monognathidae (Pisces, Anguilliformes). *Steenstrupia* 13:141–98.

Bhupathy, S. 1990. Blotch structure in individual identification of the Indian Python *Python molurus molurus* Linn. and its possible usage in population estimation. *J. Bombay Nat. Hist. Soc.* 87:399–404.

Bhupathy, S., and V. S. Vijayan. 1989. Status, distribution and general ecology of the Indian python *Python molurus molurus* Linn. in Keoladeo National Park, Bharatpur, Rajasthan. *J. Bombay Nat. Hist. Soc.* 86:381–87.

Bierregaard, R. O., Jr. 1984. Observations of the nesting biology of the Guiana Crested Eagle (*Morphnus guianensis*). *Wilson Bull.* 96:1–5.

Blackburn, D. 1985. Evolutionary origins of viviparity in Reptilia. II. Serpentes, Amphisbaenia, and Ichthyosauria. *Amphibia-Reptilia* 6:259–91.

Blair, W. F. 1960. *The rusty lizard: A population study.* Austin: Univ. Texas Press.

Blanchard, F. N. 1930. Further studies on the eggs and young of the eastern ring-neck snake, *Diadophis punctatus blanchardi. Bull. Antivenom Inst. Amer.* 4:4–10.

Bloodworth, G. 1989. Python story I. *Swara* 12 (1): 13.

Bogert, C. M. 1969. Boas: A paradoxical family. *Anim. Kingdom* 72 (4): 18–25.

Bogert, C. M., and W. G. Degenhardt. 1961. An addition to the fauna of the United States, the Chihuahuan ridge-nosed rattlesnake in New Mexico. *Amer. Mus. Novitates* 2064:1–15.

Böhme, W., ed. 1993. *Handbuch der Reptilien und Amphibien Europas.* Vol. 3/1, *Schlangen (Serpentes) I.* Wiesbaden: Aula-Verlag.

Boinski, S. 1988. Use of a club by a wild white-faced capuchin (*Cebus capucinus*) to attack a venomous snake (*Bothrops asper*). *Amer. J. Prim.* 14:177–79.

Bolaños, R. 1984. *Serpientes venenosas y ofidismo en Centroamérica.* San José: Ed. Univ. Costa Rica.

Bonta, M. 1991. Snake attack. *Bird Watcher's Digest* 13 (6): 92–96.

Bothner, R. C. 1974. Some observations on the feeding habits of the cottonmouth in southeastern Georgia. *J. Herp.* 8:257–58.

Bourgeois, M. 1965. *Contribution à la morphologie comparée du crâne des ophidiens de l'Afrique centrale.* Publ. Univ. Officielle du Congo, vol. 18. Lubumbashi: Univ. Officielle du Congo.

Bowden, C. 1991. *Desierto: Memories of the future.* New York: W. W. Norton.

Bowers, B. B., and G. B. Burghardt. 1991. The scientist and the snake: Relationships with reptiles. In *The inevitable bond,* ed. M. Davis and D. Balfour, 250–63. Cambridge: Cambridge Univ. Press.

Branch, W. R. 1978. The venomous snakes of southern Africa, Part I: Introduction and Viperidae. *Snake* 9:67–86.

———. 1986. Hemipenial morphology of African snakes: A taxonomic review, Part I: Scolecophidia and Boidae. *J. Herp.* 20:285–99.

———. 1988. *Bill Branch's field guide to the snakes and other reptiles of southern Africa.* Sanibel Island, Fla.: Ralph Curtis Books.

———. 1991a. *Everyone's guide to snakes of southern Africa.* Johannesburg: Central News Agency.

———. 1991b. *Homoroselaps lacteus,* spotted harlequin snake: Envenomation. *J. Herp. Assoc. Africa* 39:28.

Branch, W. R., and W. D. Hacke [*sic*]. 1980. A fatal attack on a young boy by an African rock python *Python sebae. J. Herp.* 14:305–7.

Brattstrom, B. H. 1964. Evolution of the pit vipers. *Trans. San Diego Soc. Nat. Hist.* 13:185–268.

Breidenbach, C. H. 1990. Thermal cues influence strikes in pitless vipers. *J. Herp.* 24:448–50.

Breithaupt, B. H., and D. Duvall. 1986. The oldest record of serpent aggregation. *Lethaia* 19:181–85.

———. 1990. Paleontological evidence of an ancient snake aggregation. In *Evolutionary paleobiology of behavior and coevolution,* ed. A. J. Boucot, 149–54. Amsterdam: Elsevier.

Broadley, D. G. 1963. Two rare fossorial reptiles in southeastern Rhodesia: Predator and prey. *J. Herp. Assoc. Rhodesia* 20:7–8.

———. 1968. The venomous snakes of central and south Africa. In *Venomous animals and their venoms,* vol. 1, *Venomous Vertebrates,* ed. W. Bücherl, E. Buckley, and V. Deulofeu, 403–35. New York: Academic Press.

———. 1974. Predation by birds on reptiles and amphibians in south-eastern Africa. *Honeyguide* 78:11–19.

———. 1978. English names for Rhodesian reptiles and amphibians. *J. Herp. Assoc. Africa* 19:25–36.

———. 1983. *FitzSimon's snakes of southern Africa.* Johannesburg: Delta Books.

———. 1987. Caudal autotomy in African snakes of the genera *Natriciteres* Loveridge and *Psammophis* Boie. *J. Herp. Assoc. Africa* 33:18–19.

———. 1994. A collection of snakes from eastern Sudan, with the description of a new species of *Telescopus* Wagler, 1830 (Reptilia: Ophidia). *J. African Zool.* 108:201–8.

Broadley, D. G., and G. Watson. 1976. A revision of the worm snakes of southeastern Africa (Serpentes: Leptotyphlopidae). *Occ. Papers Natl. Mus. Monuments Rhodesia* B5:465–510.

Brodie, E. D., Jr., R. A. Nussbaum, and R. M. Storm. 1969. An egg-laying aggregation of five species of Oregon reptiles. *Herpetologica* 25:223–27.

Brodie, E. D., III. 1989. Genetic correlations between morphology and antipredator behaviour in natural populations of the garter snake *Thamnophis ordinoides. Nature* 342:542–43.

———. 1993a. Consistency of individual differences in anti-predator behaviour and colour pattern in the garter snake, *Thamnophis ordinoides. Anim. Behav.* 45:851–61.

———. 1993b. Differential avoidance of coral snake banded patterns by free-ranging avian predators in Costa Rica. *Evolution* 47:227–35.

Brodie, E. D., III, and E. D. Brodie Jr. 1990. Tetrodotoxin resistance in garter snakes: An evolutionary response of predators to dangerous prey. *Evolution* 44:651–59.

Brodmann, P. 1987. *Die Giftschlangen Europas und die Gattung* Vipera *in Afrika und Asien.* Bern: Kümmerly + Frey.

Brooks, D. R., and D. A. McLennan. 1991. *Phylogeny, ecology, and behavior: A research program in comparative biology.* Chicago: Univ. Chicago Press.

Brown, D. G. 1990. Observations of a Prairie Rattlesnake (*Crotalus viridis viridis*) consuming neonatal cottontail rabbits (*Sylvilagus nuttalli*), with defense of the young cottontails by adult conspecifics. *Bull. Chicago Herp. Soc.* 25:24–26.

Brown, W. S. 1993. Biology, status, and management of the timber rattlesnake (*Crotalus horridus*): A guide for conservation. *Soc. Study Amph. Rept., Herp. Circ.* 22:1–72.

Bullock, R. E. 1981. Tree climbing bullsnakes. *Blue Jay* 39:139–40.

Bullock, T. H., and R. B. Cowles. 1952. Physiology of an infrared receptor: The facial pit of pit vipers. *Science* 115:541–43.

Burdick, A. 1994. It's not the only alien invader. *New York Times Mag.,* Nov. 13, 49–55, 78, 80–81, 86–87.

Burger, J., and R. T. Zappalorti. 1986. Nest site selection by pine snakes, *Pituophis melanoleucus,* in the New Jersey Pine Barrens. *Copeia* 1986:116–21.

Burghardt, G. M. 1970. Chemical perception in reptiles. In *Communication by chemical signals,* ed. J. W. Johnson, D. G. Moulton, and A. Turk, 241–308. New York: Appleton-Century-Crofts.

———. 1978. Behavioral ontogeny in reptiles: Whence, whither, and why? In *The development of behavior: Comparative and evolutionary aspects,* ed. G. M. Burghardt and M. Bekoff, 149–74. New York: Garland STPM.

———. 1985. Animal awareness: Current perceptions and historical perspective. *Amer. Psych.* 40:905–19.

———. 1988. Precocity, play, and the ectotherm-endotherm transition: Profound reorganization or superficial adaptation? In *Handbook of behavioral neurobiology,* vol. 9, ed. E. M. Blass, 107–48. New York: Plenum Press.

———. 1991. Cognitive ethology and critical anthropomorphism: A snake with two heads and hognose snakes that play dead. In *Cognitive ethology: The minds of other animals,* ed. C. A. Ristau, 53–90. Hillsdale, N.J.: Lawrence Erlbaum Assoc.

———. 1996. Amending Tinbergen: A fifth aim for ethology. In *Anthropomorphism, anecdotes, and animals: The emperor's new clothes,* ed. R. W. Mitchell, N. S. Thompson, and H. L. Miles, 254–76. Lincoln: Univ. Nebraska Press.

Burghardt, G. M., and H. W. Greene. 1988. Predator simulation and duration of death feigning in neonate hognose snakes. *Anim. Behav.* 36:1842–44.

Burghardt, G. M., R. O. Hietala, and M. R. Pelton. 1972. Knowledge and attitudes concerning black bears by users of the Great Smoky Mountains National Park. In *Bears: Their biology and management,* ed. S. Herrero, 255–73. Morges, Switzerland: Int. Union Conserv. Nat. Natur. Res.

Burton, R. W. 1950. The record hamadryad or king cobra [*Naja hannah* (Cantor)] and lengths and weights of large specimens. *J. Bombay Nat. Hist. Soc.* 49:561–62.

Bushey, C. L. 1985. Man's effect upon a colony of *Sistrurus c. catenatus* (Raf.) in northeastern Illinois (1834–1975). *Bull. Chicago Herp. Soc.* 20:1–12.

Bustard, H. R. 1969. Defensive behavior and locomotion of the Pacific boa, *Candoia aspera,* with a brief review of head concealment in snakes. *Herpetologica* 25:164–70.

Butler, J. A., and E. Shitu. 1985. Uses of some reptiles by the Yoruba people of Nigeria. *Herp. Rev.* 16:15–16.

Butler, J. A., T. W. Hull, and R. Franz. 1995. Neonate aggregations and maternal attendance of young in the Eastern Diamondback Rattlesnake, *Crotalus adamanteus. Copeia* 1995:196–98.

Butynski, T. M., and J. Kalina. 1994. Uganda's Impenetrable Forest: A birding hot spot. *East African Nat. Hist. Soc. Bull.* 24:50–54.

Cadle, J. E. 1982. Problems and approaches in the interpretation of the evolutionary history of venomous snakes. *Mem. Inst. Butantan.* 46:255–74.

———. 1987. Geographic distribution: Problems in phylogeny and zoogeography. In *Snakes: Ecology and evolutionary biology,* ed. R. A. Seigel, J. T. Collins, and S. S. Novak, 77–105. New York: Macmillan.

———. 1992. Phylogenetic relationships among vipers: Immunological evidence. In *Biology of the pitvipers,* ed. J. A. Campbell and E. D. Brodie Jr., 41–48. Tyler, Tex.: Selva.

———. 1994. The colubrid radiation in Africa (Serpentes: Colubridae): Phylogenetic relationships and evolutionary patterns based on immunological data. *Zool. J. Linn. Soc.* 110:103–40.

Cadle, J. E., and P. Chuna M. 1995. A new lizard of the genus *Macropholidus* (Teiidae) from a relictual humid forest of northwestern Peru, and notes on *Macropholidus ruthveni* Noble. *Breviora* 501:1–39.

Cadle, J. E., and H. W. Greene. 1993. Phylogenetic patterns, biogeography, and the ecological structure of neotropical snake assemblages. In *Species diversity in ecological communities: Historical and geographical perspectives,* ed. R. E. Ricklefs and D. Schluter, 281–93. Chicago: Univ. Chicago Press.

Cagle, F. R. 1946. *Typhlops braminus* in the Mariana Islands. *Copeia* 1946:101.

Caldwell, G. S., and R. W. Rubinoff. 1983. Avoidance of venomous sea snakes by naive herons and egrets. *Auk* 100:195–98.

Camilleri, C., and R. Shine. 1990. Sexual dimorphism and dietary divergence: Differences in trophic morphology between male and female snakes. *Copeia* 1990:649–58.

Campbell, J. A., and E. D. Brodie Jr., eds. 1992. *Biology of the pitvipers.* Tyler, Tex.: Selva.

Campbell, J. A., and J. L. Camarillo R. 1992. The Oaxacan Dwarf Boa *Exiliboa placata* (Serpentes: Tropidophiidae): Descriptive notes and life history. *Carib. J. Sci.* 28:17–20.

Campbell, J. A., and W. W. Lamar. 1989. *The venomous reptiles of Latin America.* Ithaca: Cornell Univ. Press.

———. 1992. Taxonomic status of miscellaneous neotropical viperids, with the description of a new genus. *Occ. Papers Mus. Texas Tech Univ.* 153:1–31.

Carpenter, C. C. 1977. Communication and displays in snakes. *Amer. Zool.* 17:217–23.

———. 1984. Dominance in snakes. *Spec. Publ. Univ. Kansas Mus. Nat. Hist.* 10:195–202.

Carr, A. F. 1940. *A contribution to the herpetology of Florida.* Univ. Florida Publ., Biol. Sci. Ser., vol. 3, no. 1.

Gainesville: Univ. Florida.

———. 1982. Armadillo dilemma. *Anim. Kingdom* 85 (5): 40–43.

———. 1994. *A naturalist in Florida: A celebration of Eden.* Ed. M. H. Carr. New Haven: Yale Univ. Press.

Case, T. J., D. T. Bolger, and A. D. Richman. 1992. Reptilian extinctions: The last ten thousand years. In *Conservation biology: The theory and practice of nature preservation and management,* ed. P. L. Fielder and S. K. Jain, 91–125. New York: Chapman and Hall.

Cervantes-Reza, F. A. 1981. Some predators of the zacatuche (*Romerolagus diazi*). *J. Mamm.* 62:850–51.

Chapman, C. A. 1986. *Boa constrictor* predation and group response in white-faced *Cebus* monkeys. *Biotropica* 18: 171–72.

Charland, M. B. 1995. Thermal consequences of reptilian viviparity: Thermoregulation in gravid and nongravid garter snakes (*Thamnophis*). *J. Herp.* 29: 383–90.

Cheney, D. L., and R. M. Seyfarth. 1990. *How monkeys see the world: Inside the mind of another species.* Chicago: Univ. Chicago Press.

Chiszar, D., D. Boyer, R. Lee, J. B. Murphy, and C. W. Radcliffe. 1990. Caudal luring in the southern death adder, *Acanthophis antarcticus. J. Herp.* 24:253–60.

Chiszar, D., H. M. Smith, C. M. Bogert, and J. Vidaurri. 1991. A chemical sense of self in timber and prairie rattlesnakes. *Bull. Psychonom. Soc.* 29:153–54.

Chiszar, D., R. K. K. Lee, H. M. Smith, and C. W. Radcliffe. 1992. Searching behavior by rattlesnakes following predatory strikes. In *Biology of the pitvipers,* ed. J. A. Campbell and E. D. Brodie Jr., 369–82. Tyler, Tex.: Selva.

Chiszar, D., J. B. Murphy, and H. M. Smith. 1993. In search of zoo-academic collaborations: A research agenda for the 1990's. *Herpetologica* 49:488–500.

Clark, D. R. 1970. Ecological study of the worm snake *Carphophis vermis* (Kennicott). *Univ. Kansas Publ. Mus. Nat. Hist.* 19:85–194.

Cobb, V. A. 1990. Reproductive notes on the eggs and offspring of *Tantilla gracilis* (Serpentes: Colubridae), with evidence of communal nesting. *Southw. Nat.* 35:222–24.

Cock Buning, T. 1983. Thermal sensitivity as a specialization for prey capture and feeding in snakes. *Amer. Zool.* 23: 363–75.

Cogger, H. G. 1992. *Reptiles and amphibians of Australia.* Ithaca: Cornell Univ. Press.

Cogger, H., and R. G. Zweifel, eds., 1992. *Reptiles and amphibians.* New York: Smithmark Publ.

Cogger, H., H. Heatwole, Y. Ishikawa, M. McCoy, N. Tamiya, and T. Teruuchi. 1987. The status and natural history of the Rennell Island sea krait, *Laticauda crockeri* (Serpentes: Laticaudidae). *J. Herp.* 21:255–66.

Coleman, K., L. A. Rothfuss, H. Ota, and K. V. Kardong. 1993. Kinematics of egg-eating in the specialized Taiwan snake *Oligodon formosus* (Colubridae). *J. Herp.* 27:320–27.

Collins, J. T. 1990. Standard common and current scientific names for North American amphibians and reptiles. 3d ed. *Soc. Study Amph. Rept., Herp. Circ.* 19:1–41.

Conant, R., and J. T. Collins. 1991. *A field guide to reptiles and amphibians of eastern and central North America.* Rev. ed. Boston: Houghton Mifflin Co.

Cooper, W. E., and N. Greenberg. 1992. Reptilian coloration and behavior. In *Biology of the Reptilia,* vol. 18, *Physiology E: Hormones, brain, and behavior,* ed. C. Gans and D. Crews, 289–422. Chicago: Univ. Chicago Press.

Cooper, W. E., C. S. Deperno, and J. Arnett. 1994. Prolonged poststrike elevation in tongue-flicking rate with rapid onset in Gila monster, *Heloderma suspectum:* Relation to diet and foraging and implications for evolution of chemosensory searching. *J. Chem. Ecol.* 20: 2867–81.

Corkhill, N. L., and R. Kirk. 1954. Poisoning by the Sudan mole viper *Atractaspis microlepidota* Günther. *Trans. Roy. Soc. Trop. Med. Hyg.* 48:376–84.

Corn, M. J. 1974. Report on the first certain collection of *Ungaliophis panamensis* from Costa Rica. *Carib. J. Sci.* 14: 167–75.

Cox, M. J. 1991. *The snakes of Thailand and their husbandry.* Malabar, Fla.: Krieger.

Crother, B. I., J. A. Campbell, and D. M. Hillis. 1992. Phylogeny and historical biogeography of the palm-pitvipers, genus *Bothriechis:* Biochemical and morphological evidence. In *Biology of the pitvipers,* ed. J. A. Campbell and E. D. Brodie Jr., 1–19. Tyler, Tex.: Selva.

Cundall, D. 1987. Functional morphology. In *Snakes: Ecology and evolutionary biology,* ed. R. A. Seigel, J. T. Collins, and S. S. Novak, 106–40. New York: Macmillan.

———. 1995. Feeding behavior in *Cylindrophis* and its bearing on the evolution of alethinophidian snakes. *J. Zool.* 237:353–76.

Cundall, D., and F. J. Irish. 1986. Aspects of locomotor and feeding behavior in the Round Island boa *Casarea dussumieri. Dodo: J. Jersey Wildl. Preserv. Trust* 23:108–11.

———. 1989. The function of the intramaxillary joint in the Round Island boa, *Casarea dussumieri. J. Zool.* 217:569–98.

Cundall, D., and D. A. Rossman. 1993. Cephalic anatomy of the rare Indonesian snake *Anomochilus weberi. Zool. J. Linn. Soc.* 109:235–73.

Cundall, D., V. Wallach, and D. A. Rossman. 1993. The systematic relationships of the snake genus *Anomochilus. Zool. J. Linn. Soc.* 109:275–99.

Cunha, O. R. da, and F. P. do Nascimento. 1978. *Ofidios da Amazônia, X: As cobras da região leste do Pará.* Publ. Avulsas, no. 31. Belém, Brazil: Mus. Paraense Emílio Goeldi.

Daltry, J. C., W. Wüster, and R. S. Thorpe. 1996. Diet and snake venom evolution. *Nature* 379:537–40.

Daniel, J. C. 1983. *The book of Indian reptiles.* Bombay: Bombay Nat. Hist. Soc.

de Queiroz, K., and J. Gauthier. 1994. Toward a phylogenetic system of biological nomenclature. *Trends Ecol. Evol.* 9:27–31.

DeSilva, A. 1990. *Colour guide to the snakes of Sri Lanka.* Portishead, England: R & A Publ.

Dial, B. E., R. E. Gatten Jr., and S. Kamel. 1987. Energetics of concertina locomotion in *Bipes biporus* (Reptilia: Amphisbaenia). *Copeia* 1987:470–77.

Diller, L. V., and D. R. Johnson. 1982. *Ecology of reptiles in the Snake River Birds of Prey Area.* Boise, Idaho: Bureau of Land Management.

Ditmars, R. L. 1932. *Thrills of a naturalist's quest.* Garden City, N.Y.: Halcyon House.

Dixon, J. R., and C. P. Kofron. 1983. The Central and South American anomalepid snakes of the genus *Liotyphlops. Amphibia-Reptilia* 4:241–64.

Dixon, J. R., and P. Soini. 1977. The reptiles of the upper Amazon basin,

Iquitos region, Peru. II. Crocodilians, turtles, and snakes. *Contr. Biol. Geol. Milwaukee Publ. Mus.* 12:1–71.

Dmi'el, R., G. Perry, and H. Mendelssohn. 1990. Sexual dimorphism in *Walterinnesia aegyptia* (Reptilia: Ophidia: Elapidae). *Snake* 22:33–35.

Dodd, C. K., Jr. 1987. Status, conservation, and management. In *Snakes: Ecology and evolutionary biology,* ed. R. A. Seigel, J. T. Collins, and S. S. Novak, 478–513. New York: Macmillan.

———. 1988. *Drymarchon corais couperi* (eastern indigo snake): Ecdysis. *Herp. Rev.* 19:84.

———. 1993. Strategies for snake conservation. In *Snakes: Ecology and behavior,* ed. R. A. Seigel and J. T. Collins, 363–93. New York: McGraw-Hill.

Dowling, H. G., and J. M. Savage. 1960. A guide to the snake hemipenis: A survey of basic structure and systematic characteristics. *Zoologica* 45:17–30.

Drummond, H. M. 1979. Stimulus control of amphibious predation in the northern water snake (*Nerodia s. sipedon*). *Z. Tierpsych.* 50:18–44.

———. 1983. Aquatic foraging in garter snakes: A comparison of specialists and generalists. *Behaviour* 86:1–30.

———. 1985. The role of vision in the predatory behaviour of natricine snakes. *Anim. Behav.* 33:206–15.

———. 1989. Limitations of a generalist: A field comparison of foraging snakes. *Behaviour* 108:23–43.

Drummond, H., and G. W. Wolfe. 1981. An observation of a diving beetle larva (Insecta: Coleoptera: Dytiscidae) attacking and kiling a garter snake, *Thamnophis elegans* (Reptilia: Serpentes: Colubridae). *Coleop. Bull.* 35:121–24.

Duarte, M. R., and I. L. Laporta-Ferreira. 1993. Fat body, lipids, and protein changes after induced long-term fasting in the pitviper, *Bothrops jararaca* (Wied, 1824) (Serpentes, Viperidae) in captivity. *Zool. Anz.* 230:111–21.

Duarte, M. R., G. Puorto, and F. L. Franco. 1995. A biological survey of the pitviper *Bothrops insularis* Amaral (Serpentes, Viperidae): An endemic and threatened offshore island snake of southeastern Brazil. *Studies Neotrop. Fauna Envir.* 30:1–13.

Dudley, R., and P. DeVries. 1990. Tropical rain forest structure and the geographical distribution of gliding vertebrates. *Biotropica* 22:432–34.

Duellman, W. E. 1961. *The amphibians and reptiles of Michoacán, México.* Univ.

Kansas Publ. Mus. Nat. Hist., vol. 15, no. 1. Lawrence: Univ. Kansas.

———. 1963. Amphibians and reptiles of the rainforests of southern El Petén, Guatemala. *Univ. Kansas Publ. Mus. Nat. Hist.* 15:205–49.

———. 1978. *The biology of an equatorial herpetofauna in Amazonian Ecuador.* Misc. Publ. Univ. Kansas Mus. Nat. Hist., no. 65. Lawrence: Univ. Kansas.

———. 1984. Henry S. Fitch in perspective. *Spec. Publ. Univ. Kansas Mus. Nat. Hist.* 10:3.

Dunn, E. R., and J. A. Tihen. 1944. The skeletal anatomy of *Liotyphlops albirostris. J. Morph.* 74:287–95.

Dunson, W. A., ed. 1975. *The biology of seasnakes.* Baltimore: Univ. Park Press.

Dunson, W. A., and G. W. Ehlert. 1971. Effects of temperature, salinity, and surface water flow on distribution of the sea snake *Pelamis. Limnol. Oceanogr.* 16:845–53.

Dunson, W. A., and S. A. Minton. 1978. Diversity, distribution, and ecology of Philippine marine snakes (Reptilia, Serpentes). *J. Herp.* 12:281–86.

Dunson, W. A., M. K. Dunson, and A. D. Keith. 1978. The nasal gland of the Montpellier snake *Malpolon monspessulanus:* Fine structure, secretion composition, and a possible role in reduction of dermal water loss. *J. Exp. Zool.* 203:461–74.

Duvall, D., K. M. Scudder, and D. Chiszar. 1980. Rattlesnake predatory behavior: Mediation of prey discrimination and release of swallowing by cues arising from envenomated mice. *Anim. Behav.* 28:674–83.

Duvall, D., M. B. King, and K. J. Gutwiller. 1985. Behavioral ecology and ethology of the prairie rattlesnake. *Natl. Geogr. Res.* 1:80–111.

Duvall, D., M. J. Goode, W. K. Hayes, J. K. Leonhardt, and D. G. Brown. 1990. Prairie rattlesnake vernal migration: Field experimental analyses and survival value. *Natl. Geogr. Res.* 6:457–69.

Duvall, D., G. W. Schuett, and S. J. Arnold. 1993. Ecology and evolution of snake mating systems. In *Snakes: Ecology and behavior,* ed. R. A. Seigel and J. T. Collins, 165–200. New York: McGraw-Hill.

Dyer, B. 1990. Unusual snake aggregations in South Africa. *J. Herp. Assoc. Africa* 37:48.

Edgren, R. A. 1953. Copulatory adjustment in snakes and its evolutionary implications. *Copeia* 1953:162–64.

Edwards, J. L. 1985. Terrestrial locomotion without appendages. In *Functional vertebrate morphology,* ed. M. Hildebrand, D. M. Bramble, K. F. Liem, and D. B. Wake, 159–72. Cambridge: Harvard Univ. Press.

Eichholz, M. W., and W. D. Koenig. 1992. Gopher snake attraction to bird's nests. *Southwest. Nat.* 37:293–298.

Elliott, W. B. 1978. Chemistry and immunology of reptilian venoms. In *Biology of the Reptilia,* vol. 8, *Physiology B,* ed. C. Gans and K. A. Gans, 163–436. New York: Academic Press.

Emerson, S. B., H. W. Greene, and E. L. Charnov. 1994. Allometric aspects of predator-prey interactions. In *Ecological morphology: Integrative organismal biology,* ed. P. C. Wainwright and S. M. Reilly, 123–39. Chicago: Univ. Chicago Press.

Ernst, C. H. 1993. *Venomous reptiles of North America.* Washington, D.C.: Smithsonian Inst. Press.

Ernst, C. H., and R. W. Barbour. 1989. *Snakes of eastern North America.* Fairfax, Va.: George Mason Univ. Press.

Estes, R. 1983. The fossil record and early distribution of lizards. In *Studies in herpetology and evolutionary biology: Essays in honor of Ernest Edward Williams,* ed. A. Rhodin and K. Miyata, 365–98. Cambridge, Mass.: Mus. Comp. Zool.

Estes, R., and A. Báez. 1985. Herpetofauna of North and South America during the Late Cretaceous and Cenozoic: Evidence for interchange? In *The great American biotic interchange,* ed. F. G. Stehli and S. D. Webb, 139–97. New York: Plenum Press.

Estes, R., T. H. Frazzetta, and E. E. Williams. 1970. Studies on the fossil snake *Dinilysia patagonica* Woodward: I. Cranial morphology. *Bull. Mus. Comp. Zool.* 140:25–73.

Evans, G. H. 1906. Breeding of the banded krait (*Bungarus fasciatus*) in Burma. *J. Bombay Nat. Hist. Soc.* 16:519–20.

Felger, R. S., and M. B. Moser. 1985. *People of the desert and the sea: Ethnobotany of the Seri Indians.* Tucson: Univ. Arizona Press.

Ferguson, R. K., P. H. Vlasses, and L. J. Riley. 1984. Captopril in the treatment of hypertension and congestive heart failure. *Res. Staff Physician* 30 (11): PC24–28.

Ferrarezzi, H. 1993. Nota sobre o gênero *Phalotris* com revisão do grupo *nasutus* e descrição de três espécies (Serpentes,

Greer, A. E. 1983. On the adaptive significance of the reptilian spectacle: The evidence from scincid, teiid, and lacertid lizards. In *Studies in herpetology and evolutionary biology: Essays in honor of Ernest Edward Williams,* ed. A. Rhodin and K. Miyata, 213–21. Cambridge, Mass.: Mus. Comp. Zool.

Gregory, P. T. 1975. Arboreal mating behavior in the red-sided garter snake. *Can. Field-Nat.* 89:461–62.

———. 1978. Feeding habits and diet overlap of three species of garter snakes (*Thamnophis*) on Vancouver Island. *Can. J. Zool.* 56:1967–74.

Gregory, P. T., and K. W. Stewart. 1975. Long-distance dispersal and feeding strategy of the red-sided garter snake (*Thamnophis sirtalis parietalis*) in the interlakes of Manitoba. *Can. J. Zool.* 53:238–45.

Gregory, P. T., J. M. Macartney, and D. H. Rivard. 1980. Small mammal predation and prey handling behavior by the garter snake *Thamnophis elegans. Herpetologica* 36:87–93.

Gregory, P. T., J. M. Macartney, and K. W. Larsen. 1987. Spatial patterns and movements. In *Snakes: Ecology and evolutionary biology,* ed. R. A. Seigel, J. T. Collins, and S. S. Novak, 366–95. New York: Macmillan.

Grismer, L. L., J. A. McGuire, and B. D. Hollingsworth. 1994. A report on the herpetofauna of the Vizcaíno Peninsula, Baja California, México, with a discussion of its biogeographic and taxonomic implications. *Bull. S. California Acad. Sci.* 93:45–80.

Gritis, P., and H. K. Voris. 1990. Variability and significance of parietal and ventral scales in the marine snakes of the genus *Lapemis* (Serpentes: Hydrophiidae), with comments on the occurrence of spiny scales in the genus. *Fieldiana (Zool.),* n.s., 56:1–13.

Groombridge, B. C. 1979. A previously unreported throat muscle in the Scolecophidia (Reptilia: Serpentes), with comments on other scolecophidian throat muscles. *J. Nat. Hist.* 13:661–80.

———. 1986. Phyletic relationships among viperine snakes. In *Studies in herpetology: Proc. Third Ord. Gen. Mtg. S.E.H.,* ed. Z. Rocek, 219–22. Prague: Charles Univ. for Societas Europaea Herpetologica.

Groves, F. 1973. Reproduction and venom in Blanding's tree snake. *Int. Zoo Yrbk.* 13:106–8.

Guillette, L. J., Jr. 1993. The evolution of viviparity in lizards. *BioScience* 43:742–51.

Gumbart, T. C., and K. A. Sullivan. 1990. Predation on yellow-eyed junco nestlings by twin-spotted rattlesnakes. *Southwest. Nat.* 35:367–68.

Gutiérrez, J. M., F. Chaves, and R. Bolaños. 1980. Estudio comparativo de venenos de ejemplares recién nacidos y adultos de *Bothrops asper. Rev. Biol. Trop.* 28:341–51.

Gutiérrez, J. M., C. Avila, Z. Camacho, and B. Lomote. 1990. Ontogenetic changes in the venom of the snake *Lachesis muta stenophrys* (bushmaster) from Costa Rica. *Toxicon* 28:419–26.

Gutiérrez, J. M., G. Rojas, N. J. da Silva Jr., and J. Núñez. 1992. Experimental myonecrosis induced by the venoms of South American *Micrurus* (coral snakes). *Toxicon* 30:1299–1302.

Guyer, C., and M. A. Donnelly. 1990. Length-mass relationships among an assemblage of tropical snakes in Costa Rica. *J. Trop. Ecol.* 6:65–76.

Gyi, K. K. 1970. A revision of the colubrid snakes of the subfamily Homalopsinae. *Univ. Kansas Publ. Mus. Nat. Hist.* 20:47–223.

Haacke, W. D. 1982. "Boy bites attacking python to death." *J. Herp. Assoc. Africa* 28:8–10.

Haagner, G. V. 1991. *Aspidelaps scutatus,* shield-nosed snake: Diet and reproduction. *J. Herp. Assoc. Africa* 39:26.

Hailey, A., and P. M. C. Davies. 1986a. Diet and foraging behaviour of *Natrix maura. Herp. J.* 1:53–61.

———. 1986b. Effects of size, sex, temperature and condition on activity metabolism and defence behaviour of the viperine snake, *Natrix maura. J. Zool., London (A)* 208:541–58.

Hall, P. M., and A. J. Meier. 1993. Reproduction and behavior of western mud snakes (*Farancia abacura*) in American alligator nests. *Copeia* 1993:219–22.

Hall, R. J., G. V. Haagner, and W. R. Branch. 1994. *Psammophylax rhombeatus rhombeatus,* Spotted Skaapsteker: Reproduction and grooming. *African Herp. News* 21:20–21.

Halloy, M., and G. M. Burghardt. 1990. Ontogeny of fish capture and ingestion in four species of garter snakes (*Thamnophis*). *Behaviour* 112:299–318.

Halpern, M. 1992. Nasal chemical senses in reptiles: Structure and function. In *Biology of the Reptilia,* vol. 18, *Physiology E: Hormones, brain, and behavior,* ed. C. Gans and D. Crews, 423–523. Chicago: Univ. Chicago Press.

Hardy, D. L. [Sr.] 1987. Fatal rattlesnake envenomation in Arizona: 1969–1984. *Clin. Tox.* 24:1–10.

———. 1992. A review of first aid measures for pitviper bite in North America, with an appraisal of Extractor™ and stun gun electroshock. In *Biology of the pitvipers,* ed. J. A. Campbell and E. D. Brodie Jr., 405–14. Tyler, Tex.: Selva.

———. 1994a. *Bothrops asper* (Viperidae) snakebite and field researchers in Middle America. *Biotropica* 26:198–207.

———. 1994b. A re-evaluation of suffocation as the cause of death during constriction by snakes. *Herp. Rev.* 25:45–47.

Harlow, P., and R. Shine. 1992. Food habits and reproductive biology of the Pacific Island boas (*Candoia*). *J. Herp.* 26:60–66.

Hartline, P. H., L. Kass, and M. S. Loop. 1978. Merging of modalities in the optic tectum: Infrared and visual integration in rattlesnakes. *Science* 199:1225–29.

Hasegawa, M., and M. Moriguchi. 1989. Geographic variation in food habits, body size and life history traits of the snakes on the Izu Islands. In *Current herpetology in East Asia,* ed. M. Matui, T. Hikida, and R. C. Goris, 414–32. Kyoto: Herp. Soc. Japan.

Hastings, J. R., and R. M. Turner. 1965. *The changing mile: An ecological study of vegetation change with time in the lower mile of an arid and semiarid region.* Tucson: Univ. Arizona Press.

Haverschmidt, F. 1970. Wattled jacana caught by an anaconda. *Condor* 72:364.

Hay, P. W., and P. W. Martin. 1966. Python predation on Uganda kob. *East African Wildl. J.* 4:151–52.

Hayes, W. K. 1995. Venom metering by juvenile prairie rattlesnakes, *Crotalus v. viridis:* Effects of prey size and experience. *Anim. Behav.* 50:33–40.

Hayes, W. K., and J. G. Galuska. 1984. Effects of rattlesnake (*Crotalus viridis oreganus*) envenomation upon mobility of male wild and laboratory mice (*Mus musculus*). *Bull. Maryland Herp. Soc.* 20:135–44.

Hayes, W. K., I. I. Kaiser, and D. Duvall. 1992a. The mass of venom expended by prairie rattlesnakes when feeding on rodent prey. In *Biology of the pitvipers,* ed. J. A. Campbell and E. D. Brodie Jr., 383–88. Tyler, Tex.: Selva.

Hayes, W. K., D. Duvall, and G. W. Schuett. 1992b. A preliminary report on the courtship behavior of free-ranging prairie rattlesnakes, *Crotalus viridis viridis* (Rafinesque), in south-central Wyoming. *Contr. Herp. Greater Cincinnati Herp. Soc.* 1992:45–48.

Hayes, W. K., P. Lavín-Murcio, and K. V. Kardong. 1993. Delivery of Duvernoy's secretion into prey by the brown tree snake, *Boiga irregularis* (Serpentes: Colubridae). *Toxicon* 31:881–87.

Heatwole, H. 1987. *Sea snakes.* Kensington, Australia: New South Wales Univ. Press.

Heatwole, H., S. A. Minton Jr., R. Taylor, and V. Taylor. 1978. Underwater observations on sea snake behaviour. *Rec. Austr. Mus.* 31:737–61.

Hecht, M. K., C. Kropach, and B. M. Hecht. 1974. Distribution of the yellow-bellied sea snake, *Pelamis platurus,* and its significance in relation to the fossil record. *Herpetologica* 30:387–96.

Hedges, S. B., C. A. Hass, and T. K. Maugel. 1989. Physiological color change in snakes. *J. Herp.* 23:450–55.

Hedges, S. B., C. A. Hass, and L. R. Maxson. 1992. Caribbean biogeography: Molecular evidence for dispersal in West Indian terrestrial vertebrates. *Proc. Natl. Acad. Sci. USA* 89:1909–13.

Henderson, R. W. 1994. A splendid quintet: The widespread boas of South America. *Lore* 44 (4): 2–9.

Henderson, R. W., and M. H. Binder. 1980. The ecology and behavior of vine snakes (*Ahaetulla, Oxybelis, Thelotornis, Uromacer*): A review. *Contr. Biol. Geol. Milwaukee Publ. Mus.* 37:1–38.

Henderson, R. W., J. R. Dixon, and P. Soini. 1978. On the seasonal incidence of tropical snakes. *Contr. Biol. Geol. Milwaukee Publ. Mus.* 17:1–15.

Henderson, R. W., T. A. Noeske-Hallin, J. A. Ottenwalder, and A. Schwartz. 1987. On the diet of the boa *Epicrates striatus* on Hispaniola, with notes on *E. fordi* and *E. gracilis. Amphibia-Reptilia* 8:251–58.

Henderson, R. W., T. Waller, P. Micucci, G. Puorto, and R. W. Bourgeois. 1995. Ecological correlates and patterns in the distribution of neotropical boines (Serpentes: Boidae): A preliminary assessment. *Herp. Nat. Hist.* 3:15–27.

Herrmann, H.-W., U. Joger, and G. Nilson. 1992a. Phylogeny and systematics of viperine snakes. I. General phylogeny of European vipers (*Vipera* sensu stricto). In *Proc. Sixth Ord. Gen. Mtg. S.E.H.,* ed. Z. Korsos and I. Kiss, 219–34. Budapest: Societas Europaea Herpetologica.

———. 1992b. Phylogeny and systematics of viperine snakes. III. Resurrection of the genus *Macrovipera* (Reuss, 1927) as suggested by biochemical evidence. *Amphibia-Reptilia* 13:375–92.

Hersek, M. J., R. G. Coss, and D. F. Hennessy. 1992. Combat between rattlesnakes (*Crotalus viridis oreganus*) in the field. *J. Herp.* 26:105–7.

Herzog, H. A., Jr., and G. M. Burghardt. 1988. Development of antipredator responses in snakes. III. Stability of individual and litter differences over the first year of life. *Ethology* 77:250–58.

Heyer, W. R., and S. Pongsapipatana. 1970. Gliding speeds of *Ptychozoon lionatum* (Reptilia: Gekkonidae) and *Chrysopelea ornata* (Reptilia: Serpentes). *Herpetologica* 26:317–19.

Heymann, E. W. 1987. A field observation of predation on a moustached tamarin (*Saguinus mystax*) by an anaconda. *Int. J. Primatol.* 8:193–95.

Hibbard, C. W. 1964. A brooding colony of the blind snake, *Leptotyphlops dulcis dissectus* Cope. *Copeia* 1964:222.

Hisaw, F. L., and H. K. Gloyd. 1926. The bull snake as a natural enemy of injurious rodents. *J. Mamm.* 7:200–205.

Hoffmann, L. A. C. 1988. Notes on the ecology of the horned adder *Bitis caudalis* (A. Smith) from Gobabeb, Namib-Naukluft Park. *J. Herp. Assoc. Africa* 35:33–34.

Holmstrom, W. F., Jr. 1981. Post-parturient behavior of the common anaconda, *Eunectes murinus. Zool. Gart.,* n.s., 51:353–56.

Huang, M. 1989. Studies on *Agkistrodon shedaoensis* ecology. In *Current herpetology in East Asia,* ed. M. Matui, T. Hikida, and R. C. Goris, 381–83. Kyoto: Herp. Soc. Japan.

Inger, R. F., and H. Marx. 1962. Variation of hemipenis and cloaca in the colubrid snake *Calamaria lumbricoidea. Syst. Zool.* 11:32–38.

———. 1965. The systematics and evolution of the oriental colubrid snakes of the genus *Calamaria. Fieldiana (Zool.)* 49:1–304.

Inglish, H. 1991. *Larry Hatteberg's Kansas people.* Wichita: Jular Publ.

Ionides, C. J. P., and C. R. S. Pitman. 1965. Notes on three East African venomous snake populations. *Puku* 3:87–95.

Irish, F. J., and P. Alberch. 1989. Heterochrony in the evolution of bolyeriid snakes. *Fortschr. Zool.* 35:205.

Iverson, J. B. 1986. Notes on the natural history of the Caicos Islands Dwarf Boa, *Tropidophis greenwayi. Carib. J. Sci.* 22:191–98.

Jackson, J. A. 1974. Gray rat snakes versus red-cockaded woodpeckers: Predator-prey adaptations. *Auk* 91:342–47.

Jackson, J. F. 1979. Effects of some ophidian tail displays on the predatory behavior of grison (*Galictis* sp.). *Copeia* 1979:169–72.

Jackson, J. F., W. Ingram III, and H. W. Campbell. 1976. The dorsal pigmentation pattern of snakes as an antipredator strategy: A multivariate approach. *Amer. Nat.* 110:1029–53.

Jackson, K., and T. H. Fritts. 1995. Evidence from tooth morphology for a posterior maxillary origin of the proteroglyph fang. *Amphibia-Reptilia* 16:273–88.

Jackson, M. K., and H. W. Reno. 1975. Comparative skin structure of some fossorial and subfossorial leptotyphlopid and colubrid snakes. *Herpetologica* 31:350–59.

Jaensch, M. 1988. *Aspidelaps lubricus infuscatus,* western coral snake: Distribution, behaviour and feeding. *J. Herp. Assoc. Africa* 34:46.

Jansen, D. W. 1983. A possible function of the secretion of Duvernoy's gland. *Copeia* 1983:262–64.

Jansen, D. W., and R. C. Foehring. 1983. The mechanism of venom secretion from Duvernoy's gland of the snake *Thamnophis sirtalis. J. Morph.* 175:271–77.

Jayne, B. C. 1982. Comparative morphology of the semispinalis-spinalis muscle of snakes and correlations with locomotion and constriction. *J. Morph.* 172:83–96.

———. 1986. Kinematics of terrestrial snake locomotion. *Copeia* 1986:915–27.

———. 1988. Muscular mechanisms of snake locomotion: An electromyographic study of lateral undulation of the Florida banded water snake (*Nerodia fasciata*) and the yellow rat snake (*Elaphe obsoleta*). *J. Morph.* 197:159–81.

Jayne, B. C., H. K. Voris, and H. B. Heang. 1988. Diet, feeding behavior,